SILICON-BASED HYBRID NANOPARTICLES: FUNDAMENTALS, PROPERTIES, AND APPLICATIONS

SILICON-BASED HYBRID NANOPARTICLES: FUNDAMENTALS, PROPERTIES, AND APPLICATIONS

Edited by

SABU THOMAS
Polymer Science and Engineering, School of Chemical Sciences, and Centre for Neuroscience and Nanotechnology, Mahatma Gandhi University, Kottayam, India

TUAN ANH NGUYEN
Institute for Tropical Technology, Vietnam Academy of Science and Technology, Hanoi, Vietnam

MAZAHER AHMADI
Department of Analytical Chemistry, Bu-Ali Sina University, Hamedan, Iran

GHULAM YASIN
Institute for Advanced Study, College of Physics and Optoelectronic Engineering, Shenzhen University, Shenzhen, P.R. China

NIRAV JOSHI
Physics Department, Sao Carlos Institute of Physics at the University of Sao Paulo, Sao Paulo, Brazil

Elsevier
Radarweg 29, PO Box 211, 1000 AE Amsterdam, Netherlands
The Boulevard, Langford Lane, Kidlington, Oxford OX5 1GB, United Kingdom
50 Hampshire Street, 5th Floor, Cambridge, MA 02139, United States

Copyright © 2022 Elsevier Inc. All rights reserved.

No part of this publication may be reproduced or transmitted in any form or by any means, electronic or mechanical, including photocopying, recording, or any information storage and retrieval system, without permission in writing from the publisher. Details on how to seek permission, further information about the Publisher's permissions policies and our arrangements with organizations such as the Copyright Clearance Center and the Copyright Licensing Agency, can be found at our website: www.elsevier.com/permissions.

This book and the individual contributions contained in it are protected under copyright by the Publisher (other than as may be noted herein).

Notices
Knowledge and best practice in this field are constantly changing. As new research and experience broaden our understanding, changes in research methods, professional practices, or medical treatment may become necessary.

Practitioners and researchers must always rely on their own experience and knowledge in evaluating and using any information, methods, compounds, or experiments described herein. In using such information or methods they should be mindful of their own safety and the safety of others, including parties for whom they have a professional responsibility.

To the fullest extent of the law, neither the Publisher nor the authors, contributors, or editors, assume any liability for any injury and/or damage to persons or property as a matter of products liability, negligence or otherwise, or from any use or operation of any methods, products, instructions, or ideas contained in the material herein.

British Library Cataloguing-in-Publication Data
A catalogue record for this book is available from the British Library

Library of Congress Cataloging-in-Publication Data
A catalog record for this book is available from the Library of Congress

ISBN: 978-0-12-824007-6

For Information on all Elsevier publications
visit our website at https://www.elsevier.com/books-and-journals

Publisher: Matthew Deans
Acquisitions Editor: Simon Holt
Editorial Project Manager: Isabella C. Silva
Production Project Manager: Debasish Ghosh
Cover Designer: Greg Harris

Typeset by MPS Limited, Chennai, India

Contents

List of contributors xiii
Preface xvii

1 Silicon-based hybrid nanoparticles: An introduction 1
Surbhi Sharma and Soumen Basu
1.1 Introduction 1
References 5

**2 Synthesis and characterization of silicon-based
hybrid nanoparticles 11**
Gautam M. Patel, Gaurang J. Bhatt and Pradeep T. Deota
2.1 Introduction 11
2.2 Synthesis 12
    2.2.1 Fabrication process 12
    2.2.2 Sol–gel process 15
    2.2.3 Polymer grafting 17
    2.2.4 Encapsulation process 20
    2.2.5 Drug loading 21
    2.2.6 Colloidal system 23
    2.2.7 Solution blending 25
    2.2.8 Gas-phase synthesis 26
    2.2.9 Coating with magnetic nanoparticles 26
    2.2.10 Catalyst 27
2.3 Characterization 27
    2.3.1 Fourier-transform infrared spectroscopy (FTIR) 27
    2.3.2 X-ray photoelectron spectroscopy (XPS) 29
    2.3.3 Dynamic light scattering (DLS) 31
    2.3.4 Transmission electron microscopy (TEM) 33
2.4 Conclusion 34
References 35

**3 Properties of silicon–carbon (CNTs/graphene)
hybrid nanoparticles 45**
Xinyi Chen, Deng Long and Jingqin Cui
3.1 Introduction 45

3.2 Design of Si/C nanoparticles 46
    3.2.1 Mixed Si/C nanoparticles 47
    3.2.2 Finely configured Si/C hybrid nanoparticles 47
    3.2.3 Cross-dimensional Si/C hybrids 48
3.3 Electrical properties of Si/C nanoparticles 49
    3.3.1 Influence of morphology and structure 50
    3.3.2 Influence of chemical components 52
    3.3.3 In situ characterization on of Si/C hybrids' properties 53
3.4 Optical properties of ultrasmall Si/C nanoparticles 55
    3.4.1 Fluorescence of Si/C quantum dots 56
    3.4.2 Light absorption and photocarrier behavior of silicon
          quantum dots/graphene 58
3.5 Conclusion 60
References 60

4 Properties of silicon−ZnO hybrid nanoparticles 65
Adem Kocyigit
4.1 Introduction 65
4.2 Zinc oxide nanoparticles: synthesis and properties 65
4.3 Preparation of silicon−zinc oxide hybrid structures 67
    4.3.1 Anodization of silicon wafers 68
    4.3.2 Metal assisted chemically etching of silicon wafers 69
    4.3.3 Synthesis of silicon−zinc oxide hybrid structures 70
4.4 Structural and morphological properties of silicon−zinc
    oxide hybrid nanoparticles 73
    4.4.1 X-ray diffraction (XRD) patterns of the silicon−zinc
          oxide hybrid structures 73
    4.4.2 Fourier-transform infrared spectroscopy or Raman
          graph of the silicon−zinc oxide hybrid structures 75
    4.4.3 Morphological characterization of the silicon−zinc oxide
          hybrid structures 77
4.5 Electrical properties of silicon−zinc oxide
    hybrid nanoparticles 79
4.6 Optical properties of silicon−zinc oxide
    hybrid nanoparticles 80
    4.6.1 UV−Vis spectroscopy results of silicon−zinc oxide
          hybrid structures 81

  4.6.2 Photoluminescence spectroscopy results of
    silicon-zinc oxide hybrid structures 83
4.7 Applications of the silicon—zinc oxide hybrid nanoparticles 84
References 84

5 Assembly and electroluminescence of sheet-like zinc oxide/silicon light-emitting diode by a radio frequency magnetron sputtering technique 89
L. Castañeda
5.1 Introduction 89
5.2 Experimental details 92
  5.2.1 Fabrication of the samples 92
  5.2.2 Analysis of the samples 93
5.3 Results and Discussions 93
5.4 Conclusions 98
Conflicts of Interest 99
Acknowledgments 99
References 99

6 Silicon-based nanomaterials for energy storage 103
Shumaila Ibraheem, Ghulam Yasin, Rashid Iqbal, Adil Saleem, Tuan Anh Nguyen and Sehrish Ibrahim
6.1 Introduction 103
6.2 General background and progress 105
  6.2.1 Structure—property relationship 106
6.3 Si-based nanomaterials for lithium storage 108
  6.3.1 Si/carbon-based nanomaterials for Li storage 109
  6.3.2 Si/metals-based nanomaterials for Li storage 112
  6.3.3 Si/other materials-based nanostructures
    for Li storage 113
6.4 Si-based nanomaterials for supercapacitors 115
6.5 Conclusion 118
References 118

7 Application of silicon-based composite in batteries 125
Runwei Mo
7.1 Introduction 125

7.2 Capacity failure mechanism of Si electrode in LIB 126
7.3 Classification of Si-based composites 127
    7.3.1 Si/carbon composites 127
    7.3.2 Si /metal composites 130
    7.3.3 Si/metal oxide composites 133
    7.3.4 Si/polymer composites 135
7.4 Conclusions 137
7.5 Outlook 138
References 141

8 Nano silicon carbon hybrid particles and composites for batteries: Fundamentals, properties and applications 145
Yohan Oudart, Rudy Guicheteau, Jean-Francois Perrin, Raphael Janot, Mathieu Morcrette, Mariana Gutierrez, Laure Monconduit and Nicolas Louvain

8.1 Introduction 145
8.2 Nanosilicon for batteries 146
    8.2.1 Silicon generalities 146
    8.2.2 Nanosizing 146
    8.2.3 Various forms of nanosilicon 148
    8.2.4 General behavior and SEI 149
8.3 Carbon 151
    8.3.1 Carbon forms 152
    8.3.2 Pitch as precursor 155
8.4 Carbon coating of silicon 157
    8.4.1 Mechanical milling 157
    8.4.2 Gas-phase synthesis 158
    8.4.3 Liquid process 159
    8.4.4 Si/C composites 161
8.5 Conclusion 163
References 164

9 Nanostructured silicon for energy applications 169
Tenzin Ingsel and Ram K. Gupta

9.1 Introduction 169
9.2 Silicon for energy applications 171
    9.2.1 Applications of silicon in photoelectrochemical and photovoltaic devices 171

9.2.2 Silicon for lithium-ion batteries 177
9.2.3 Supercapacitors based on nanostructured silicon 183
9.2.4 Silicon as electrocatalyst for hydrogen
evolution reaction 186
9.3 Conclusion 194
References 194

10 Application of silicon-based hybrid nanoparticles in catalysis 199
Pratibha and Jaspreet Kaur Rajput
10.1 Introduction 199
10.2 Types of silicon-based nanoparticles 201
10.3 Silicon-based hybrid nanoparticles as catalyst for
organic conversions 203
10.3.1 Coupling reactions 203
10.3.2 Oxidation reactions 217
10.3.3 Reduction reactions 220
10.3.4 Multicomponent reactions 222
10.3.5 Carbon dioxide conversion 228
10.3.6 Addition reactions 233
10.3.7 Miscellaneous reaction 235
10.4 Conclusion and future scope 239
References 239

11 Silicon-based biosensor 247
Sandeep Arya, Anoop Singh, Asha Sharma and Vinay Gupta
11.1 Introduction 247
11.2 Fabrication of silicon-based hybrid nanoparticles 248
11.3 Attachment of bioreceptor on silicon-based materials 255
11.3.1 Chemical surface modification 255
11.4 Applications of biosensor 257
11.4.1 DNA detection 258
11.4.2 Enzyme detection 259
11.4.3 Antibody detection 260
11.4.4 Cell detection 260
11.4.5 Virus detection 261
11.4.6 Protein detection 262

        11.4.7 Small analyte detection 262
    11.5 Conclusion 264
    Reference 265

12 **Graphene-based field effect transistor (GFET) as nanobiosensors...269**
    Homa Farmani, Ali Farmani and Tuan Anh Nguyen
    12.1 Introduction 269
        12.1.1 Immunological assays 270
        12.1.2 Amplification method 271
        12.1.3 Nanobiosensors 271
    12.2 Graphene-based field effect transistor (FET) as biosensors 272
    12.3 Conclusion 273
    References 274

13 **Biomedical applications 277**
    Jih-Hsing Chang, Narendhar Chandrasekar, Shan-Yi Shen,
    Mohd. Shkir and Mohanraj Kumar
    13.1 Introduction 277
    13.2 Why silica nanoparticles are most suited for
         biomedical applications? 285
    13.3 Pharmaceutical applications of mesoporous silica
         nanoparticles 287
    13.4 Drug loading for MSNs 290
        13.4.1 Drug encapsulation mechanisms 290
    13.5 Gene delivery 295
    13.6 Protein absorption and separation 299
    13.7 Nucleic acid detection and purification 305
    13.8 Remarkable trends in mesoporous silica nanoparticles
         toward cancer therapeutic applications 307
    13.9 Biosafety of mesoporous silica nanoparticles 312
    13.10 Summary 314
    References 316

14 **Application in hyperthermia treatment 325**
    Sabrina A. Camacho, J.J. Hernández-Sarria, Josino Villela S. Neto,
    M. Montañez-Molina, F. Muñoz-Muñoz, H. Tiznado,
    J. López-Medina, O.N. Oliveira Jr and J.R. Mejía-Salazar
    14.1 Introduction 325

14.2 Nanoparticle heating: fundamentals 326
            14.2.1 LSPP-based heating mechanism 327
            14.2.2 Magnetic-based heating mechanism 331
        14.3 Synthesis of hyperthermia agents 331
            14.3.1 Plasmonic hyperthermia nanoagents 332
            14.3.2 Magnetic hyperthermia nanoagents 334
        14.4 Hyperthermia applications 339
            14.4.1 Plasmonic hyperthermia applications 340
            14.4.2 Magnetic hyperthermia applications 341
        14.5 Final remarks 343
        14.6 Acknowledgments 345
            14.6.1 Funding 345
        References 345

15 Silicon—metal hybrid nanoparticles as nanofluid scale inhibitors in oil/gas applications 353
Yasser Mahmoud A. Mohamed and Mohamed F. Mady
    15.1 Introduction 353
    15.2 Conclusions and future perspectives 358
    References 358

Index 363

List of contributors

Sandeep Arya Department of Physics, University of Jammu, Jammu, India

Soumen Basu School of Chemistry and Biochemistry, Affiliate Faculty—TIET-Virginia Tech Center of Excellence in Emerging Materials, Thapar Institute of Engineering & Technology, Patiala, India

Gaurang J. Bhatt Applied Chemistry Department, Faculty of Technology & Engineering, The MS University of Baroda, Vadodara, India

Sabrina A. Camacho Instituto de Física de São Carlos, Universidade de São Paulo, CP 369, São Carlos, Brasil; São Paulo State University (UNESP), School of Sciences, Humanities and Languages, Assis, Brazil

L. Castañeda Sección de Estudios de Posgrado e Investigación de la Escuela Superior de Medicina, Instituto Politécnico Nacional, Mexico City, México

Narendhar Chandrasekar Department of Nanoscience and Technology, Sri Ramakrishna Engineering College, Coimbatore, India

Jih-Hsing Chang Department of Environmental Engineering and Management, Chaoyang University of Technology, Taichung City, Taiwan

Xinyi Chen Pen-Tung Sah Institute of Micro-Nano Science and Technology, Xiamen University, Xiamen, P.R. China

Jingqin Cui Pen-Tung Sah Institute of Micro-Nano Science and Technology, Xiamen University, Xiamen, P.R. China

Pradeep T. Deota Applied Chemistry Department, Faculty of Technology & Engineering, The MS University of Baroda, Vadodara, India

Ali Farmani School of Electrical Engineering, Lorestan University, Khoramabad, Iran

Homa Farmani School of Electrical Engineering, Lorestan University, Khoramabad, Iran

Rudy Guicheteau Nanomakers, Rambouillet, France

Ram K. Gupta Department of Chemistry, Kansas Polymer Research Center, Pittsburg State University, Pittsburg, KS, United States

Vinay Gupta Department of Mechanical Engineering, Khalifa University of Science and Technology, Abu Dhabi, United Arab Emirates

Mariana Gutierrez Nanomakers, Rambouillet, France; Laboratoire de Réactivité et Chimie des Solides, Université de Picardie Jules Verne, Hub de l'énergie, Amiens, France

J.J. Hernández-Sarria Instituto de Física de São Carlos, Universidade de São Paulo, CP 369, São Carlos, Brasil

Shumaila Ibraheem Institute for Advanced Study, Shenzhen University, Shenzhen, P.R. China; College of Physics and Optoelectronic Engineering, Shenzhen University, Shenzhen, P.R. China

Sehrish Ibrahim College of Life Science and Technology, Beijing University of Chemical Technology, Beijing, China

Tenzin Ingsel Department of Chemistry, Kansas Polymer Research Center, Pittsburg State University, Pittsburg, KS, United States

Rashid Iqbal Institute for Advanced Study, Shenzhen University, Shenzhen, P.R. China; College of Physics and Optoelectronic Engineering, Shenzhen University, Shenzhen, P.R. China

Raphael Janot Laboratoire de Réactivité et Chimie des Solides, Université de Picardie Jules Verne, Hub de l'énergie, Amiens, France; Réseau sur le Stockage Electrochimique de l'Energie (RS2E), Hub de l'Energie, Amiens, France

Adem Kocyigit Department of Electrical and Electronics Engineering, Faculty of Engineering, Igdir University, Igdir, Turkey; Department of Electronics and Automation, Vocational High School, Bilecik Şeyh Edebali University, Bilecik, Turkey

Mohanraj Kumar Department of Environmental Engineering and Management, Chaoyang University of Technology, Taichung City, Taiwan

Deng Long Pen-Tung Sah Institute of Micro-Nano Science and Technology, Xiamen University, Xiamen, P.R. China

J. López-Medina CONACYT – Centro de Nanociencias y Nanotecnología, UNAM, Ensenada, México

Nicolas Louvain Réseau sur le Stockage Electrochimique de l'Energie (RS2E), Hub de l'Energie, Amiens, France; ICGM, The University of Montpellier, CNRS, ENSCM, Montpellier, France

Mohamed F. Mady Green Chemistry Department, National Research Center, Cairo, Egypt; Chemistry Department, Bioscience and Environmental Engineering, Faculty of Science and Technology, University of Stavanger, Stavanger, Norway

J.R. Mejía-Salazar Instituto Nacional de Telecomunicações (Inatel), Santa Rita do Sapucaí, Brasil

Runwei Mo School of Mechanical and Power Engineering, East China University of Science and Technology, Shanghai, P.R. China

Yasser Mahmoud A. Mohamed Photochemistry Department, National Research Center, Cairo, Egypt

Laure Monconduit Réseau sur le Stockage Electrochimique de l'Energie (RS2E), Hub de l'Energie, Amiens, France; ICGM, The University of Montpellier, CNRS, ENSCM, Montpellier, France

M. Montañez-Molina Centro de Investigación Científica y Educación Superior de Ensenada-CICESE, Ensenada, México

Mathieu Morcrette Laboratoire de Réactivité et Chimie des Solides, Université de Picardie Jules Verne, Hub de l'énergie, Amiens, France; Réseau sur le Stockage Electrochimique de l'Energie (RS2E), Hub de l'Energie, Amiens, France

F. Muñoz-Muñoz Facultad de Ingeniería, Arquitectura y Diseño, Universidad Autonoma de Baja California (UABC), Ensenada BC, Mexico

Josino Villela S. Neto Instituto Nacional de Telecomunicações (Inatel), Santa Rita do Sapucaí, Brasil

Tuan Anh Nguyen Institute for Tropical Technology, Vietnam Academy of Science and Technology, Hanoi, Vietnam

O.N. Oliveira Jr Instituto de Física de São Carlos, Universidade de São Paulo, CP 369, São Carlos, Brasil

Yohan Oudart Nanomakers, Rambouillet, France

Gautam M. Patel Department of Chemistry, School of Sciences, ITM (SLS) Baroda University, Vadodara, India

Jean-Francois Perrin Nanomakers, Rambouillet, France

Pratibha Department of Chemistry, Dr. B.R Ambedkar National Institute of Technology, Jalandhar, India

Jaspreet Kaur Rajput Department of Chemistry, Dr. B.R Ambedkar National Institute of Technology, Jalandhar, India

Adil Saleem Institute for Advanced Study, Shenzhen University, Shenzhen, P.R. China; College of Physics and Optoelectronic Engineering, Shenzhen University, Shenzhen, P.R. China

Asha Sharma Department of Physics, University of Jammu, Jammu, India

Surbhi Sharma School of Chemistry and Biochemistry, Affiliate Faculty—TIET-Virginia Tech Center of Excellence in Emerging Materials, Thapar Institute of Engineering & Technology, Patiala, India

Shan-Yi Shen Department of Environmental Engineering and Management, Chaoyang University of Technology, Taichung City, Taiwan

Mohd. Shkir Advanced Functional Materials & Optoelectronics Laboratory (AFMOL), Department of Physics, Faculty of Science, King Khalid University, Abha, Saudi Arabia

Anoop Singh Department of Physics, University of Jammu, Jammu, India

H. Tiznado Centro de Nanociencias y Nanotecnología, Universidad Nacional Autónoma de México, Ensenada BC, México

Ghulam Yasin Institute for Advanced Study, Shenzhen University, Shenzhen, P.R. China; College of Physics and Optoelectronic Engineering, Shenzhen University, Shenzhen, P.R. China

Preface

As the dominant material of the "Silicon Age" ("Digital Age"), silicon has been widely used in semiconductor electronics/optoelectronics. Recently, with the development of nanoscience and nanotechnology, electronic/optoelectronic devices have become smarter, stronger, more durable, and smaller with lower power consumption. In this regard, nanostructured semiconductors are at the heart of these advanced nanodevices.

The hybridization of silicon nanoparticles with other semiconductor/metal/metal oxide nanoparticles may show superior features, compared to the lone individual nanoparticles. These new hybrid approaches can overcome the limits of single components, improve properties/achieve new properties, and achieve multiple functionalities for single nanoparticles. In addition, these nanohybrids can be synthesized in situ directly on the silicon wafers, producing the Lab-on-a-chip platforms.

For example, silicon nanohybrid-based SERS chips (with silver/gold nanoparticles) could serve as a multifunctional platform, not only for broad-range/high sensitive/simultaneous quantification of heavy metals in the environment but also can enable simultaneous capture/discrimination/inactivation of bacteria. In the case of the anode materials in lithium-ion batteries, hybridization of silicon nanoparticles with nanocarbon (CNTs/graphene) could enhance significantly their cycling performance and rate capabilities, with higher reversible lithium storage capacities. Besides, the silicon/iron oxide hybrid nanoparticles possess excellent fluorescence, superparamagnetism, and biocompatibility that can be used effectively for the diagnostic imaging system in vivo. Similarly, gold–silicon nanohybrids could be used as highly efficient near-infrared hyperthermia agents for cancer cell destruction.

These interesting findings of these nanohybrids open up promising applications with the synergistic effects of individual nanoparticles.

In the first part of the book, the chapters focus on the rational design, controlled synthesis (Chapter 2), and in-depth understanding of structure-property relationships, especially for silicon-carbon (Chapter 3) and silicon-ZnO (Chapters 4 and 5). The second part emphasizes the promising applications of these hybrid nanoparticles in energy storage/conversion

(Chapters 6-9), catalysis (Chapter 10), sensors (Chapters 11 and 12), biomedicine (Chapter 13), hyperthermia treatment (Chapter 14), and oil/gas applications (Chapter 15).

Tuan Anh Nguyen
Institute for Tropical Technology, Vietnam Academy of Science and Technology, Hanoi, Vietnam

Silicon-based hybrid nanoparticles: An introduction

Surbhi Sharma and Soumen Basu
School of Chemistry and Biochemistry, Affiliate Faculty—TIET-Virginia Tech Center of Excellence in Emerging Materials, Thapar Institute of Engineering & Technology, Patiala, India

1.1 Introduction

Silicon (Si) is the second most abundant element on earth and is readily accessible. Moreover, it is potentially inexpensive, nontoxic, and eco-friendly (Zuo, Zhu, Müller-Buschbaum, & Cheng, 2017). Si nanoparticles (NPs) demonstrate supremely effectual photoemission in the visible and near IR ranges (Kutrovskaya et al., 2017). They have gained immense interest owing to their versatile applications in comparison to the bulk material, due to their physical and chemical properties like photostability (not only in aqueous medium but also in ambient air), high surface area, good biocompatibility, and tunable optical properties (Mariani et al., 2019; Sato et al., 2011). Porous Si NPs have been known to be biodegradable as they are degraded easily into a nontoxic compound, orthosilicic acid (Peng et al., 2014; Xia et al., 2019). The low cytotoxicities of Si NPs are enticing even after the illumination of light (Sato et al., 2011). Nonetheless, as compared to lone Si NPs, hybrids with other semiconductors/metals/nonmetals/metal-oxides display superior properties and, hence, are important to be explored. Doping of Si NPs with different heteroatoms or metal ions can give rise to some new and appealing features. The Si-based hybrid materials have huge applications in magnetic resonance imaging (MRI), drug delivery, sensing, photovoltaic applications, diodes, battery electrodes, and catalysis.

The chief benefit of silicon-based hybrid NPs is an excellent performance as sensors in comparison to the individual components regarding dual-mode detection, enhanced signal stability, and higher sensitivity (Arshavsky-Graham, Massad-Ivanir, Segal, & Weiss, 2019). The doping of Si NPs with boron and phosphorus has been reported to provide tunable IR plasmon resonances

(Kramer, Schramke, & Kortshagen, 2015; Zhou et al., 2015). Phosphorus-doped Si nanocrystals display a localized surface plasmon resonance (LSPR) which rapidly disappeared on oxidation. Boron-doped Si nanocrystals manifest no LSPR as-produced, however, they developed plasmonic response after oxidation (Kramer, Schramke, & Kortshagen, 2015). Metals doped in Si NPs produce a strong magnetic resonance signal which is substantial for the doped Si NPs to act as a dual-modality clinic probe. This probe finds application in highly sensitive fluorescence imaging as well as MRI (Li, Wang, He, Li, & Zhang, 2018). The gadolinium (Gd)-doped Si NPs fabricated by the hydrothermal route are quite stable, water-soluble, and exhibit low biotoxicity. The hybrid NPs have enormous potential in the domain of biomedical application, offering good spatial resolution besides high sensitivity (Li, Wang, He, Li, & Zhang, 2018).

For drug delivery applications, the large surface area and extensively flexible surface chemistries of porous Si carriers facilitate better drug-loading capabilities and controlled release of the bioactive agents (Zhu et al., 2016). The porous Si particles do not lead to any inflammatory responses in animal models. The hybridization of Si and polymers is known to augment the properties of porous Si materials as drug-delivery platforms (Zhang, Esser, Vasani, Thissen, & Voelcker, 2019). A versatile nanohybrid based on acetylated dextran, Si NPs, and gold (Au) NPs manifested huge prospects for the diagnosis as well as treatment of acute liver failure (Liu et al., 2018). Encapsulation of alkyl terminated Si nanocrystals within this amphiphilic polymer shell (poly(maleic anhydride)) renders the hybrid particles water-dispersible, luminescent, and stable over a wide pH range. These are an appropriate choice for in vivo biological imaging applications (Hessel et al., 2010). The Si-Au nanohybrids have been utilized as hyperthermia agents for the destruction of cancer cells in vitro, and thermally ablate tumors in vivo (Zhu et al., 2016). Au NPs trapped in the pores of porous Si result in an improved photothermal effect owing to dipole–dipole coupling in addition, to extended retention time at target sites (Xia et al., 2019). In addition, to drug delivery, Au-Si hybrids are significant in optical integrated circuits. Covering the Si NPs manifests an enhanced near-field emission due to the nanoantenna action of Au NPs (Kutrovskaya et al., 2017). The magnetic hybrid nanomaterials based on Si and magnetic iron oxides magnetite (Fe_3O_4) and maghemite (γ-Fe_2O_3) are significant in MRI diagnosis and targeted drug delivery systems. These magnetic hybrid NPs are quite crucial for contrast agents in vivo (MRI monitoring; fluorescence optical imaging) (Sato et al., 2011). The hybrid NPs with superparamagnetism of 3.0 nm can magnetically function only after the application of an external magnetic field. The redispersion occurs immediately as the magnetic field is removed (Sato et al., 2011).

Moreover, Si nanohybrids also play a key role in surface-enhanced Raman scattering (SERS), a technique that is used for ultrasensitive analyses as well as detection (Meng et al., 2018). The Si nanohybrids display higher values of enhancement factor (EF) because of the effective coupling of highly efficient SERS hotspots through Si substrates. It considerably encourages the improvement of EF values in addition, to the detection limit. The metal NPs tend to be properly immobilized on the Si substrate and thereby escaping the aggregation of free NPs (Jiang et al., 2012; Wang et al., 2014). The high values of EF could be accredited to a combined electromagnetic enhancement as well as a charge transfer mechanism (Harraz et al., 2015). The Si hybrid NPs (Ag/Au) act as a high-quality probe for the detection of chemical and biological entities promising specificity, reproducibility, and reliability (Zhang et al., 2016).

Silicon−metal oxide hybrid nanostructures have been utilized for diverse applications. Hybrid Si/MoO_3 nanostructures have been utilized for the sensing of CO_2 (Thomas et al., 2021). Porous Si ameliorates the concentration limit besides the quality of gas detection to a large degree due to the boosted surface reactions on the porous Si having a large surface to volume ratio. Amino-capped Si quantum dots find application as 2,4,6-trinitrotoluene (TNT) detectors (Ban, Zheng, & Zhang, 2015). Functionalization of Si nanowires with conducting polymer polypyrrole boosts NH_3-sensing (Qin, Cui, Zhang, & Liu, 2018) and NO_2-sensing performance at room temperature (Qin, Cui, Wen, & Bai, 2019). Silica-based composites also find huge application in the field of catalysis and environmental remediation (Sharma & Basu, 2020; Sharma & Basu, 2021; Wu et al., 2015).

The Si NPs-based hybrids have been investigated most extensively for lithium-ion batteries (LIB) that have a substantial role in the development of electric vehicles as the latter hold a huge potential concerning climate protection goals in the transport sector (Kwon et al., 2020). It is imperative to explore alternative energy sources for addressing climate change problems since the relentless consumption of conventional resources of energy has caused an upsurge in energy demand (Sharma, Basu, Shetti, & Aminabhavi, 2020; Sharma, Basu, Shetti, & Kamali, et al., 2020) and grievous environmental challenges (Sharma, Kundu, Basu, Shetti, & Aminabhavi, 2020). In addition, to electrical vehicles, LIB has gained immense interest in consumer electronics and energy storage. LIB comprises the anode, cathode, electrolyte, and a separator (Yang et al., 2020).

Silicon has come across as a capable anode contender for the replacement of commercially used graphite attributing to features, that is theoretical capacity of 3579 mAh g^{-1} in a fully alloyed form of $Li_{15}Si_4$, comparatively low redox voltage (0.5 V vs Li/Li$^+$), and environmental friendliness (Li, Hwang, & Sun, 2019). Nevertheless, significant volume expansion upon lithiation and consequent poor cyclic stability is a noteworthy shortcoming of Si anodes. In this regard, Si-C hybrid nanomaterials are quite promising as there is a combined high electron conductivity plus the stability of carbon in addition, to the high specific capacity of Si (Shen et al., 2018). The combination of several allotropes of carbon, such as graphite, graphene, and carbon nanotubes (CNTs) attains higher battery performance. The carbon layers encasing Si NPs supply various electrical contact points as well as reduce the electron transport paths. This favors the enhancement of electrical conductivity of the anodes by assuaging electrical insulation (Yang et al., 2020). Silicon nanohybrids with metals such as Ag (Yoo, Lee, Ko, & Park, 2013), Co (Jin, Zhu, Lu, Liu, & Zhu, 2017), Cu (Kwon, Kim, Kim, & Hong, 2019), etc., have a better electrical conductivity as well as mechanical strength as compared to Si-C nanohybrids. Metals assist the acceleration in the transport of electrons providing extra conductive paths inside the Si NPs, which eventually facilitates reaction kinetics during the cycles. Additionally, the combination of metal oxides like TiO_2 (Yang et al., 2017), CuO (Rangasamy, Hwang, & Choi, 2014), Cu_2O (Rangasamy, Hwang, & Choi, 2014), MoO_3 (Martinez-Garcia et al., 2015), Al_2O_3 (Li et al., 2014), etc. with Si NPs help in improving the electrochemical activity of the battery. The Si@Fe_2O_3/C NPs with a core−shell structure can also be an effective anode material. The activated Si hybrid NPs are constructed by using Si@MIL-88-Fe as a precursor (Fig. 1.1). The SEM and TEM images of the Si@Fe_2O_3/C are presented in Fig. 1.2. The nanocomposite performs better in comparison to lone Si NPs, manifesting an improved cycle life and specific capacity (Wang et al., 2019).

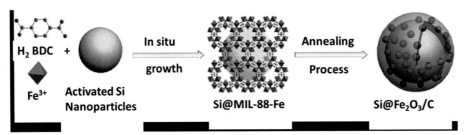

Figure 1.1 Representation of the synthesis of Si@Fe_2O_3/C nanocomposite. Reproduced from Wang, Q., Guo, C., He, J., Yang, S., Liu, Z., Wang, Q. (2019). Fe_2O_3/C-modified Si nanoparticles as anode material for high-performance lithium-ion batteries. *Journal of Alloys and Compounds*; copyright Elsevier. *795*, 284−290. https://doi.org/10.1016/j.jallcom.2019.050.038.

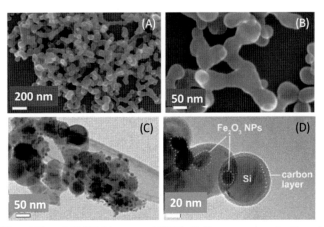

Figure 1.2 (A and B) SEM and (C and D) TEM images of Si@Fe$_2$O$_3$/C nanocomposite. Reproduced from Wang, Q., Guo, C., He, J., Yang, S., Liu, Z., Wang, Q. (2019). Fe$_2$O$_3$/C-modified Si nanoparticles as anode material for high-performance lithium-ion batteries. *Journal of Alloys and Compounds, 795*, 284−290; copyright Elsevier.

Similarly, nanohybrids of Si NPs with conducting polymers, namely, polyaniline (Wu et al., 2013), polypyrrole (Luo et al., 2015), and polyethylene dioxythiophene (McGraw et al., 2016), offer better electrical conductivity as well as mechanical flexibility which helps in enhancing the electrochemical performance. Thus silicon-based hybrid NPs are a suitable choice for anode materials in LIB.

Therefore the silicon-based hybrid materials hold paramount importance in biological applications, magnetic resonance imaging, sensing, photovoltaic applications, battery electrodes, diodes, and catalysis. The presence of silicon in profuse amounts along with the versatility of these nanohybrids guarantees that in the years to come, silicon-based nanomaterials will continue to play an important role in day-to-day life.

References

Arshavsky-Graham, S., Massad-Ivanir, N., Segal, E., & Weiss, S. (2019). Porous silicon-based photonic biosensors: Current status and emerging applications. *Analytic Chemistry, 91*, 441−467. Available from https://doi.org/10.1021/acs.analchem.8b05028.

Ban, R., Zheng, F., & Zhang, J. (2015). A highly sensitive fluorescence assay for 2,4,6-trinitrotoluene using amine-capped silicon quantum dots as a probe. *Analytic Methods, 7*, 1732−1737. Available from https://doi.org/10.1039/C4AY02729A.

Harraz, F. A., Ismail, A. A., Bouzid, H., Al-Sayari, S. A., Al-Hajry, A., & Al-Assiri, M. S. (2015). Surface-enhanced Raman scattering (SERS)-active substrates from silver plated-porous silicon for detection of crystal violet. *Applied*

Surface Science, 331, 241−247. Available from https://doi.org/10.1016/j.apsusc.2015.010.042.

Hessel, C. M., Rasch, M. R., Hueso, J. L., Goodfellow, B. W., Akhavan, V. A., Puvanakrishnan, P., . . . Korgel, B. A. (2010). Alkyl passivation and amphiphilic polymer coating of silicon nanocrystals for diagnostic imaging. *Small, 6*, 2026−2034. Available from https://doi.org/10.1002/smll.201000825.

Jiang, Z. Y., Jiang, X. X., Su, S., Wei, X. P., Lee, S. T., & He, Y. (2012). Silicon-based reproducible and active surface-enhanced Raman scattering substrates for sensitive, specific, and multiplex DNA detection. *Applied Physics Letters, 100*, 203104. Available from https://doi.org/10.1063/1.3701731.

Jin, Y., Zhu, B., Lu, Z., Liu, N., & Zhu, J. (2017). Challenges and recent progress in the development of Si anodes for lithium-ion battery. *Advanced Energy Materials, 7*, 1700715. Available from https://doi.org/10.1002/aenm.201700715.

Kramer, N. J., Schramke, K. S., & Kortshagen, U. R. (2015). Plasmonic properties of silicon nanocrystals doped with boron and phosphorus. *Nano Letters, 15*, 5597−5603. Available from https://doi.org/10.1021/acs.nanolett.5b02287.

Kutrovskaya, S., Arakelian, S., Kucherik, A., Osipov, A., Evlyukhin, A., & Kavokin, A. V. (2017). The synthesis of hybrid gold-silicon nano particles in a liquid. *Scientific Reports., 7*, 10284. Available from https://doi.org/10.1038/s41598-017-09634-y.

Kwon, H. J., Hwang, J.-Y., Shin, H.-J., Jeong, M.-G., Chung, K. Y., Sun, Y.-K., & Jung, H.-G. (2020). Nano/microstructured silicon−carbon hybrid composite particles fabricated with corn starch biowaste as anode materials for li-ion batteries. *Nano Letters, 20*, 625−635. Available from https://doi.org/10.1021/acs.nanolett.9b04395.

Kwon, S., Kim, K.-H., Kim, W.-S., & Hong, S.-H. (2019). Mesoporous Si−Cu nanocomposite anode for a lithium ion battery produced by magnesiothermic reduction and electroless deposition. *Nanotechnology, 30*, 405401. Available from https://doi.org/10.1088/1361-6528/ab2dd2.

Li, P., Hwang, J. Y., & Sun, Y. K. (2019). Nano/microstructured silicon-graphite composite anode for high-energy-density li-ion battery. *ACS Nano, 13*, 2624−2633. Available from https://doi.org/10.1021/acsnano.9b00169.

Li, S., Wang, F., He, X.-W., Li, W.-Y., & Zhang, Y.-K. (2018). One-pot hydrothermal preparation of gadolinium-doped silicon nanoparticles as a dual-modal probe for multicolor fluorescence and magnetic resonance imaging. *Journal of Materials Chemistry B., 6*, 3358−3365. Available from https://doi.org/10.1039/C8TB00415C.

Li, Y., Sun, Y., Xu, G., Lu, Y., Zhang, S., Xue, L., . . . Zhang, X. (2014). Tuning electrochemical performance of Si-based anodes for lithium-ion batteries by employing atomic layer deposition alumina coating. *Journal of Materials Chemistry A*. Available from https://doi.org/10.1039/c4ta01562b.

Liu, Z., Li, Y., Li, W., Xiao, C., Liu, D., Dong, C., . . . Santos, H. A. (2018). Multifunctional nanohybrid based on porous silicon nanoparticles, gold nanoparticles, and acetalated Dextran for liver regeneration and acute liver failure theranostics. *Advanced Materials, 30*, 1703393. Available from https://doi.org/10.1002/adma.201703393.

Luo, L., Zhao, P., Yang, H., Liu, B., Zhang, J.-G., Cui, Y., . . . Wang, C.-M. (2015). Surface coating constraint induced self-discharging of silicon nanoparticles as anodes for lithium ion batteries. *Nano Letters, 15*, 7016−7022. Available from https://doi.org/10.1021/acs.nanolett.5b03047.

Mariani, S., Paghi, A., La Mattina, A. A., Debrassi, A., Dähne, L., & Barillaro, G. (2019). Decoration of porous silicon with gold nanoparticles via layer-by-layer nanoassembly for interferometric and hybrid photonic/plasmonic (bio) sensing. *ACS Applied Materials & Interfaces.*, *11*, 43731−43740. Available from https://doi.org/10.1021/acsami.9b15737.

Martinez-Garcia, A., Thapa, A. K., Dharmadasa, R., Nguyen, T. Q., Jasinski, J., Druffel, T. L., & Sunkara, M. K. (2015). High rate and durable, binder free anode based on silicon loaded MoO_3 nanoplatelets. *Scientific Reports*, *5*, 10530. Available from https://doi.org/10.1038/srep10530.

McGraw, M., Kolla, P., Yao, B., Cook, R., Quiao, Q., Wu, J., & Smirnova, A. (2016). One-step solid-state in-situ thermal polymerization of silicon-PEDOT nanocomposites for the application in lithium-ion battery anodes. *Polymer (Guildf)*, *99*, 488−495. Available from https://doi.org/10.1016/j.polymer.2016.050.044.

Meng, X., Wang, H., Chen, N., Ding, P., Shi, H., Zhai, X., ... He, Y. (2018). A graphene−silver nanoparticle−silicon sandwich SERS chip for quantitative detection of molecules and capture, discrimination, and inactivation of bacteria. *Analytic Chemistry*, *90*, 5646−5653. Available from https://doi.org/10.1021/acs.analchem.7b05139.

Peng, F., Su, Y., Zhong, Y., Fan, C., Lee, S.-T., & He, Y. (2014). Silicon nanomaterials platform for bioimaging, biosensing, and cancer therapy. *Accounts of Chemical Research*, *47*, 612−623. Available from https://doi.org/10.1021/ar400221g.

Qin, Y., Cui, Z., Wen, Z., & Bai, Y. (2019). Highly sensitive NO_2 sensors based on core-shell array of silicon nanowires/polypyrrole and new insight into gas sensing mechanism of organic/inorganic hetero-contact. *Polymer Composites*, *40*, 3275−3284. Available from https://doi.org/10.1002/pc.25183.

Qin, Y., Cui, Z., Zhang, T., & Liu, D. (2018). Polypyrrole shell (nanoparticles)-functionalized silicon nanowires array with enhanced NH_3-sensing response. *Sensors and Actuators B: Chemical*, *258*, 246−254. Available from https://doi.org/10.1016/j.snb.2017.110.089.

Rangasamy, B., Hwang, J. Y., & Choi, W. (2014). Multi layered Si−CuO quantum dots wrapped by graphene for high-performance anode material in lithium-ion battery. *Carbon N. Y*, *77*, 1065−1072. Available from https://doi.org/10.1016/j.carbon.2014.060.022.

Sato, K., Yokosuka, S., Takigami, Y., Hirakuri, K., Fujioka, K., Manome, Y., ... Fukata, N. (2011). Size-tunable silicon/iron oxide hybrid nanoparticles with fluorescence, superparamagnetism, and biocompatibility. *Journal of the American Chemical Society*, *133*, 18626−18633. Available from https://doi.org/10.1021/ja202466m.

Sharma, S., & Basu, S. (2020). Highly reusable visible light active hierarchical porous WO_3/SiO_2 monolith in centimeter length scale for enhanced photocatalytic degradation of toxic pollutants. *Separation and Purification Technology*, *231*, 115916. Available from https://doi.org/10.1016/j.seppur.2019.115916.

Sharma, S., & Basu, S. (2021). Fabrication of centimeter-sized Sb_2S_3/SiO_2 monolithic mimosa pudica nanoflowers for remediation of hazardous pollutants from industrial wastewater. *Journal of Cleaner Production*, *280*, 124525. Available from https://doi.org/10.1016/j.jclepro.2020.124525.

Sharma, S., Basu, S., Shetti, N. P., & Aminabhavi, T. M. (2020). Waste-to-energy nexus for circular economy and environmental protection: Recent trends in hydrogen energy. *Science of the Total Environment*, *713*, 136633. Available from https://doi.org/10.1016/j.scitotenv.2020.136633.

Sharma, S., Basu, S., Shetti, N. P., Kamali, M., Walvekar, P., & Aminabhavi, T. M. (2020). Waste-to-energy nexus: A sustainable development. *Environmental*

Pollution, 267, 115501. Available from https://doi.org/10.1016/j.envpol.2020.115501.

Sharma, S., Kundu, A., Basu, S., Shetti, N. P., & Aminabhavi, T. M. (2020). Sustainable environmental management and related biofuel technologies. *Journal of Environmental Management, 273*, 111096. Available from https://doi.org/10.1016/j.jenvman.2020.111096.

Shen, X., Tian, Z., Fan, R., Shao, L., Zhang, D., Cao, G., ... Bai, Y. (2018). Research progress on silicon/carbon composite anode materials for lithium-ion battery. *Journal of Energy Chemistry, 27*, 1067–1090. Available from https://doi.org/10.1016/j.jechem.2017.120.012.

Thomas, T., Kumar, Y., Ramos Ramón, J. A., Agarwal, V., Sepúlveda Guzmán, S., Pushpan, R. R. S., ... Sanal, K. C. (2021). Porous silicon/α-MoO$_3$ nanohybrid based fast and highly sensitive CO$_2$ gas sensors. *Vacuum, 184*, 109983. Available from https://doi.org/10.1016/j.vacuum.2020.109983.

Wang, Q., Guo, C., He, J., Yang, S., Liu, Z., & Wang, Q. (2019). Fe$_2$O$_3$/C-modified Si nanoparticles as anode material for high-performance lithium-ion batteries. *Journal of Alloys and Compounds, 795*, 284–290. Available from https://doi.org/10.1016/j.jallcom.2019.050.038.

Wang, S.-Y., Jiang, X.-X., Xu, T.-T., Wei, X.-P., Lee, S.-T., & He, Y. (2014). Reactive ion etching-assisted surface-enhanced Raman scattering measurements on the single nanoparticle level. *Applied Physics Letters, 104*, 243104. Available from https://doi.org/10.1063/1.4884060.

Wu, G., Li, J., Fang, Z., Lan, L., Wang, R., Lin, T., ... Chen, Y. (2015). Effectively enhance catalytic performance by adjusting pH during the synthesis of active components over FeVO$_4$/TiO$_2$-WO$_3$-SiO$_2$ monolith catalysts. *Chemical Engineering Journal, 271*, 1–13. Available from https://doi.org/10.1016/j.cej.2015.020.012.

Wu, H., Yu, G., Pan, L., Liu, N., McDowell, M. T., Bao, Z., & Cui, Y. (2013). Stable li-ion battery anodes by in-situ polymerization of conducting hydrogel to conformally coat silicon nanoparticles. *Nature Communications, 4*, 1943. Available from https://doi.org/10.1038/ncomms2941.

Xia, B., Zhang, W., Tong, H., Li, J., Chen, Z., & Shi, J. (2019). Multifunctional chitosan/porous silicon@Au nanocomposite hydrogels for long-term and repeatedly localized combinatorial therapy of cancer via a single injection. *ACS Biomaterials Science & Engineering, 5*, 1857–1867. Available from https://doi.org/10.1021/acsbiomaterials.8b01533.

Yang, J., Wang, Y., Li, W., Wang, L., Fan, Y., Jiang, W., ... Zhao, D. (2017). Amorphous TiO$_2$ shells: A vital elastic buffering layer on silicon nanoparticles for high-performance and safe lithium storage. *Advanced Materials, 29*, 1700523. Available from https://doi.org/10.1002/adma.201700523.

Yang, Y., Yuan, W., Kang, W., Ye, Y., Yuan, Y., Qiu, Z., ... Tang, Y. (2020). Silicon-nanoparticle-based composites for advanced lithium-ion battery anodes. *Nanoscale, 12*, 7461–7484. Available from https://doi.org/10.1039/C9NR10652A.

Yoo, S., Lee, J.-I., Ko, S., & Park, S. (2013). Highly dispersive and electrically conductive silver-coated Si anodes synthesized via a simple chemical reduction process. *Nano Energy, 2*, 1271–1278. Available from https://doi.org/10.1016/j.nanoen.2013.060.006.

Zhang, C., Jiang, S. Z., Yang, C., Li, C. H., Huo, Y. Y., Liu, X. Y., ... Man, B. Y. (2016). Gold@silver bimetal nanoparticles/pyramidal silicon 3D substrate with high reproducibility for high-performance SERS. *Scientific Reports, 6*, 25243. Available from https://doi.org/10.1038/srep25243.

Zhang, D.-X., Esser, L., Vasani, R. B., Thissen, H., & Voelcker, N. H. (2019). Porous silicon nanomaterials: Recent advances in surface engineering for controlled drug-delivery applications. *Nanomedicine, 14,* 3213–3230. Available from https://doi.org/10.2217/nnm-2019-0167.

Zhou, S., Pi, X., Ni, Z., Ding, Y., Jiang, Y., Jin, C., ... Nozaki, T. (2015). Comparative study on the localized surface plasmon resonance of boron- and phosphorus-doped silicon nanocrystals. *ACS Nano, 9,* 378–386. Available from https://doi.org/10.1021/nn505416r.

Zhu, G., Liu, J.-T., Wang, Y., Zhang, D., Guo, Y., Tasciotti, E., ... Liu, X. (2016). In Situ reductive synthesis of structural supported gold nanorods in porous silicon particles for multifunctional nanovectors. *ACS Applied Materials & Interfaces, 8,* 11881–11891. Available from https://doi.org/10.1021/acsami.6b03008.

Zuo, X., Zhu, J., Müller-Buschbaum, P., & Cheng, Y.-J. (2017). Silicon based lithium-ion battery anodes: A chronicle perspective review. *Nano Energy, 31,* 113–143. Available from https://doi.org/10.1016/j.nanoen.2016.110.013.

Synthesis and characterization of silicon-based hybrid nanoparticles

Gautam M. Patel[1], Gaurang J. Bhatt[2] and Pradeep T. Deota[2]

[1]Department of Chemistry, School of Sciences, ITM (SLS) Baroda University, Vadodara, India [2]Applied Chemistry Department, Faculty of Technology & Engineering, The MS University of Baroda, Vadodara, India

2.1 Introduction

One of the most active fields of science today is the area of nanostructured materials such as nanowires, nanorods, nanospheres, and nanobelts (Joshi et al., 2018; Patel, Pillai, Bhatt, & Mohammad, 2020; Patel, Shah, Bhatt, & Deota, 2021; Patel, Vora, & Pillai, 2019; Wu et al., 2020). Specifically, silicon nanoparticles (SiNPs) are a booming and fascinating area of science with major technical implications. SiNPs demonstrate noncytotoxic properties (Park et al., 2009; Sato et al., 2009) and persistent multicolor luminescence of blue light with a reduction of the mean diameter at room temperature (Sato, Kishimoto, & Hirakuri, 2007). The latest interest came nearly a decade after Canham's exciting 1990 discovery (Canham, 1990). SiNPs have a number of significant advantages, such as low toxicity, emissions from a single material across the entire visible spectrum, and silicon's ability to form covalent carbon bonds and thus combine inorganic and organic components at the molecular scale. An additional feature expanding their usage manifold as an extraordinary support material is the possibility of functionalizing both the external and internal pore surfaces with several organic moieties. A number of investigations were conducted to change the surface of SiNPs using alkyl chains in order to make Silicon Nanocrystals (SiNCs) stable in different solutions (Anderson et al., 2012; Báñez-Redín, Joshi, & do Nascimento, 2020; Dasog & Veinot, 2012; Dasog et al., 2013; Joshi et al., 2016; Li & Ruckenstein, 2004; Li, Swihart, & Ruckenstein, 2004; Panthani et al., 2012; Singh et al., 2013). Further practical changes to the SiNPs are also desirable for applications such as optics, sensors, biology, medicine, membranes, and microelectronics. Nanocomposites generated by SiNCs/polymer blends have lately received a lot of attention due to the improvement of their mechanical and thermal properties (Bao, Yang, Shi, & Ma, 2014; Deubel, Steenackers, Garrido, Stutzmann, & Jordan, 2013; Liu, Nayfeh, & Yau, 2011; Sato et al., 2010). In addition, long alkyl chains passivate the NC surface to maintain colloidal stability and avoid surface oxidation of SiNCs (Hua, Swihart, & Ruckenstein, 2005; Mangolini & Kortshagen, 2007; Mastronardi et al., 2012). These surface modifiers facilitate the formation of nanomaterials with a wide range of significant features. In this chapter we will confine our discussion of their structure, synthesis, and characterization to a broad range of silicon-based hybrid nanoparticles (Si-HNPs). We will also provide an outlook for some potential clinical applications of Si-HNPs.

2.2 Synthesis

Several processes for the synthesis of silicon-based hybrid nanoparticles have been developed. These include the fabrication process, sol−gel process, polymer graphing, encapsulation process, drug loading, colloidal system, solution blending, gas-phase synthesis, and combination with magnetic NPs. We will discuss the synthesis and characterization with some applications of Si-HNPs.

2.2.1 Fabrication process

In recent years, the design and manufacture of multicomponent heterostructured NPs have drawn substantial scientific interest (Cagnani, Joshi, & Shimizu, 2019; Malik et al., 2018; Mishra, Malik, Tomer, & Joshi, 2020). A new hybrid device can be manufactured at room temperature which might possess cost effective silicon−organic hybrid moieties. Using this method, individuals illustrated a hybrid memory system employing AuNPs as charging storage elements deposited by a self-assembly chemical process. The NPs are isolated by the Langmuir−Blodget technique at room temperature from the silicone channel (Kolliopoulou et al., 2003). The Al−insulator−semiconductor memory structures were fabricated with AuNPs as the charging storage elements. Using spin-coating thin organic insulating layers of 40-nm thicknesses were produced. The organic insulator was then deposited, Al top contacts were removed by heating and Al/SiO$_2$/Si and Al/isolator/SiO$_2$/Si control devices were prepared (Mabrook, Pearson, Kolb, Zeze, & Petty, 2008). TiO_2 NPs were integrated into vertical silicon nanowires (SiNWs) arrays to create hybrid devices due to the dielectric properties of TiO_2 nanostructures. Using the coprecipitation method, TiO_2 NPs were prepared and SiNWs formed by metal assisted electronless chemical etching. Then the TiO_2 NPs precipitates were poured on the SiNWs' surface. After that, 50 nm of chromium followed by 300 nm of Au layer was fabricated on both sides of the substrate to create electrical contacts at the top and bottom. These hybrid devices have great potential for use as in transistors, photodetectors, solar cells, memory devices, inverters, and bistable optical switches (Rasool, Rafiq, Ahmad, Imran, & Hasan, 2012). Confining SiNPs into porous CNF (Si/PCNF) increased the electrochemical efficiency of silicon-based electrodes. The Si/PCNF hybrid was produced using a stable suspension of SiNPs in a solution of polyacrylonitrile/DMF through simple electrospinning, successive calcination, and then etching with 10% hydrogen fluoride to finish the fabrication of Si/PCNF. The hybrid NPs were designed and prepared for high-performance catalysts, electrode materials and sensors for Li/Na-based batteries and supercapacitors (Zhou, Wan, & Guo, 2013). To synthesize a lightweight, high-performance metal matrix hybrid nanocomposite, some researchers presented the tendency of combining the nano silicon carbide (SiC) with aluminum alloy 6061. In this fabrication method, the novel ultrasonic cavitation process was used. At first, alloy melt was mechanically stirred to obtain the homogeneous mixture, after which the reinforcement mixture of SiC with nano boron carbide (BC) was slowly applied in the melt to achieve better dispersion. The fabricated SiC and BC nanoparticles increase the hardness that give the soft matrix its inherent hardness properties (Poovazhagan, Kalaichelvan, Rajadurai, & Senthilvelan, 2013). It has been reported that ternary biocompatible multicore magnetoplasmonic nanoparticles are easily synthesized in single phase. NPs were produced simultaneously by sputtering high-purity targets of Ag, Fe, and Silicon using an upgraded magnetron-sputter inert-gas condensation device. Because of the large free positive energy of Fe and Ag mix, segregation of the components which form a core or dumbbell framework is required due to their strength, which is not enough to induce their absolute coalescence. Si fills the Fe-Ag NPs surface resulting in a core structure. This manufacture of ternary NPs blends plasmonic and magnetic functionality, including strong stability of colloidal suspension, making it desirable to be used in the field of biomedicine and nanotechnology (Benelmekki et al., 2014). The Si−Ag system is fascinating since Si is photoluminescent, whereas Ag is plasmonic and, quite notably, since Si and Ag are immiscible in solid state, they form eutectic systems. Studies explored the coating process of freshly nucleated SiNPs with Ag atoms with MD computer simulations in this guided preparation for multicomponent heterostructured metal-semiconductor NPs. The amount of Ag nanoclusters that decorate the surface of SiNPs is altered by changing the energy provided to the sputter source Ag magnetron. Ag first land as discrete molecules mostly on surface of SiNPs, nucleating to form nanoclusters. Ag can be anticipated to segregate from the metastable silicide at room temperature toward the end of the simulation (Cassidy et al., 2013). Silver nanoparticles (AgNPs) provide a possible solution to boost the power conversion efficiency (PCE) of organic solar cells, demonstrating effective local field improvements across the AgNPs that could improve light dispersion and absorption in organic film. An effective way for improving the PCE of SiNW/poly(3-hexylthiophene) (P3HT) hybrid solar cells by decorating AgNPs on the SiNW surface has been reported. Metal-assisted chemical etching has prepared vertically arranged SiNW arrays. The synthesized SiNW arrays were then submerged in a mixture of HF-AgNO$_3$ for depositing AgNPs on SiNWs. Using spin-coating with a weight ratio of 1:1, the polymer, P3HT:PCBM, was deposited onto SiNWs. After the manufacture of the hybrid solar cells, the entire substrate is baked for 20 min at 110°C in nitrogen. This hybrid tallied up to 58% of the short-circuit current (Liu, Qu, Zhang, Tan, & Wang, 2013). An amorphous SiNPs backbone of graphene nanocomposite (aSBG) anode was formed in which aSiNPs possess elastic behavior with relaxation of strains and strong diffusivity of Li ions. This hybrid design offers many additional advantages, such as accommodating volume variations without losing SiNPs from the graphene matrix, a large surface reaction area among the active components, strong electrical conductivity of a total electrode the electrolyte due to nanometric effects, and an increase of the Si mass loading in the composite. The nonaggregated 3D porous graphene synthesized by freezing plays a key function in securing a simple pathway to SiH$_4$ gas penetration during the fabrication process. An ideal electrode structure with a clear distribution of aSiNPs islands was offered by aSBG hybrid on both sides of graphene sheets. This novel aSBG fabrication approach tackled the boosted kinetics and cycling stability problems of silicon anodes (Ko, Chae, Jeong, Oh, & Cho, 2014). For binder-free electrodes, a simple fabrication of SiNPs/porous carbon nanofiber (Si/PCNF) hybrids has been reported for the conductive carbon nanofiber (CNF) matrix and porous structure introduction. Precursor electrospinning solutions were prepared by dispersing adequate amounts of silicon to achieve homogeneous suspension, polyacrylonitrile (PAN), and template (F127) (1:2:2) in DMF.

The Si/PCNF displays a high reversible capacity, a rate capability in the absence of binders and additives, stable cycling performance, and excellent electrochemical performance (Wang, Song, Wang, & Fan, 2015). Another process can fabricate simultaneously two forms of graphene-coated porous alloy anode. Homogeneous Si (yin) and tin (yang) blend along with graphene layers formed a yin-yang hybrid using a Chinese traditional process. Porous tin–graphene composite can alleviate the stress produced by etching to eliminate the Si or tin component due to the void cavity in the structure. This kind of cost-effective hybrid is beneficial for thermoelectric and optoelectronic uses (Jin et al., 2017). As per schematic diagram shown in Figure 2.1, aggregated SiNPs/Ionic liquid hybrid can be synthesized (Tchalala et al., 2020) (Fig. 2.1).

2.2.2 Sol–gel process

The sol–gel method has been used since the 1970s to depose in situ inorganic minerals with the organic polymer matrix (Landry, Coltrain, & Brady, 1992; Noell, Wikes, Mohanty, & Mcgrath, 1990).

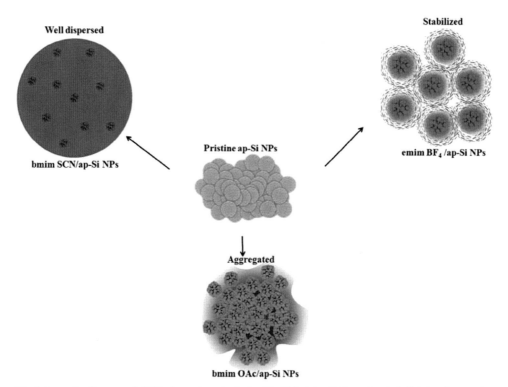

Figure 2.1 Schematic diagram of SiNPs in various ionic liquids (Tchalala, El-Demellawi, & Kalakonda, 2020).

Sol—gel is one of many processes for the manufacture of fibers, including the production of silica fibers (Sakka, Tomozawa, & Doremus, 1982). Furthermore, organic—inorganic hybrids developed via the sol—gel process have been shown to serve as precursors of ceramic powder of nonoxides (Wei, Kennedy, & Harris, 1984). The ease of producing homogeneous multicomponent materials is another benefit of this process. The introduction of Ti into silicon carbide (SiC) fibers enhances resistance to oxidation and compatibility with metals and plastics when manufacturing composites, and shows various electrical properties. The synthetic procedure for SiC-TiC fibers investigated preparations of hybrid SiO_2-TiO-phenolic resin fibers precursor by sol—gel processing, and then the conversion of the hybrid using a carbothermal reduction into SiC-TiC hybrid fibers. Tetraethoxysilane (TEOS) and 2-propanol solution of Ti tetrakis (2,4-pentanedionate) were added to the EtOH and then HCl was added. The viscosity of the liquid enhanced substantially, and the preparation was obtained just before the solutions were gelated. Hybrid SiO_2/TiO_2/phenolic resin fibers with Si:Ti ratios down to 10.4 can be easily formed using an additional Ti source (Hasegawa, Nakamura, & Kajiwara, 1996). A further hybrid SiO_2/ZrO_2/phenolic resin was developed by the same group. The conditions of process for forming ZrO_2/phenolic hybrid resin fibers were first investigated. These hybrid fibers have been synthesized by dissolving zirconium tetra-kis(2,4-pentanedionate) (ZTP) and phenolic resins in 2,4-pentanedione in EtOH, accompanied by H_2SO_4. With time, the viscosity of liquid is enhanced, producing viscous sols. The fibers formed through the sol—gel process have been sticky, which is due to solvent inclusion. They were allowed to dry overnight at room temperature to prevent coalescence of the fibers (Hasegawa, Fukuda, & Kajiwara, 1999). Sol—gel processes can obtain nanospheric silica particles during the reaction of tetraethoxysilane (TEOS) that is catalyzed using alkalis. Development of NPs includes the hydrolysis of alkoxide and the expansion of particles through condensation. By a modified process, the compounds could be dissolved in hydrocarbons using an alkyl silicone terminal (Pedroso, Dias, Azuma, & Mothé, 2000). Surface modification makes the silica nanospheres highly compatible in hydrocarbons, thus producing homogeneous dispersion, based upon the requirements of the copolymerization of TEOS. TEOS copolymerization using dimethyldiethoxysilane (DMDEOS) resulted in a high organophilic silica conversion. However, after the reaction ended, a wax-like gel was produced while drying by vacuum. Alkylalkoxysilane hydrolysis becomes prevalent and an organic-rich layer develops. This hybrid structure is defined by the presence of two such layers with unique structure (Dias, Pedroso, Mothé, & Azuma, 2004). Inorganic/organic nanocomposites could be formed by sol–gel processing using TEOS to form an Si-polymer framework. Cationic polyurethane (PU) was provided in a microemulsion form to minimize the surface energy of nanosilica in the formation of SiO_2/PU hybrid. After prepolymer formation, a certain amount of TEOS with PU solution follows. So the aqueous emulsions were formed of SiO_2/PU NPs (Zhu & Sun, 2004). Poly(vinyl alcohol) (PVA) is extensively studied due to its excellent film-forming and physical properties. PVA/SiO_2/TiO_2 hybrids can be formed using a sol–gel coating process. In H_2O, PVA was dissolved with TEOS, HCl was added continuously, thus producing reactive silica sol. Then, tetrabutyl titanate was added with stirring in acetylacetone. Thus the hybrid PVA/SiO_2-TiO_2 sols were prepared. The hybrid fibers adequately prevent UV radiation and reduce the aging of the hybrid (Ma, Shi, & Song, 2014). Polyacrylate/SiO_2 composites were synthesized by adding colloidal silica to a polyacrylate emulsion (MPS). The development of Si-O-Si polymer cross-link networks in the formation of a film by the addition of a cosolvent in the sol–gel process enhances the hardness, thermal stability, and solvent resistance of the composite films. The composite latexes FS-PSA/SiO_2 were formed by mixing colloidal silica in EtOH at room temperature. The sol–gel processes enhance the thermal stability. Particles of Si and the silanol groups on the latex surface improve the roughness and hydrophobicity of films (Ye et al., 2019).

2.2.3 Polymer grafting

Researchers have always been keen to research the improvements in the different characteristics of polymers with the inclusion of additives. When combined using the right amount of NPs as reinforcement, polymers possess substantial enhancement (Jin, Rafiq, Gill, & Song, 2013; Lin, Chen, Chan, & Wu, 2011; Sarita et al., 2013). Reversible addition–fragmentation chain transfer (RAFT) provides a method for synthesizing a diverse range of polymers with relatively small distributions. It is accomplished via RAFT reagent carrying a thiocarbonyl group which binds chain radicals reversibly and thus facilitates polymerization. This approach has functionalized a broad range of materials including carbon nanotubes, CdSe nanoparticles, cellulose, cotton, silica particles, and gold nanorods. By comparison, SiNCs/polymer hybrids are still in their initial phases of progress, largely due to the poor supply of well-established SiNCs. Physical component blending offers an excellent methodology to procuring hybrids. Liu et al. used this technique to combine hydrides to terminate SiNCs with regioregular P3HT and studied these in prototype photovoltaic systems. SiNCs processed through silane dissociation in a radio frequency (rf) plasma were collected on mesh filters. SiNCs/P3HT on indium tin oxide coated glass substrates were fabricated as solar cells using a spin-coating process. This spin-coating blend created a good device with efficiency improvements of up to 1.15% (Liu, Holman, & Kortshagen, 2009). To enhance material homogeneity Mitra et al. used a technique for coating SiNCs with PEDOT: PSS polymer using atmospheric pressure microplasma. This supplied a homogeneous coating of polymer and enhanced the dispersibility of solvents, photoluminescence, and stability. The SiNCs in ethanol were added to a PEDOT:PSS filtered solution in distilled water to produce a SiNCs/PEDOT:PSS/water colloid. The properties of the SiNCs/polymer colloid were analyzed to enhance the comprehension of surface engineering (Mitra et al., 2013). Hessel and coworkers optimized semiconductor quantum dots and demonstrated the synthesis of SiNCs/poly(maleic anhydride) hybrids. Within this amphiphilic polymer shell, encapsulating alkyl terminated SiNCs produced the dispersible NPs which are convenient for studying bioimaging in vivo. They even indicated how HNPs were luminescent, and broad enough to circumvent the renal system, and stay for prolonged periods in the circulatory system. The mixture varies from yellow to brown due to a reduction of NCs' size during the process of oxide removal. The SiNCs are passivated with a dodecane monolayer which is covalently bonded. In amphiphilic polymer micelles, the NCs were encapsulated using a dry film. Then fabricated NCs were centrifuged and added to a saline phosphate buffer (Hessel et al., 2010; Messaddeq et al., 2019). The creation of covalent links between the SiNCs and polymers offers one approach for evaluating long-term stability, charge transport, and homogeneity. A photoluminescent, solution-processable, homogeneous, and chemical-resistant SiNCs/polystyrene (PS) was reported by Yang. Styrene radicals react with free styrene to produce SiNCs/PS hybrids. The technique for the preparation of these SiNCs/PS hybrids is outlined. The SiNCs were formulated using a known technique that exposes the thermally induced disproportion of hydrogen silsesquioxane (HSQ). The resulting solids were grounded and etched to release hydride-terminated SiNCs. In an argon atmosphere, styrene in toluene was mixed with SiNCs and reflux. After 15 h of heating, an orange solution precipitated an amber solid upon the addition of ethanol (Yang et al., 2014) (Fig. 2.2).

Dung et al. reported a designed SiNCs/PS hybrid that is a helpful charge trapping component in a metal−insulator−semiconductor

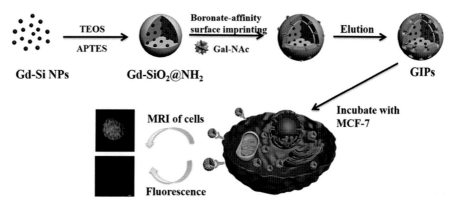

Figure 2.2 Synthetic diagram of hybrid polymer coated SiNPs via polymerization (Ren et al., 2021).

prototype system (Dung, Choi, & Jeong, 2013). Choi et al. synthesized hybrids of SiNCs/PS with different ratios of the NCs and polymer components (Choi, Dung, & Jeong, 2014). With the SiNCs concentration the refractive index of spin-coated films increased. Recently, Kehrle et al. developed a multistage surface-initiated group-transfer polymerization (SIGTP) process for improved SiNCs. The surfaces of hydride-terminated SiNCs were initially updated using ethyleneglycol dimethacrylate with photoinduced hydrosilylation. Such a unique surface has acted as a catalyst anchor which readily polymerizes diethylvinylphosphonate (DEVP) to produce SiNCs/PDEVP. This result is water soluble and luminescent (Kehrle et al., 2014). Sato et al. stated the reaction of SiNCs/silicone polymer hybrids which are flexible, extendible, transparent, and luminescent. Silicone elastomers were connected effectively with the surface of the SiNC via siloxane links. This photostable hybrid was crafted to form a range of shapes. The SiNPs with luminescence color were well dispersed in EtOH by ultrasonics and spun in ethanol for 30 min with a centrifugation filtration process to extract the supernatant containing uneven SiNPs. The silicone elastomer was then applied with a micropipette to SiNPs, and further blended at room temperature. The SiNPs carrying silicone elastomer were subsequently cooled down to obtain the SiNP/polymer hybrid. These specially formed SiNPs/polymer composites can be extensively used, such as in diagnostic imaging systems, electroluminescent displays, and surgical bypasses in various organs (Sato et al., 2010). An alternative method to interface polymers with SiNCs was reported by Hçhlein. To develop luminescent SiNCs/polymer hybrids, they demonstrated surface-mediated

RAFT polymerization with SiNCs and methyl methacrylate, styrene, N-isopropylacrylamide, 4-vinylbenzyl chloride, and hexyl acrylate. It formed reactive chlorosilane surface SiNCs that react with nucleophiles under mild conditions. In a two-step reaction, the RAFT reagent 6-hydroxyhexyl 3-(methylthio)-2-phenyl-3-thioxopropanoate (HMT) was prepared. HMT was mixed with the toluene dispersion of SiNC-SiCl to stabilize the RAFT reagent and then stirred overnight. The HMT functionalized SiNCs were purified through precipitation from CH_3CN/toluene (Höhlein, Werz, Veinot, & Rieger, 2015). Kassiba et al. reported hybrid core–shell nanocomposites doped with camphor sulfonic acid (CSA) based on silicon carbide NPs (SiC) and polyaniline (PANI). These hybrids were prepared by polymerization in CSA with SiC NPs with a 1.23 wt% solvent ratio. The resulting SiC/PANI/CSA hybrids were dialyzed using distilled water via a cellophane membrane to eliminate residues and impurities. Then this reaction was accomplished by water evaporation followed by vacuum drying to isolate the final product (Kassiba et al., 2007). Vijaya et al. produced SiC/Si_3N_4/Nylon-6 nanocomposite fibers. Initially, SiC NPs were coated with a sonochemical method on Si_3N_4 nanorods. Si_3N_4, SiC NPs, and CTAB in ethanol were added to an ultrasonic cell jacket and the reaction was kept under nitrogen at 10°C for 3 h. The resulting dry hybrid SiC/Si_3N_4 NPs were used for Nylon-6 fibers reinforcement in a mechanical blender at 5°C (Rangari, Yousuf, & Jeelani, 2013).

2.2.4 Encapsulation process

In recent times, Liu et al. reported a yolk–shell framework of carbon-encapsulated silicon with good coulombic efficiency above 99%. It has SiNPs as "yolk" and amorphous carbon as the "shell." They developed a method using a room temperature solution to coat SiNPs first with SiO_2, then with polydopamine, and then with nitrogen-doped carbon coating. This hybrid can be a next generation for Li-ion batteries to increase coulombic efficiency and cycle life by introducing it to various high-capacity alloy-type anode materials (Liu et al., 2012). Then Luo et al. reported another method to encapsulate SiNPs using shells with reduced graphene oxide (rGO). The easy one-step method includes an rGO shell to permit SiNPs to broaden without shattering the graphene shell and hence protect Si particles efficiently. The SiNPs and GO sheet colloidal solution was then injected to develop droplets that were pumped at 600°C through a preheated tube furnace under N_2 gas. Crumpled graphene shell encapsulation improved Si performance significantly in Li ion battery anode, including cycling stability, coulombic strength, and rate capability (Luo et al., 2012). Kong et al. encapsulated SiNPs in polydopamine (PDA)-derived hollow graphitized CNFs using an electrospinning process to prepare a novel hybrid system. Weak electrical conduction induced by the high insulation Si content may be balanced by the good electrical conductivity of the C-PDA shell. Initialy, PAN—SiNFs were prepared through electrospinning. The suspension of dopamine hydrochloride and PAN-SiNFs was then applied to tris(hydroxymethyl) aminomethane to initialize polymerization by constant air bubbling. After that, to leach out PAN, it was heated at 60°C in DMF. To prepare the final hybrid SiNPs encapsulated hollow hybrid nanofibers (SiNFs/C-PDA), the dried PDA—SiNFs were then annealed at 700°C under Ar gas for 3 h. This hybrid showed greatly enhanced electrochemical characteristics, like better high-rate capacity and higher cycling stability (Kong et al., 2013). By varying the surface properties of various NPs and modulating the composition of the organic phase, Liu et al. efficiently encapsulated porous silicon and AuNPs into the polymer matrix simultaneously. Firstly, using dextran in anhydrous DMSO under N_2, acetylated dextran (AcDEX) was prepared. The Murray place exchange method was performed to functionalize the AuNPs with the desired ligand to further conjugate AuNPs on AcDEX. Finally, dextranylated porous Si and Au were dispersed into the AcDEX in $EtOH/CH_3CN$ to prepare the AuNPs-encapsulated nanohybrids (Liu et al., 2018).

2.2.5 Drug loading

PSiNPs have shown great success in cancer diagnosis and therapy as multifunctional nanosystems (Park et al., 2009; Tasciotti et al., 2008; Tzur-Balter, Shatsberg, Beckerman, Segal, & Artzi, 2015; Xia et al., 2015; Xu et al., 2016). PSiNPs have a flexible loading functionality due to its controllable porosity, huge surface area, customized surface functionalization of various organic and biomolecules, or even NPs (Anglin, Cheng, Freeman, & Sailor, 2008; Godin, Tasciotti, Liu, Serda, & Ferrari, 2011; Santos, Mäkilä, Airaksinen, Bimbo, & Hirovnen, 2014). In contrast, tunable photoluminescence (PL) may also be shown by PSiNPs, resulting in simple monitoring of its biodistribution in the body using fluorescence signals. All these benefits make PSi an excellent alternative for the delivery of drugs and biomedical processes. Zhang et al. combined PMVEMA with thermally hydrocarbonized PSiNPs modified by carboxylic acid using polyethyleneimine (PEI) as a linker polymer. Positively charged amine could help in strengthening the cellular interactions of formed particles. The microfluidic procedure was consequently used to develop multistage TDDSs by encapsulating the PSi/PEI/PMVEMA. This hybrid composite increases the effect of inhibition on cell proliferation and strengthens the permeability of the drug (Zhang et al., 2014). A near-infrared light-triggered device based on PSiNPs in conjunction with IR820 dyes has been designed to supervise the release of doxorubicin hydrochloride (DOX) into cancer cells to combine chemophotothermal therapy. IR820 molecules may be adsorbed onto NH_2-PSiNPs with positive zeta potential surfaces via electrostatic attraction in aqueous medium. To construct hybrid composites, hydrophilic DOX was loaded into IR820/NH_2-PSiNPs that could be effectively internalized into cancer cells. Intracellular DOX discharge from DOX/IR820/NH_2-PSiNPs can be progressively noticed in situ after cellular absorption by localized NIR laser photostimulation. The results indicated that in physiological conditions, DOX/IR820/NH_2-PSiNPs had outstanding biodegradability. The therapeutic method based on nanocomposite of PSiNPs integrated with IR820 and DOX molecules, therefore might have tremendous capability for NIR light-stimulated cancer treatment (Xia, Wang, Chen, Zhang, & Shi, 2016). A novel approach of integrating high DOX loading into styrene-linked PSiNPs (SPSiNPs) through π-stacking to produce DOX/SPSiNPs has also been published by the same group. To form SPSiNPs, firstly divinylbenzene was grafted through microwave-assisted hydrosilylation into hydrogen-terminated PSiNPs, and SPSiNPs were then fully distributed by ultrasonication in DOX aqueous solution. DOX/SPSiNPs in the neutral buffer remained highly stable. DOX is efficiently released from DOX/SPSiNPs within Hela cells after cellular uptake, contributing to an outstanding anticancer effect (Xia, Wang, Zhang, & Shi, 2016). NIR dye compounds were readily released from PSiNPs and quickly decomposed under NIR laser irradiation in the aqueous solution, restricting their possible in vivo applications. Hence, biodegradable nanosystems with a stable photothermal effect have been developed by the same group. Polyaniline was covalently fabricated via polymerization into PSiNPs as shown in Fig. 2.3. PSi were sonicated with 4-vinylaniline, and then incubated for 60 min at 100°C by microwave heating to prepare a vinylaniline-terminated hybrid composite (VANi-PSiNPs). The oxidative polymerization of aniline into VANi-PSiNPs was performed in an acid medium containing VANi-PSiNPs, aniline, and the oxidant. The resulting PANi-PSiNPs composite demonstrated outstanding biodegradability and biocompatibility (Xia et al., 2017).

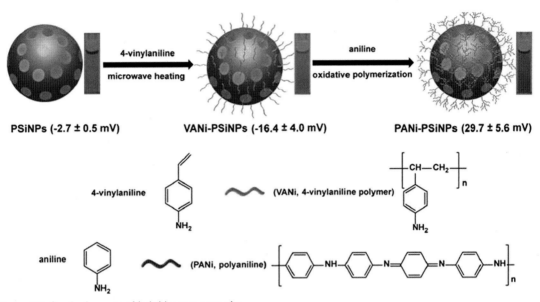

Figure 2.3 Synthetic route of hybrid nanocomposite.

2.2.6 Colloidal system

Possessing unique multifunctional synergistic characteristics, colloidal hybrid NPs have newly drawn significant interest. Specifically, in photocatalysis, bioimaging, and solar energy conversion, semiconductor–metal HNPs have been investigated extensively (Lee, Shevchenko, & Talapin, 2008; Lee, Bodnarchuk, Shevchenko, & Talapin, 2010; Li et al., 2015; Zhang, Tang, Lee, & Ouyang, 2010). A variety of requirements are being sought for discrete luminescent quantum dots of semiconductors, very prominently CdS and CdSe, from bioimaging to hybrid light emitters, for which good NCs are easily generated as colloidal dispersions. Hua et al. proposed steps toward the objective of manufacturing excellent SiNPs in macroscopic proportions as colloidal dispersions (Hua et al., 2005). Since then, a unique approach has been reported to synthesize luminescent SiNPs in which NPs are treated with CO_2 laser heating of SiH_4/H_2/He and after sonication of SiNPs in MeOH, etched with acidic medium to minimize the size and passivate the surface (Li, He, Talukdar, & Swihart, 2003). Washing of particles in a membrane filter eliminates the acid mixture with excessive MeOH. The particles displayed bright, clear photoluminescence as a result. Mastronardi et al. have effectively isolated colloidally stable allylbenzene-capped SiNCs using size-selective precipitation into many visible emitting monodisperse fractions, as shown in Fig. 2.4 (Mastronardi et al., 2012). In order to prepare colloidally stable SiNCs, this easy and efficient method has been used to reduce the size distribution of passivated SiNCs. A reductive disproportionate nucleation was triggered by the thermally processed hydridosilicate glasses, developing a SiO_2 matrix with SiNCs. The etching method was used to eliminate SiNCs from an encapsulated matrix and develop hydride terminated SiNCs. The hydrophobic SiNCs were collected in a nonpolar solvent, and then a hydrosilylation reaction started thermally to passivate the NCs. The study of HRTEM and SAED reveals that most of the tiny particles in the sample could be amorphous or nanoclusters.

Polydisprse alkyl-capped SiNCs were prepared by the same group using organic density gradient ultracentrifugation by size-separation. The alkyl chain passivated SiNCs were fractionated using 2,4,6-tribromotoluene mixed with chlorobenzene. This high degree of alkyl surface cover offers incredible colloidal stability for existing SiNCs (Mastronardi et al., 2011). A new and effective self-limiting synthesis of colloidally stable semiconductor metal such as Au, Ag, and Pt hybrid NPs-based SiNCs was stated by Sugimoto et al. Instant formation of HNPs containing metal cores and SiNC shells occurred by simply blending of colloidal SiNCs with metal salts. SiNCs and $HAuCl_4$ colloidal dispersion solutions are prepared for the formation of Si/Au hybrid NPs. Hybrid Si/AgNPs were formed by combining SiNC and $AgNO_3$. Likewise, hybrid Si/PtNPs were produced using SiNCs in a solution of H_2PtCl_6 and stirred for 6 h (Sugimoto, Fujii, & Imakita, 2016). Kutrovskaya et al. demonstrated that the process of laser ablation can be used effectively to form SiNPs in which the colloidal solution formed by laser ablation has been settled from a single layer composed of SiNPs in EtOH to prevent the oxidation. As a result, the formation of hybrid Si-AuNPs is noted by laser action on a mixed colloidal

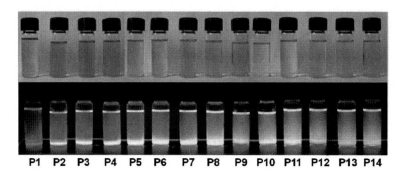

Figure 2.4 Top fractions in visible light and bottom in photoexcitation at 365 nm.

solution. The rise in laser irradiation power led to a broader distribution of 100 nm diameter NP sizes. The resulting hybrid Si-AuNPs are shown in Fig. 2.5 (Kutrovskaya et al., 2017).

2.2.7 Solution blending

To enhance the properties of nitrile rubber (NBR), graphene oxide (GO) hybridized SiNPs have been developed by Zhang et al. On the surface and between layers of GO, SiNPs were homogeneously distributed, and a better thermally stable hybrid was formed. These hybrid nanocomposites were prepared by the solution blending method to analyze the mechanical properties. Alkaline solution of SiNPs and ethyl orthosilicate were mixed. The mixture was kept for 3 h at 55°C. A certain amount of NH_4OH was added into the GO suspension. Then, SiNPs were mixed with the GO suspension at 80°C for 3 h. The formed precipitates were washed with DW and stored in a cooled place. For the formation of final $GO/SiO_2/NBR$ nanocomposites, a dispersed solution of GO/SiNPs in DMF was mixed with NBR in DMF suspension with vortex stirring. This hybrid showed outstanding tensile strength and elongation (Zhang, He, Wang, Rodrigues, & Zhang, 2018).

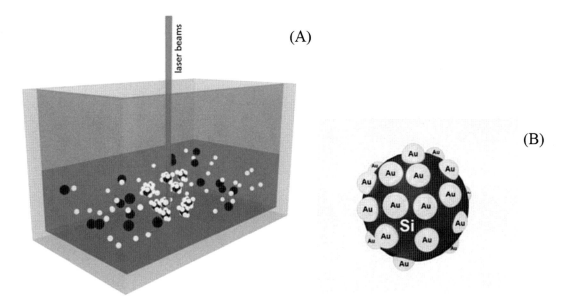

Figure 2.5 (A) Laser irradiation in colloidal solution. (B) Formation of hybrid Si-AuNPs.

2.2.8 Gas-phase synthesis

The efficiency of hot wall reactor (HWR) gas-phase synthesis to generate Ge on Si anisotropic particles was displayed by Mehringer et al. For anisotropic hybrid Si-GeNPs, they proposed a successive gas-phase synthesis method that delivers sufficient command over patch number and size solely. GeH_4 from the HWR-2 was added into HWR-1 with SiNPs, leading to Ge accumulation on the surface of the SiNPs. Synthesis at 600–700°C led to the formation of Janus-like particles. These NPs display a high degree of crystallinity and connection of an epitaxial orientation among the patches and particles of the Si core (Mehringer et al., 2014).

2.2.9 Coating with magnetic nanoparticles

Magnetic hybrid composites are extensively studied for their outstanding application and characteristics. Core–shell is the standard framework of magnetic hybrid composite, with the shell as inorganics or organics and the core as inorganic magnetic NPs such as Fe, Ni, Co, etc. Due to the nontoxicity of the magnetic Fe oxides, they are suitable for synthesis. Biotechnological and photocatalytic applications are offered by the deposition of magnetic Fe oxide into Ti. However, the photocatalytic efficiency of TiO_2 is reduced after the introduction of Fe_3O_4 as the cores of the hybrid Fe_3O_4/TiO_2 due to photodissolution. To solve this problem, Wang et al. employed SiO_2 because of its nontoxicity as coatings and barrier layers among the magnetic cores and Ti shells. First, sequentially aqueous ammonia and tetraethyl orthosilicate were mixed to the dispersed solution of magnetic NPs and stirred for 6 h. Using the magnetic separation method, Fe_3O_4/SiO_2 particles were isolated and washed excessively with EtOH/water and dried. After that this particles were again suspended in a hexane:water solution, and subsequently tetrabutyl titanate was added. The mixture was moved with γ-Fe_2O_3 planes to the autoclave for the calcination. This hybrid provides photocatalytic ability and is also a strong applicant for magnetic separation to purify polluted water (Wang et al., 2009). A simpler technique to synthesize new hybrid NPs combining Si and Fe_3O_4 consisting of γ-Fe_2O_3 was reported by Sato et al. By reducing the particle size, this synthetic method can regulate optical and magnetic behavior, with the hybrid displaying green fluorescence and superparamagnetic nature. Si atoms and Fe_3O_4 were systematically inserted into the oxide matrix through codeposition during the sputtering process. The precipitation of Si and Fe_3O_4 was subsequently undertaken during the annealing step in a thick oxide matrix at 1100°C for 60 min in Ar atmosphere, leading to the formation of the hybrid. After the extraction of the hybrid from the matrix using HF, they were centrifuged repeatedly to eliminate the uneven NPs and HF from the surface. These hybrid NPs show significant capabilities for in vivo MRI screening and optical fluorescence imaging (Sato et al., 2011).

2.2.10 Catalyst

The development of strong and reusable catalysts is one of the most key factors for pharmaceutical applications as well as for organic synthesis. For catalytic transformation exciting nanodevices are predicted to understand immediate, selective reaction systems (Astruc, 2008). For the arylation, Yamada and coworkers designed and examined a catalyst of silicon nanowire array (SiNA)-stabilized palladium NPs, along with various palladium-catalyzed methods. It would be interesting to develop hybrid SiNA-PdNPs as nanoscopic and macroscopic catalysts for good activity and recyclability. This hybrid needed to fit on a 1-cm^2 sized silicon wafer, with constrained nanosize reaction fields. For the cleaning and deployment of Si-H surface groups, the Si wafer was washed using acids. First, the H-terminated wafer reacted with $AgNO_3$ in HF/H_2O_2, forming the Si-AgNPs. AgNPs were extracted and the Si-H surface reconstruction of the SiNA was achieved with HNO_3 and HF. To achieve the hybrid SiNA-PdNPs catalyst, PdNPs immobilization was conducted with K_2PdCl_4 on SiNA via the Mizoroki−Heck reaction (Yamada, Yuyama, Sato, Fujikawa, & Uozumi, 2014).

2.3 Characterization

A number of analytical methods were used to look into the characteristics of the synthesized hybrid SiNPs in terms of understanding the functionality. It is useful to evaluate the nature of these compounds and also to detect any required changes to the NPs. In addition, the reliability of the chemical processes used to construct NPs must be checked by various characterization methods.

2.3.1 Fourier-transform infrared spectroscopy (FTIR)

FTIR operates with the electromagnetic spectrum region of IR radiation. It is a very effective method for detecting various kinds of chemical bonds through the infrared absorption spectrum. At certain frequencies, every functional group acquired the radiation and thus generated structural data. Hua et al. reported that vinyl acetate and ethyl undecylenate grafted silicon can be hydrolyzed to obtain −OH, −COOH functionality. ATR FTIR spectra for SiNPs during photoinitiated hydrosilylation with numerous double bond terminated components are shown in Fig. 2.6A. The absence of the peak of the terminal double bond is shown by FTIR, confirming that organic molecules are covalently connected with the Si surface (Hua et al., 2005). Dispersed SiNCs:decyl by DGU, described by Mastronardi et al. can be identified by FTIR. These recorded C-H stretches, which are correlated with uneven packing of hydrocarbons, are near to the 2922 and 2850 cm^{-1} antisymmetric and symmetric vibrations found in Fig. 2.6B (Mastronardi et al., 2011).

FTIR can detect the chemical composition of PSiNPs-based nanocomposites reported by Xia et al. The C=N and C=C stretching vibrations of the quinoid and benzenoid polyaniline rings in PANi-PSiNPs have been allocated at 1584 and 1504 cm^{-1}. With the successively grafted polymerization of aniline and 4-vinylaniline, the two bands at 2983 and 803 cm^{-1} have progressively expanded. In the meantime the band increased steadily at 1302 cm^{-1} because of the C-N stretching for the aromatic 2° amine that informed polymerization on PSiNPs (Xia et al., 2017). FTIR verified the effective modification of DPSi and DAu on hybrid. The IR spectrum compilation illustrates the encapsulation process (Liu et al., 2018).

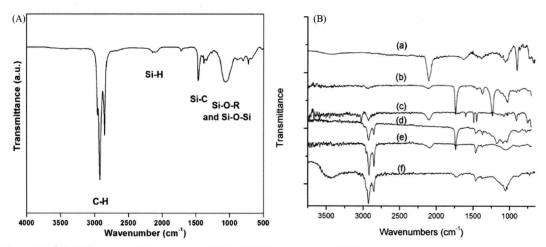

Figure 2.6 (A) FTIR of hydrosilylated hybrid SiNPs. (B) Alkyl passivated SiNCs.

Alkyl passivation polymer coating of SiNCs reported by Hessel's group was characterized by ATR-FTIR. The FTIR demonstrates strong bands at 2108 and 898 cm^{-1}, proving that the NCs are hydride-coated. At 2108 and 898 cm^{-1} the Si-H peaks vanish, and aliphatic C-H strong stretching at 2853, 2925, 2957 cm^{-1} confirmed the hydrosilylation onto the surface of the NCs proceeded as planned (Hessel et al., 2010; Messaddeq et al., 2019). In PEDOT:PSS with SiNCs reported by Mitra et al. peaks at 2200 to 2250 cm^{-1} are due to Si-H bonds with an oxygen backbone. The Si-O band at 1000–1100 cm^{-1} is associated with PEDOT:PSS absorption. In FTIR, the Si-CH$_2$ and Si-C bonds appear at 1258 and 1273 cm^{-1}, respectively (Mitra et al., 2013). SiNCs/polystyrene hybrids were analyzed by FTIR with a peak of C-H bond at 3000 cm^{-1} of the phenyl ring and the aliphatic polymer backbone of 2650–2900 cm^{-1}.

At 2000 and 1650 cm^{-1}, poor absorption indicative of overtone peaks resulting from monosubstituted aromatic rings is noted. The mixture of phenyl group stretching C=C and aliphatic bending C-H has also been related to the strong peak at 1450 cm^{-1} (Ren et al., 2021; Yang et al., 2014). FTIR measurements were performed by Zhang et al. to investigate the contact between GO and SiNPs. The major peaks of this component have been recorded, including several groups, like C-O, C=C, C=O and −OH, and at 1255, 1623, 1744 and 3432 cm^{-1} respectively. Si-O-C bonds can be prompted by two new bands at 1114 and 661 cm^{-1} (Zhang et al., 2018).

2.3.2 X-ray photoelectron spectroscopy (XPS)

XPS method is a surface analysis, that is employed for chemical investigation. In particular, XPS calculates the electronic states, composition of elements, empirical and chemical formula within a substance. This approach is intended to establish the local chemical environment of a specific atom, in contrast to the surface chemical composition. By bombarding a substance with X-ray photons, XPS spectra are obtained when the kinetic energy and the sum of electrons released from the sample are measured even at the same time.

XPS defined alkyl passivation of SiNCs. The typical peaks of Si element and SiO$_2$ at 99.4 eV and 103.5 eV, respectively, are XPS of the oxide embedded NCs. There is low suboxide available, proposing that the SiNCs and the SiO$_2$ matrix have an abrupt interface. A single XPS peak at 99.2 eV after acid etching is due to Si, confirming the whole SiO$_2$ matrix was clearly extracted (Hessel et al., 2010; Messaddeq et al., 2019). XPS

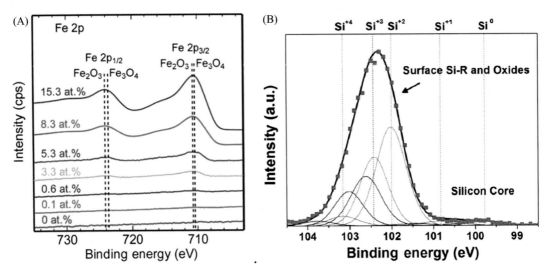

Figure 2.7 (A) XPS spectrum of hybrid SiNPs/Magnetic particles. (B) Allylbenzene capped SiNCs.

characterized hybrid SiNPs/magnetic particles. The hybrids in which specific NPs with varying Fe content were systematically revealed on the substrate surface. Because of the spin-orbit coupling, Fe 2p core level peaks with splitting components of Fe $2p_{1/2}$ and Fe $2p_{3/2}$ are detected from all hybrid NPs, while no certain peaks are observed for the SiNPs. In addition, the Fe $2p_{1/2}$ and Fe $2p_{3/2}$ line shapes show two peaks at the binding energies of 723.6 and 724.1 eV, whereas the Fe $2p_{3/2}$ line shows the 710.4 and 710.7 eV peaks (Fig. 2.7A). The peaks at 710.4 and 723.6 eV are well in line with those recorded (Sato et al., 2011). Colloidally stable allylbenzene capped SiNCs was confirmed by XPS, in which a strong peak was noted at 102.3 eV, and also a weak peak at 99.9 eV. As shown in Fig. 2.7B, a peak of 99.9 eV in the NC core is related to elemental Si and a wide peak of 102.3 eV is associated with a mixture of the $Si^{\delta+}$-$C^{\delta-}$ and $Si^{\delta+}$-$O^{\delta-}$ groups (Mastronardi et al., 2012). SiNCs and PEDOT:PSS polymer-coated hybrid NPs were characterized by Si 2p XPS spectrum as per Fig. 2.8. It was equipped with respective Si $2p_{3/2}$ and Si $2p_{1/2}$ doublets isolated with 0.6 eV.

The existence of Si-Si bonds connected to SiNCs is strongly demonstrated by peaks at 98.8 eV of Si $2p_{3/2}$, 99.3 eV of Si $2p_{1/2}$, 102.4 eV of Si $2p_{3/2}$, and 103 eV of Si $2p_{1/2}$. The peaks of Si $2p_{3/2}$ at 101 eV and of Si $2p_{3/2}$ at 101.6 eV relate to suboxide (Mitra et al., 2013). DOX/SPSiNPs could be analyzed by the XPS method to identify atomic concentrations of elemental components. The

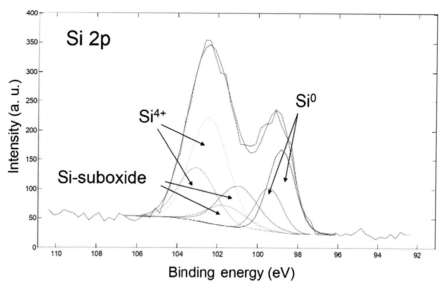

Figure 2.8 XPS analysis of SiNCs/PEDOT:PSS.

signals for C 1 s, O 1 s, Si 2p, and Si 2 s at 285, 532, 103, and 154 eV, respectively, were recorded in the XPS spectrum, which was reliable with the integration of organic substances into PSiNPs. C-C and C-H were designated to the 284.4 eV peak, and C-carboxyl was designated to the 288.5 eV peak, that was attributed to carboxyl groups in DOX molecules (Xia et al., 2015; Wang, et al., 2015). FS-PSA/SiO$_2$ hydrophobic composites can be described by XPS. The peaks at 285 and 532 eV relating to C_{1S} and O_{1S} are distinctly identified, whereas the peak at 689 eV refers to F_{1S}, originating from DFMA copolymerization. There are Si_{2P} peaks referring to Si-O at 102 eV and two types of Si-O bonds, with peaks at 101.59 and 101.08 eV corresponding to the Si-O-Si group (Ye et al., 2019).

2.3.3 Dynamic light scattering (DLS)

A technique for classifying the size of NPs (1 nm diameter) usually in the submicron region is DLS, which is also defined as photon correlation spectroscopy (PCS). The DLS computed diameter is bound to a value which indicates how a particle diffuses inside a mixture and thus it is attributed as a hydrodynamic diameter. DLS tests the scattered light from a monochromatic beam of light including a laser which travels via a colloidal solution and evaluates the distributed light intensity modulation as a time function. SiNC alkyl passivation polymer coating can be detected using the DLS test, where SiNCs had 17.8 ± 0.4 nm (hydrodynamic diameter) in the PBS solution as determined by DLS and a −31 mV zeta potential. The DLS results demonstrate the passivation of NCs using an amphiphilic polymer of low-molecular-weight are an efficient approach to have a hydrophilic grafting without drastically increasing the total hydrodynamic diameter (Hessel et al., 2010). Sandra et al. have efficiently synthesized silicon-TiO_2 using a sol−gel process, in which they found Si−O−Ti bond formation in the hybrid matrix. They conducted DLS analysis to evaluate the particle size distribution as a function of the concentration of TiO_2 NPs. This approach calculates and compares the Brownian movement of the particles to the size.

It should be observed that 0.02 and 0.04 M TiO_2 generate a narrow peak in the size distribution with a mean diameter at 44 nm, as per Fig. 2.9. An often broader peak suggests the convergence of particles in the hybrid matrix as the concentration of TiO_2 rises (Messaddeq et al., 2019). Moraes et al. effectively grafted polystyrene onto the SiNPs, confirmed by the improvement in particle size by DLS from 130 to 330 nm. DLS suggests that although comparatively low PDI is 0.047 and polydispersity at 0.007, the particles exhibit a significant increase in size, associated with the presence of disulfide bridges (Moraes, Ohno, Maschmeyer, & Perrier, 2015). DLS can be classify the mixture of PDEVP and photoluminescent SiNCs, where they are founded with the hydrodynamic radius of 177 nm. The thermal evaluation by DLS of the hydrodynamic radii of the particle−polymer mixture showed the thermoresponsive nature of the hybrid

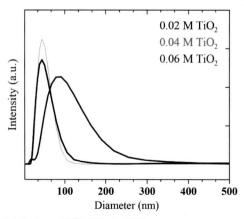

Figure 2.9 DLS analysis of size distribution of TiO_2 NPs in HNPs.

(Ren et al., 2021; Yang et al., 2014). DLS analysis characterized hybrid DOX/SPSiNPs. The tests revealed that the high negative zeta potential of DOX/SPSiNPs was -23.1 ± 5.8 mV and mean hydrodynamic size was around 257 nm, resulting in their excellent aqueous solution dispersibility and stability (Xia et al., 2015; Wang et al., 2015). DLS analyses of THF dispersions were carried out to evaluate the size of the hybrid SiNC-gPS. A significant increase after 12 h of polymerization in hydrodynamic diameter to 37 nm from 5.6 nm was detected. Its dispersity was observed to be greater with shorter reaction times, which may be due to particle interactions (Höhlein et al., 2015).

2.3.4 Transmission electron microscopy (TEM)

TEM is often used to analyze compounds through an electron-based approach. TEM is effective and has a better resolution under a high vacuum that can evaluate the internal structure of the compound giving information of the morphology. An electron beam interacts intensely with sample, boosting the potential to use contrast for identification. The size of particles serves a crucial factor in the properties included in SiNPs, thus the calculation of the particle size, crystalline stricture of Si core, and size distribution of SiNPs is essential. Hybrid Si-Au-Ag NPs were recorded for TEM analyses by the Sugimoto group, by dropping hybrid NP solutions onto carbon-coated copper grids. NCs created a monolayer in the absence of 3D agglomeration, a round NC with lattice fringes corresponding to the Si crystal, and hybrid NPs in solution were shown in the TEM image. STEM-EDS element mapping supports a development of the hybrid Si-AuNPs core–shell. Synthesized hybrid γ-Fe_2O_3 SiNPs were characterized by TEM as per Fig. 2.10. The spherical products were roughly about 100 nm in diameter. The nature of the structure of the NPs are represented in Fig. 2.10B. Hybrid γ-Fe_2O_3 SiNPs with 15 nm mean size and superparamagnetism nature were described in Fig. 2.10C (Wang et al., 2009). Zhou et al. reported hybrid Si/PCNF was characterized by TEM and HRTEM analysis. TEM images demonstrate that SiNPs are well-enclosed in the porous CNF and there is void space within SiNPs and the carbon shell. The HRTEM analysis shows that a 30-nm SiNP with a single crystalline structure and also free of SiO on their surface was encapsulated with a thin carbon film. The Si/PCNF has mesopores of about 3.7 nm and a diameter of 31.4 nm, which is associated with the TEM results (Zhou et al., 2013).

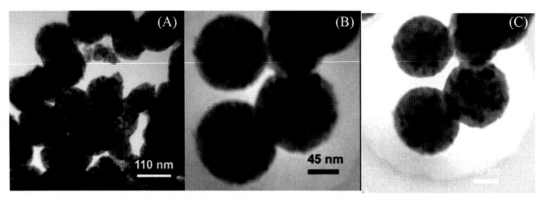

Figure 2.10 TEM image of hybrid γ-Fe$_2$O$_3$ SiNPs. (A) Low magnification TEM image (B) High magnification TEM image (C) TEM image of as-prepared Hybrid γ-Fe$_2$O$_3$ SiNPs.

Bright-field TEM evaluated the homogeneity of SiNC/PS hybrids. Investigations indicated marginal SiNC clustering irrespective of size, showing that NCs are evenly distributed across the polymer (Ren et al., 2021; Yang et al., 2014).

FS-PSA/SiO$_2$ hybrids were characterized using TEM analysis. TEM images showed that FS-PSA latexes had a clean core–shell structure, in which the dark regions serve the phase of PSA core and the light region represents FS containing the PSA shell. The results show the particle size and distribution of hybrid composites. Furthermore, the mean diameter of the hybrid composite was enhanced marginally, which was due to subsequent hydrolysis and condensation (Ye et al., 2019).

2.4 Conclusion

In this chapter, we have presented effective approaches for the synthesis of hybrid SiNPs, which show considerable interests for diverse applications. Overall, silicon-based hybrid nanoparticles were successfully produced by various methods such as fabrication, polymer grafting, sol–gel process, colloidal solution, and gas-phase synthesis. Al/insulator/SiO$_2$/Si, SiNWs/TIO$_2$, Si/PCNF, SiNWs/P3HT, and amorphous silicon-backed graphene were synthesized using the fabrication process to enhance their electrical, dielectric, and photodetection properties for the preparation of hybrid memory devices. PVA/SiO$_2$-TiO$_2$, SiO$_2$/PU, FS-PSA/SiO$_2$, polyacrylate/SiO$_2$ were produced using the sol−gel method to form hybrid membrane, UV radiation shield fibers, and to improve thermal stability. Using the polymer grafting method, various hybrids were prepared, such as SiNCs/PEDOT:PSS/water colloid, SiNC/poly(maleic anhydride), SiNC/polystyrene, and SiNCs-g-PEGDM, for applications such as hybrid solar cells and photoluminescence for us in in *vivo* imaging. Drug loading on silicon enhances the therapeutic efficacy and bioavailability. Various hybrids like PANi-PSiNPs, DOX/SPSiNPs, and PSi-PEI-PMVEMA are used in photothermal therapy and chemotherapy of cancer cells, whether in vitro or in vivo. SiNA-Pd hybrid material can be used as a catalyst in the Mizoroki−Heck reaction. To enhance the magnetic properties to disinfect waste water using magnetic extraction, silicon, magnetic iron, and titanium oxides functional hybrid NPs consisting of Fe_3O_4 and γ-Fe_2O_3 were prepared by researchers. These can be useful also in diagnostic imaging system in vivo. The formation of hybrid SiNPs was analyzed by various spectroscopic characterization such as FTIR. XPS, DLS, and TEM analysis. With a wide variety of applications, these silicon-based hybrid nanoparticles will undoubtedly have an extremely beneficial impact in the future.

References

Anderson, I. E., Shircliff, R. A., Macauley, C., Smith, D. K., Lee, B. G., Agarwal, S., ... Collins, R. T. (2012). Silanization of low-temperature-plasma synthesized silicon quantum dots for production of a tunable, stable, colloidal solution. *The Journal of Physical Chemistry C, 116*, 3979−3987. Available from https://doi.org/10.1021/jp211569a.

Anglin, E. J., Cheng, L. Y., Freeman, W. R., & Sailor, M. J. (2008). Porous silicon in drug delivery devices and materials. *Advanced Drug Delivery Review, 60*, 1266−1277. Available from https://doi.org/10.1016/j.addr.2008.03.017.

Astruc, D. (2008). *Nanoparticles and catalysis*. Weinheim: Wiley-VCH. Available from http://doi.org/10.1002/9783527621323.

báñez-Redín, G., Joshi, N., do Nascimento, G. F., et al. (2020). Determination of p53 biomarker using an electrochemical immunoassay based on layer-by-layer films with $NiFe_2O_4$ nanoparticles. *Microchimica Acta, 187*, 619. Available from https://doi.org/10.1007/s00604-020-04594-z.

Bao, Y., Yang, Y. Q., Shi, C. H., & Ma, J. Z. (2014). Fabrication of hollow silica spheres and their application in polyacrylate film forming agent. *Journal of Materials Science, 49*(24), 8215−8225. Available from https://doi.org/10.1007/s10853-014-8530-7.

Benelmekki, M., Bohra, M., Kim, J.-H., Diaz, R. E., Vernieres, J., Grammatikopoulos, P., & Sowwan, M. (2014). A facile single-step synthesis of ternary multicore magneto-plasmonic nanoparticles. *Nanoscale, 6*, 3532. Available from https://doi.org/10.1039/C3NR06114K.

Cagnani, G. R., Joshi, N., & Shimizu, F. M. (2019). Carbon nanotubes-based nanocomposite as photoanode. *Interfacial Engineering in Functional Materials for Dye-Sensitized Solar Cells*. Available from https://doi.org/10.1002/9781119557401.ch10.

Canham, L. T. (1990). Silicon quantum wire array fabrication by electrochemical and chemical dissolution of wafers. *Applied Physics Letters, 57*, 1046. Available from https://doi.org/10.1063/1.103561.

Cassidy, C., Singh, V., Grammatikopoulos, P., Djurabekova, F., Nordlund, K., & Sowwan, M. (2013). Inoculation of silicon nanoparticles with silver atoms. *Scientific Reports, 3*, 3083. Available from https://doi.org/10.1038/srep03083.

Choi, J. K., Dung, M. X., & Jeong, H. D. (2014). Novel synthesis of covalently linked silicon quantum dot–polystyrene hybrid materials: Silicon quantum dot–polystyrene polymers of tunable refractive index. *Materials Chemistry and Physics, 148*, 463–472. Available from https://doi.org/10.1016/j.matchemphys.2014.08.016.

Dasog, M., & Veinot, J. G. C. (2012). Size independent blue luminescence in nitrogen passivated silicon nanocrystals. *Physica Status Solidi A, 209*, 1844–1846. Available from https://doi.org/10.1002/pssa.201200273.

Dasog, M., Yang, Z., Regli, S., Atkins, T. M., Faramus, A., Singh, M. P., ... Veinot, J. G. (2013). Chemical insight into the origin of red and blue photoluminescence arising from freestanding silicon nanocrystals. *ACS Nano, 7*, 2676–2685. Available from https://doi.org/10.1021/nn4000644.

Deubel, F., Steenackers, M., Garrido, J. A., Stutzmann, M., & Jordan, R. (2013). Macromol. Semiconductor/polymer nanocomposites of acrylates and nanocrystalline silicon by laser-induced thermal polymerization. *Materials Science and Engineering*. Available from https://doi.org/10.1002/mame.201200392.

Dias, M. L., Pedroso, M. A. S., Mothé, C. G., & Azuma, C. (2004). Core shell silica-silicon hybrid nanoparticles: Synthesis and characterization. *Journal of Metastable Nanocrystalline Materials, 22*, 83. Available from https://doi.org/10.4028/www.scientific.net/JMNM.22.83.

Dung, M. X., Choi, J. K., & Jeong, H. D. (2013). Newly synthesized silicon quantum dot–polystyrene nanocomposite having thermally robust positive charge trapping. *ACS Applied Materials & Interfaces, 5*, 2400–2409. Available from https://doi.org/10.1021/am400356r.

Godin, B., Tasciotti, E., Liu, X., Serda, R. E., & Ferrari, M. (2011). Multistage nanovectors: From concept to novel imaging contrast agents and therapeutics. *Accounts of Chemical Research, 44*, 979–989. Available from https://doi.org/10.1021/ar200077p.

Hasegawa, I., Fukuda, Y., & Kajiwara, M. (1999). Inorganic–organic hybrid route to synthesis of ZrC and Si–Zr–C fibers. *Ceramics International, 25*, 523–527. Available from https://doi.org/10.1016/S0272-8842(97)00089-8.

Hasegawa, I., Nakamura, T., & Kajiwara, M. (1996). Synthesis of continuous silicon carbide-titanium carbide hybrid fibers through sol-gel processing. *Materials Research Bulletin, 31*, 869. Available from https://doi.org/10.1016/0025-5408(96)00063-3.

Hessel, C. M., Rasch, M. R., Hueso, J. L., Goodfellow, B. W., Akhavan, V. A., Puvanakrishnan, P., ... Korgel, B. A. (2010). Alkyl passivation and amphiphilic polymer coating of silicon nanocrystals for diagnostic imaging. *Small (Weinheim an der Bergstrasse, Germany), 6*, 2026–2034. Available from https://doi.org/10.1002/smll.201000825.

Höhlein, I. M. D., Werz, P. D. L., Veinot, J. G. C., & Rieger, B. (2015). Photoluminescent silicon nanocrystal-polymer hybrid materials via surface initiated reversible addition−fragmentation chain transfer (RAFT) polymerization. *Nanoscale*, *7*, 7811−7818. Available from https://doi.org/10.1039/C5NR00561B.

Hua, F., Swihart, M. T., & Ruckenstein, E. (2005). Efficient surface grafting of luminescent silicon quantum dots by photoinitiated hydrosilylation. *Langmuir: the ACS Journal of Surfaces and Colloids*, *21*, 6054−6062. Available from https://doi.org/10.1021/la0509394.

Jin, J., Rafiq, R., Gill, Y. Q., & Song, M. (2013). Preparation and characterization of high performance of graphene/nylon nanocomposites. *European Polymer Journal*, *49*(9), 2617−2626. Available from https://doi.org/10.1016/j.eurpolymj.2013.06.004.

Jin, Y., Tan, Y., Hu, X., Zhu, B., Zheng, Q., Zhang, Z., ... Zhu, J. (2017). Scalable production of the silicon−tin Yin-Yang hybrid structure with graphene coating for high performance lithium-ion battery anodes. *ACS Appied Materials Interfaces*, *9*, 15388. Available from https://doi.org/10.1021/acsami.7b00366.

Joshi, N., da Silva, L. F., Jadhav, H., M'Peko, J.-C., Millan Torres, B. B., Aguir, K., ... Oliveira, O. N., Jr. (2016). One-step approach for preparing ozone gas sensors based on hierarchical NiCo$_2$O$_4$ structures. *RSC Advances*, *6*, 92655−92662. Available from https://doi.org/10.1039/C6RA18384K.

Joshi, N., da Silva, L. F., Jadhav, H. S., Shimizu, F. M., Suman, P. H., M'Peko, J.-C., ... Oliveira, O. N., Jr. (2018). Yolk-shelled ZnCo$_2$O$_4$ microspheres: Surface properties and gas sensing application. *Sensors and Actuators B*, *257*, 906−915. Available from https://doi.org/10.1016/j.snb.2017.11.041.

Kassiba, A., Bednarski, W., Pud, A., Errien, N., Makowska-Janusik, M., Laskowski, L., ... Noskov, Y. (2007). Hybrid core − shell nanocomposites based on silicon carbide nanoparticles functionalized by conducting polyaniline: Electron paramagnetic resonance investigations. *The Journal of Physical Chemistry*, *111*, 11544. Available from https://doi.org/10.1021/jp070966y.

Kehrle, J., Hçhlein, I. M. D., Yang, Z., Jochem, A. R., Helbich, T., Kraus, T., ... Rieger, B. (2014). Thermoresponsive and photoluminescent hybrid silicon nanoparticles by surface-initiated group transfer polymerization of diethyl vinylphosphonate. *Angewandte Chemie International Edition*, *53*, 12494−12497. Available from https://doi.org/10.1002/anie.201405946.

Ko, M., Chae, S., Jeong, S., Oh, P., & Cho, J. (2014). Elastic a-silicon nanoparticle backboned graphene hybrid as a self-compacting anode for high-rate lithium ion batteries. *ACS Nano*, *8*, 8591. Available from https://doi.org/10.1021/nn503294z.

Kolliopoulou, S., Dimitrakis, P., Normand, P., Zhang, H. L., Cant, N., Evans, S. D., ... Tsoukalas, D. (2003). Hybrid silicon−organic nanoparticle memory device. *Journal of Applied Physics*, *94*, 5234. Available from https://doi.org/10.1063/1.1604962.

Kong, J., Yee, W. A., Wei, Y., Yang, L., Ang, J. M., Phua, S. L., ... Lu, X. (2013). Silicon nanoparticles encapsulated in hollow graphitized carbon nanofibers for lithium ion battery anodes. *Nanoscale*, *5*, 2967. Available from https://doi.org/10.1039/C3NR34024D.

Kutrovskaya, S., Arakelian, S., Kucherik, A., Osipov, A., Evlyukhin, A., & Kavokin, A. V. (2017). The synthesis of hybrid gold-silicon nano particles in a liquid. *Scientific Reports*, *7*, 10284. Available from https://doi.org/10.1038/s41598-017-09634-y.

Landry, C. J. T., Coltrain, B. K., & Brady, B. K. (1992). In situ polymerization of tetraethoxysilane in poly (methyl methacrylate): Morphology and dynamic mechanical properties. *Polymer, 33*, 1486. Available from https://doi.org/10.1016/0032-3861(92)90126-H.

Lee, J., Bodnarchuk, M. I., Shevchenko, E. V., & Talapin, D. V. (2010). Magnet-in-the-semiconductor FePt − PbS and FePt − PbSe nanostructures: Magnetic properties, charge transport, and magnetoresistance. *Journal of the American Chemical Society, 132*, 6382−6391. Available from https://doi.org/10.1021/ja100029s.

Lee, J. S., Shevchenko, E. V., & Talapin, D. V. (2008). Au − PbS core − shell nanocrystals: Plasmonic absorption enhancement and electrical doping via intra-particle charge transfer. *Journal of the American Chemical Society, 130*, 9673−9675. Available from https://doi.org/10.1021/ja802890f.

Li, J., Cushing, S. K., Meng, F., Senty, T. R., Bristow, A. D., & Wu, N. (2015). Plasmon-induced resonance energy transfer for solar energy conversion. *Nature Photonics, 9*, 601−607. Available from https://doi.org/10.1038/nphoton.2015.142.

Li, X., He, Y., Talukdar, S. S., & Swihart, M. T. (2003). Process for preparing macroscopic quantities of brightly photoluminescent silicon nanoparticles with emission spanning the visible spectrum. *Langmuir: the ACS Journal of Surfaces and Colloids, 19*, 8490. Available from https://doi.org/10.1021/la034487b.

Li, Z., & Ruckenstein, E. (2004). Water-soluble poly (acrylic acid) grafted luminescent silicon nanoparticles and their use as fluorescent biological staining labels. *Nano Letters, 4*, 1463−1467. Available from https://doi.org/10.1021/nl0492436.

Li, Z., Swihart, M., & Ruckenstein, E. (2004). Luminescent silicon nanoparticles capped byconductive polyaniline through the self-assembly method. *Langmuir: the ACS Journal of Surfaces and Colloids, 20*, 1963−1971. Available from https://doi.org/10.1021/la0358926.

Lin, Y., Chen, H., Chan, C.-M., & Wu, J. (2011). Effects of coating amount and particle concentration on the impact toughness of polypropylene/CaCO$_3$ nanocomposites. *European Polymer Journal, 47*(3), 294−304. Available from https://doi.org/10.1016/j.eurpolymj.2010.12.004.

Liu, C. Y., Holman, Z. C., & Kortshagen, U. R. (2009). Hybrid solar cells from P3HT and silicon nanocrystals. *Nano Letters, 9*, 449−452. Available from https://doi.org/10.1021/nl8034338.

Liu, K., Qu, S., Zhang, X., Tan, F., & Wang, Z. (2013). Improved photovoltaic performance of silicon nanowire/organic hybrid solar cells by incorporating silver nanoparticles. *Nanoscale Research Letters, 8*(1), 1−6. Available from https://doi.org/10.1186/1556-276X-8-88.

Liu, N., Wu, H., McDowell, M. T., Yao, Y., Wang, C., & Cui, Y. A. (2012). A yolk-shell design for stabilized and scalable Li-ion battery alloy anodes. *Nano Letters*. Available from https://doi.org/10.1021/nl3014814.

Liu, Q., Nayfeh, M. H., & Yau, S. T. (2011). A silicon nanoparticle-based polymeric nano-composite material for glucose sensing. *Journal of Electroanalytic Chemistry, 657*, 172−175. Available from https://doi.org/10.1016/j.jelechem.2011.03.022.

Liu, Z., Li, Y., Li, W., Xiao, C., Liu, D., Dong, C., et al. (2018). Multifunctional nanohybrid based on porous silicon nanoparticles, gold nanoparticles, and acetalated dextran for liver regeneration and acute liver failure theranostics. *Advanced Materials, 30*, 1703393. Available from https://doi.org/10.1002/adma.201703393.

Luo, J., Zhao, X., Wu, J., Jang, H. D., Kung, H. H., & Huang, J. (2012). Crumpled graphene-encapsulated Si nanoparticles for lithium ion battery anodes. *The Journal of Physical Chemistry Letters*, *3*, 1824–1829. Available from https://doi.org/10.1021/jz3006892.

Ma, H., Shi, T., & Song, Q. (2014). Synthesis and characterization of novel PVA/SiO_2-TiO_2 hybrid fibers. *Fibers*, *2*, 275–284. Available from https://doi.org/10.3390/fib2040275.

Mabrook, M. F., Pearson, C., Kolb, D., Zeze, D. A., & Petty, M. C. (2008). Memory effects in hybrid silicon-metallic nanoparticle-organic thin film structures. *Organic Electronics*, *9*, 816. Available from https://doi.org/10.1016/j.orgel.2008.05.023.

Malik, R., Tomer, V. K., Joshi, N., Dankwort, T., Lin, L. W., & Kienle, L. (2018). Au–TiO_2-loaded cubic gC_3N_4 nanohybrids for photocatalytic and volatile organic amine sensing applications. *ACS Applied Materials & Interfaces*, *10*, 34087–34097. Available from https://doi.org/10.1021/acsami.8b08091.

Mangolini, L., & Kortshagen, U. (2007). Plasma-assisted synthesis of silicon nanocrystal inks. *Advanced Materials*, *19*, 2513–2519. Available from https://doi.org/10.1002/adma.200700595.

Mastronardi, M. L., Hennrich, F., Henderson, E. J., Maier-Flaig, F., Blum, C., Reichenbach, J., ... Ozin, G. A. (2011). Preparation of monodisperse silicon nanocrystals using density gradient ultracentrifugation. *Journal of the American Chemical Society*, *133*, 11928–11931. Available from https://doi.org/10.1021/ja204865t.

Mastronardi, M. L., Maier-Flaig, F., Faulkner, D., Henderson, E. J., Kübel, C., Lemmer, U., & Ozin, G. A. (2012). Size-dependent absolute quantum yields for size-separated colloidally-stable silicon nanocrystals. *Nano Letters*, *12*, 337–342. Available from https://doi.org/10.1021/nl2036194.

Mehringer, C., Wagne, R., Jakuttis, T., Butz, B., Spiecker, E., & Peukert, W. (2014). Gas phase synthesis of anisotropic silicon germanium hybrid nanoparticles. *Journal of Aerosol Science*, *67*, 119–130. Available from https://doi.org/10.1016/j.jaerosci.2013.10.005.

Messaddeq, S. H., Bonnet, A. S., Santagnelli, S. H., Salek, G., Colmenares, Y. N., & Messaddeq, Y. (2019). Photopolymerized hybrids containing TiO_2 nanoparticles for gradient-index lens. *Materials Chemstry and Physics*, *236*, 121793. Available from https://doi.org/10.1016/j.matchemphys.2019.121793.

Mishra, P. K., Malik, R., Tomer, V. K., & Joshi, N. (2020). Hybridized graphitic carbon nitride (g-CN) as high performance VOCs sensor. *Functional Nanomaterials*.

Mitra, S., Cook, S., Švrček, V., Blackey, R. A., Zhou, W., Kovač, J., ... Mariotti, D. (2013). Improved optoelectronic properties of silicon nanocrystals/polymer nanocomposites by microplasma-induced liquid chemistry. *The Journal of Physical Chemistry C*, *117*, 23198–23207. Available from https://doi.org/10.1021/jp400938x.

Moraes, J., Ohno, K., Maschmeyer, T., & Perrier, S. (2015). Selective patterning of gold surfaces by core/shell, semisoft hybrid nanoparticles. *Small (Weinheim an der Bergstrasse, Germany)*, *11*, 482–488. Available from https://doi.org/10.1002/smll.201400345.

Noell, J. L. W., Wikes, G. L., Mohanty, D. K., & Mcgrath, J. E. (1990). The preparation and characterization of new polyether ketone-tetraethylorthosilicate hybrid glasses by the sol-gel method. *Journal of Applied Polymer Science*, *40*, 1177. Available from https://doi.org/10.1002/app.1990.070400709.

Panthani, M. G., Hessel, C. M., Reid, D. K., Asillas, G., Jose-Yacaman, M., & Korgel, B. (2012). Graphene-supported high-resolution TEM and STEM imaging of silicon nanocrystals and their capping ligands. *The Journal of Physical Chemistry C, 116*, 22463−22468. Available from https://doi.org/10.1021/jp308545q.

Park, J. H., Gu, L., Maltzahn, G. V., Ruoslahti, E., Bhatia, S. N., & Sailor, M. J. (2009). Biodegradable luminescent porous silicon nanoparticles for in vivo applications. *Nature Materials, 8*, 331−336. Available from https://doi.org/10.1038/nmat2398.

Patel, G., Pillai, V., P., Bhatt, & Mohammad, S. (2020). Application of nanosensors in the food industry. In Tuan Anh Nguyen, et al. (Eds.), *Nanosensors for Smart Cities* (1st ed.). Elsevier.

Patel, G., Shah, V., Bhatt, G., & Deota, P. (2021). Humidity nanosensors for smart manufacturing. In Sabu Thomas, Tuan Anh Nguyen, et al. (Eds.), *Nanosensors for Smart Manufacturing* (1st ed.). Elsevier.

Patel, G., Vora, M., & Pillai, V. (2019). Liquid phase exfoliation of two-dimensional material for sensors and photocatalysis - A Review. *Journal of Nanoscience & Nanotechnology, 19*, 5054−5073. Available from https://doi.org/10.1166/jnn.2019.16933.

Pedroso, M. A. S., Dias, M. L., Azuma, C., & Mothé, C. G. (2000). Synthesis of acrylic-modified sol−gel silica. *Colloid and Polymer Science, 278*, 1180. Available from https://doi.org/10.1007/s00396-005-1424-0.

Poovazhagan, L., Kalaichelvan, K., Rajadurai, A., & Senthilvelan, V. (2013). Characterization of hybrid silicon carbide and boron carbide nanoparticles-reinforced aluminum alloy composites. *Procedia Engineering, 64*, 681−689. Available from https://doi.org/10.1016/j.proeng.2013.09.143.

Rangari, V. K., Yousuf, M., & Jeelani, S. (2013). Influence of SiC/Si_3N_4 hybrid nanoparticles on polymer tensile properties. *Journal of Composites Science*. Available from https://doi.org/10.1155/2013/462914.

Rasool, K., Rafiq, M. A., Ahmad, M., Imran, Z., & Hasan, M. M. (2012). TiO_2 nanoparticles and silicon nanowires hybrid device: Role of interface on electrical, dielectric, and photodetection properties. *Applied Physics Letters, 101*, 253104. Available from https://doi.org/10.1063/1.4772068.

Ren, X.-H., Wang, H.-Y., Li, S., He, X.-W., Li, W.-Y., & Zhang, Y.-K. (2021). Preparation of glycan-oriented imprinted polymer coating Gd-doped silicon nanoparticles for targeting cancer Tn antigens and dual-modal cell imaging via boronate-affinity surface imprinting. *Talanta*. Available from https://doi.org/10.1016/j.talanta.2020.121706.

Sakka, S., Tomozawa, M., & Doremus, R. (Eds.), (1982). *Gel Method for Making Glass*. New York: Academic Press. Available from https://doi.org/10.1016/S0161-9160(13)70027-9.

Santos, H. A., Mäkilä, E., Airaksinen, A. J., Bimbo, L. M., & Hirovnen, J. (2014). Porous silicon nanoparticles for nanomedicine: Preparation and biomedical applications. *Nanomedicine: Nanotechnology, Biology, and Medicine, 9*, 535−554. Available from https://doi.org/10.2217/nnm.13.223.

Sarita, K., Susheel, K., Annamaria, C., James, N., Youssef, H., & Rajesh, K. (2013). Surface modification of inorganic nanoparticles for development of organic−inorganic nanocomposites—A review. *Progress in Polymer Science, 38* (8), 1232−1261. Available from https://doi.org/10.1016/j.progpolymsci.2013.02.003.

Sato, K., Fukata, N., Hirakuri, K., Murakami, M., Shimizu, T., & Yamauchi, Y. (2010). Flexible and transparent silicon nanoparticle/polymer composites with stable luminescence. *Chemistry, an Asian Journal, 5*, 50–55. Available from https://doi.org/10.1002/asia.200900403.

Sato, K., Hiruoka, M., Fujioka, K., Fukata, N., Hirakuri, K., & Yamamoto, K. (2009). Toxicity effect of cancer cell labeled with visible luminescent nanocrystalline silicon particles and visualization observation in vivo. *Research Society Symposium Proceedings, 1145*. Available from https://doi.org/10.1557/PROC-1145-MM04-23, 1145-MM04–23.

Sato, K., Kishimoto, N., & Hirakuri, K. (2007). White luminescence from silica glass containing red/green/blue luminescent nanocrystalline silicon particles. *Journal of Applied Physics, 102*, 104305. Available from https://doi.org/10.1063/1.2811920.

Sato, K., Yokosuka, S., Takigami, Y., Hirakuri, K., Fujioka, K., Manome, Y., ... Fukata, N. (2011). Size-tunable silicon/iron oxide hybrid nanoparticles with fluorescence, superparamagnetism, and biocompatibility. *Journal of the American Chemical Society, 133*, 18626–18633. Available from https://doi.org/10.1021/ja202466m.

Singh, A., Salmi, Z., Joshi, N., Jha, P., Kumar, A., Lecoq, H., ... Gupta, S. K. (2013). Photo-induced synthesis of polypyrrole-silver nanocomposite films on N-(3-trimethoxysilylpropyl)pyrrole-modified biaxially oriented polyethylene terephthalate flexible substrates. *RSC Advances, 3*, 5506–5523. Available from https://doi.org/10.1039/C3RA22981E.

Sugimoto, H., Fujii, M., & Imakita, K. (2016). Silicon nanocrystal-noble metal hybrid nanoparticles. *Nanoscale, 8*, 10956–10962. Available from https://doi.org/10.1039/C6NR01747A.

Tasciotti, E., Liu, X., Bhavane, R., Plant, K., Leonard, A. D., Price, B. K., ... Ferrari, M. (2008). Mesoporous silicon particles as a multistage delivery system for imaging and therapeutic applications. *Nature Nanotechnology, 3*, 151–157. Available from https://doi.org/10.1038/nnano.2008.34.

Tchalala, M. R., El-Demellawi, J. K., Kalakonda, P., et al. (2020). High thermally stable hybrid materials based on amorphous porous silicon nanoparticles and imidazolium-based ionic liquids: Structural and chemical analysis. *Materials Today: Proceedings*. Available from https://doi.org/10.1016/j.matpr.2020.03.706.

Tzur-Balter, A., Shatsberg, Z., Beckerman, M., Segal, E., & Artzi, N. (2015). Mechanism of erosion of nanostructured porous silicon drug carriers in neoplastic tissues. *Nature Communications, 6*, 6208. Available from https://doi.org/10.1038/ncomms7208.

Wang, C., Yin, L., Zhang, L., Kang, L., Wang, X., & Gao, R. (2009). Magnetic (γ-Fe_2O_3@SiO_2)$_n$@TiO_2 functional hybrid nanoparticles with actived photocatalytic ability. *The Journal of Physical Chemistry C, 113*, 4008. Available from https://doi.org/10.1021/jp809835a.

Wang, M. S., Song, W. L., Wang, J., & Fan, L. Z. (2015). Highly uniform silicon nanoparticle/porous carbon nanofiber hybrids towards free-standing high-performance anodes for lithium-ion batteries. *Carbon, 82*, 337–345. Available from https://doi.org/10.1016/j.carbon.2014.10.078.

Wei, G. C., Kennedy, C. R., & Harris, L. A. (1984). Synthesis of sinterable SiC powders by carbothermic reduction of gel-derived precursors; and pyrolysis of polycarbosilane. *Bulletin of the American Ceramic Society, 63*, 1054–1061.

Wu, Y., Joshi, N., Zhao, S., Long, H., Zhou, L., Ma, G., ... Lin, L. (2020). NO_2 gas sensors based on CVD tungsten diselenide monolayer. *Applied Surface Science*, *529*, 147110. Available from https://doi.org/10.1016/j.apsusc.2020.147110.

Xia, B., Wang, B., Chen, Z., Zhang, Q., & Shi, J. (2016). Near-infrared light-triggered intracellular delivery of anticancer drugs using porous silicon nanoparticles conjugated with IR820 dyes. *Advanced Materials. Interfaces*, *3*, 1500715. Available from https://doi.org/10.1002/admi.201500715.

Xia, B., Wang, B., Shi, J., Zhang, Y., Zhang, Q., Chen, Z., & Li. (2017). Photothermal and biodegradable polyaniline/porous silicon hybrid nanocomposites as drug carriers for combined chemo-photothermal therapy of cancer. *Journal of Acta Biomaterialia*. Available from https://doi.org/10.1016/j.actbio.2017.01.015.

Xia, B., Wang, B., Zhang, W., & Shi, J. (2015). High loading of doxorubicin into styrene-terminated porous silicon nanoparticles via π-stacking for cancer treatments in vitro. *RSC Advances*, *5*, 44660−44665. Available from https://doi.org/10.1039/C5RA04843E.

Xia, X., Mai, J., Xu, R., Perez, J. E. T., Guevara, M. L., Shen, Q., ... Shen, H. (2015). Porous silicon microparticle potentiates anti-tumor immunity by enhancing cross-presentation and inducing type I interferon response. *Cell Reports*, *11*, 957−966. Available from https://doi.org/10.1016/j.celrep.2015.04.009.

Xu, R., Zhang, G., Mai, J., Deng, X., Segura-Ibarra, V., Wu, S., ... Shen, H. (2016). An injectable nanoparticle generator enhances delivery of cancer therapeutics. *Nature Biotechnology*, *34*, 414−419. Available from https://doi.org/10.1038/nbt.3506.

Yamada, Y. M. A., Yuyama, Y., Sato, T., Fujikawa, S., & Uozumi, Y. (2014). A palladium-nanoparticle and silicon-nanowire-array hybrid: A platform for catalytic heterogeneous reactions. *Angewandte Chemie International Edition*, *53*, 127−131. Available from https://doi.org/10.1002/ange.201308541.

Yang, Z., Dasog, M., Dobbie, A. R., Lockwood, R., Zhi, Y., Meldrum, A., & Veinot, J. G. C. (2014). Highly luminescent covalently linked silicon nanocrystal/polystyrene hybrid functional materials: Synthesis, properties, and processability. *Advanced Functional Materials*, *24*, 1345−1353. Available from https://doi.org/10.1002/adfm.201302091.

Ye, L., Ma, G., Zheng, S., Huang, X., Zhao, L., Luo, H., & Liao, W. (2019). Facile fabrication of fluorine−silicon-containing poly (styrene−acrylate)/SiO_2 hydrophobic composites by combining physically mixing and sol−gel process. *Journal of Coatings Technology and Research*, *16*, 1243−1252. Available from https://doi.org/10.1007/s11998-019-00198-2.

Zhang, H., Liu, D., Shahbazi, M.-A., Mäkilä, E., Herranz-Blanco, B., Salonen, J., ... Santos, H. A. (2014). Fabrication of a multifunctional nano-in-micro drug delivery platform by microfluidic templated encapsulation of porous silicon in polymer matrix. *Advanced Materials*, *26*, 4497. Available from https://doi.org/10.1002/adma.201400953.

Zhang, J., Tang, Y., Lee, K., & Ouyang, M. (2010). Nonepitaxial growth of hybrid core-shell nanostructures with large lattice mismatches. *Science (New York, N. Y.)*, *327*, 1634−1638. Available from https://doi.org/10.1126/science.1184769.

Zhang, Z., He, X., Wang, X., Rodrigues, A. M., & Zhang, R. (2018). Reinforcement of the mechanical properties in nitrile rubber by adding graphene oxide/silicon dioxide hybrid nanoparticles. *Journal of Applied Polymer Science*, *135*, 46091−46099. Available from https://doi.org/10.1002/app.46091.

Zhou, X. S., Wan, L.-J., & Guo, Y.-G. (2013). Electrospun silicon nanoparticle/porous carbon hybrid nanofibers for lithium-ion batteries. *Small (Weinheim an der Bergstrasse, Germany), 9*, 2684–2688. Available from https://doi.org/10.1002/smll.201202071.

Zhu, Y., & Sun, D.-X. (2004). Preparation of silicon dioxide/polyurethane nanocomposites by a sol–gel process. *Journal of Applied Polymer Science, 92*, 2013–2016. Available from https://doi.org/10.1002/app.20067.

Properties of silicon−carbon (CNTs/graphene) hybrid nanoparticles

Xinyi Chen, Deng Long and Jingqin Cui

Pen-Tung Sah Institute of Micro-Nano Science and Technology, Xiamen University, Xiamen, P.R. China

3.1 Introduction

Carbon sources are abundantly present on Earth, such as fossil oil, coal, graphite, plants, etc. The rich properties and abundance have made carbon one of the most broadly used materials in modern industry. The big family of nanosized carbon, such as carbon dots, fullerene, graphene, carbon nanotubes (CNTs), and their derivatives, keep attracting great interest due to their highly tunable properties, and they have exhibited great potential in the fields of optoelectronics, catalysis, medicine, energy, and the environment. The size-induced quantum effect has significantly enhanced the electrical and thermal conductivity of nanosized carbon. The carrier mobility of graphene can exceed $10{,}000 \, \text{cm}^2 \, \text{V}^{-1} \, \text{s}^{-1}$ (Novoselov et al., 2004) with thermal conductivity of about $\sim 5000 \, \text{W m}^{-1} \, \text{K}^{-1}$ (Geim, 2009), while the mobility of CNT can be as high as $79{,}000 \, \text{cm}^2 \, \text{V}^{-1} \, \text{s}^{-1}$ (Durkop & Getty, 2003) with thermal conductivity of about $2000-6000 \, \text{W m}^{-1} \, \text{K}^{-1}$ (Han & Fina, 2011). More interestingly, the electrical properties of graphene and CNTs can be further adjusted between metal and semiconductor via physically or chemically manipulated sizes, structures, native defects, and impurities. Both located in Group IVA, carbon and silicon have the same number of valence electrons and bonding modes. Compared to silicon, the properties of carbon would be closer to metal, so the complementary properties straightforwardly facilitate the attempt to combine silicon and carbon.

The design of Si/C hybrid nanoparticles will be discussed in this chapter. Different methods, starting materials, and surface functionalization result in different morphologies and structures.

The electrical properties of Si/C hybrid nanoparticles will be discussed. The main application of silicon nanoparticles based on their electrical properties is as the next-generation anode for lithium-ion batteries (LIBs), because the theoretically predicted alloying of silicon and lithium has revealed a very high capacity of silicon for lithium storage, which makes silicon a promising candidate for high energy density anode of LIBs. However, the lithiation of silicon would cause a significant volume expansion that negatively affects the stability. Carbon, in contrast, is a widely used commercial anode with very stable performance. The weakness of the current commercial carbon anodes is their relatively low capacity, which needs to be improved to meet the demand for high density and high rate anodes required by the fast development of portable electronic devices and electric vehicles. Thus how to synergize the advantages of silicon and carbon to establish a well performed high density and high rate anode for LIBs is becoming more and more attractive. This chapter will discuss how the material design, micromorphology and chemical components influence the electrochemical performance of Si/C nanoparticles.

Furthermore, when the size of nanoparticles goes down to the quantum region, namely quantum dots (QDs), the band structure will be shifted by the quantum effect, enlarging the bandgap to exhibit extra optical properties. Moreover, the edge effect might also introduce additional optical properties for quantum-sized graphene. Thus Si/C nanoparticles at the quantum region have exhibited potential for the future of optoelectronics and medicine. This chapter will then discuss the quantum effect-induced optoelectronic properties and applications of ultrasmall Si/C nanoparticles (QDs).

3.2 Design of Si/C nanoparticles

Carbon nanomaterials such as graphene, CNTs, and carbon dots can be obtained via various physical or chemical approaches. Graphene can be achieved by mechanical or chemical exfoliation (Novoselov et al., 2004). Chemical vapor deposition (CVD) could achieve high-quality, large-area monolayer or few-layered graphene on both smooth micro-, and nanomorphological substrates (Guo et al., 2017; Zhu et al., 2010). CVD is also the dominant technique to fabricate CNTs (Baughman, Zakhidov, & Heer, 2002). In addition, ultrasmall carbon particles, such as graphene QDs, could be extracted from as-prepared graphene by further mechanical or chemical pulverization with gradual and precise purification (Liu et al., 2020). Based on various fabrication and functionalization techniques of silicon and carbon materials, a series of property-oriented designs have been developed to produce Si/C nanoparticles with controllable configurations. This section will discuss three strategies of Si/C configuration: mixing, interface definition, and cross-dimensional hybrids.

3.2.1 Mixed Si/C nanoparticles

Mixing silicon and carbon nanomaterials to obtain Si/C hybrid is the most straightforward and inexpensive method for mass production. However, the rough mixing of prepared silicon and carbon nanomaterials usually results in an incompact interface. That random and incompact interface would inhibit the interaction between silicon and carbon, which makes it difficult to exhibit an optimized hybrid property. Thus mixing proper precursors of silicon and carbon before mechanical milling or chemical reaction is a usual approach for building strongly interfacial-combined Si/C hybrid nanoparticles.

The mechanical milling of mixed silicon nanoparticles and organic carbon precursors has been widely applied to prepare Si/C nanoparticles, followed by pyrolysis to obtain a highly compact material (Terranova, Orlanducci, Tamburri, Guglielmotti, & Rossi, 2014). It is found that the introduction of fluorin could etch the surface of silicon during pyrolysis, increasing the surface area of silicon, that is the interface between silicon and carbon (Liu et al., 2009). Comparably, precursors for graphene and CNTs can be also introduced into mechanical milling to achieve Si/graphene and Si/CNTs hybrids (Wang & Wang, 2005; Zuo, Yin, & Ma, 2007).

Assembly based on interfaces in liquid phase is another effective approach for achieving compact Si/C hybrid nanoparticles. An emulsion environment of water and oil with the precursors of silicon nanoparticles, carbon black, and graphene oxide, would help the assembly of silicon nanoparticles and graphene (Fig. 3.1) (Chen et al., 2014; Yun et al., 2016). Using more than one carbon precursor would sometimes favor the compact formation of Si/C hybrids, because different carbon precursors could stack together to build up a three-dimensional (3D) network for the settlement of silicon and carbon (Zhou, Yin, Cao, Wan, & Guo, 2012).

3.2.2 Finely configured Si/C hybrid nanoparticles

Although an intensive mixing of silicon and carbon has resulted in compactly mixed Si/C nanoparticles, the interfaces between silicon and carbon are randomly formed. Thus interface engineering has been developed to define the configuration

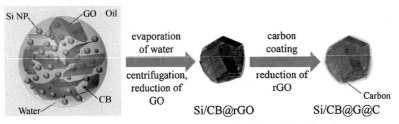

Figure 3.1 Schematic of the preparation process of Si/carbon blackB@graphene@carbon. Reprinted with permission from Yun, Q., Qin, X., He, Y.-B., Lv, W., Kaneti, Y.V., Li, B., ... Kang, F. (2016). Micron-sized spherical Si/C hybrids assembled via water/oil system for high-performance lithium ion battery. *Electrochimica Acta*, *211*, 982–988. Copyright 2016, Elsevier.

of Si/C nanoparticles, forming an adjustable interface between silicon and carbon to manipulate the interaction.

The porous Si/C hybrid is a kind of finely configured structure that can significantly increase the surface area as well as the uniformity of Si/C. The etching derived porous silicon and 3D carbon network are the usual starting materials, followed by dispersion or coating of the complementary material for Si/C (Liu et al., 2013; Tian, Tan, Xin, Wang, & Han, 2015). The introduction of a third template or coupling material would be also applied to achieve porous Si/C (Woo, Park, Hwang, & Whang, 2012; Yin, Xin, Wan, Li, & Guo, 2011).

The core–shell structure can define a very clear interface for Si/C hybrid nanoparticles. The Si/C is a commonly proposed structure, starting with silicon nanoparticles. The carbon outer layer can be usually achieved by vapor deposition or carbonization (Dimov, Fukuda, Umeno, Kugino, & Yoshio, 2003; Si et al., 2009). This core–shell structure can be also applied to silicon nanowires (Huang, Fan, Shen, & Zhu, 2009). In addition, the use of doped carbon material or an interlayer between silicon and carbon has also been found to be positive for electrical properties (He, Xu, Wang, Xu, & Zhang, 2018; Hu et al., 2008).

Moreover, a further developed core–shell structure, that is Si–void–C, namely yolk–shell structure, has been proposed aiming at the volume expansion issue in a Si-based LIB (Fig. 3.2) (Xie et al., 2017).

3.2.3 Cross-dimensional Si/C hybrids

For fully exhibiting the low-dimensional properties of graphene and CNTs, and facilitating the fabrication of freestanding devices, the cross-dimensional (0D/1D, 0D/2D) hybridization for silicon and carbon has attracted in-depth investigation.

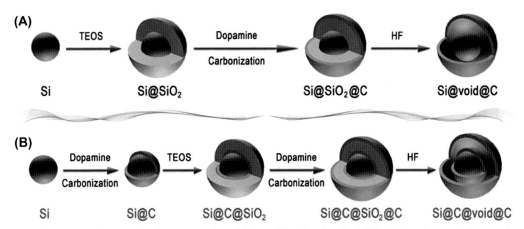

Figure 3.2 Schematic diagram for the preparation of (A) traditional yolk−shell Si@void@carbon and (B) novel core−shell yolk−shell Si@carbon@void@carbon. Reprinted with permission from Xie, J., Tong, L., Su, L., Xu, Y., Wang, L., & Wang, Y. (2017). Core-shell yolk-shell Si@C@Void@C nanohybrids as advanced lithium ion battery anodes with good electronic conductivity and corrosion resistance. *Journal of Power Sources, 342*, 529−536. Copyright 2017, Elsevier.

The 0D/1D Si/CNTs structure can start with the growth of CNTs, followed by deposition of silicon nanoparticles (Wang & Kumta, 2010). Comparably, a typical 0D/2D Si/graphene structure is a silicon nanoparticle embedded graphene sheet, while the edge of graphene can be functionalized to build a widely connected 3D graphene network (Evanoff, Magasinski, Yang, & Yushin, 2011). A multilayered 0D/2D hybrid, such as a sandwich structure, that is graphene/Si/graphene has been also developed (Fig. 3.3) (Feng et al., 2020). More complicatedly, a core−shell structured Si/C nanoparticle can further act as the 0D part for a 0D/2D Si/C hybrid (Wang, Hu, Mao, & Mao, 2020).

3.3 Electrical properties of Si/C nanoparticles

The theoretically predicted alloying of silicon and lithium has implied the very promising potential of silicon for high energy density anodes of LIBs. The biggest challenge in this field is how to solve the lithiation induced volume expansion issue, to maintain the structure of the anode for guaranteeing long-term stable performance. The Si/C hybrid nanoparticles, combining the high energy density of silicon and stability of carbon, have been widely investigated for the solution to silicon-based anodes. The structure−activity relationship of Si/C anodes, that is influence of morphology, structure, and chemical components would dominate the practical performance.

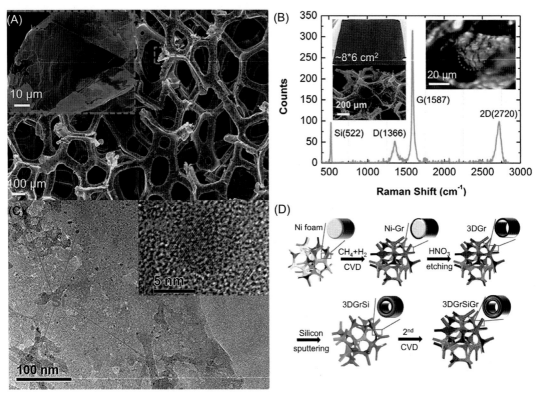

Figure 3.3 (A) Scanning electron microscopy (SEM) image. (B) Raman spectrum. (C) Transmission electron microscopy (TEM) image. (D) X-ray diffraction (XRD) of 3D graphene/Si/graphene (3DGrSiGr). The inset of (B) shows the distribution of C (red in the color version) and Si (green in the color version) via EDS mapping and optical image. The inset of (C) shows the high-resolution TEM of Si nanoparticle with distinct (220) facets of Si diamond structure. (D) Schematic 3DGrSiGr fabrication process. Reprinted with permission from Feng, Z., Huang, C., Fu, A., Chen, L., Pei, F., He, Y., ..., Cui, J. (2020). A three-dimensional network of graphene/silicon/graphene sandwich sheets as anode for Li-ion battery. *Thin Solid Films*, *693*, 137702. Copyright 2020, Elsevier.

3.3.1 Influence of morphology and structure

A pure silicon film as the anode of LIBs would cause significant volume change and structural destruction (Fig. 3.4) (Yang et al., 2018) and that is why Si/C nanoparticles have been proposed (Fig. 3.4) (Yang et al., 2018).

The size, shape, orientation, and spatial distribution of Si/C anodes would be essential to their electrical properties, while being crystalline or amorphous has a slight effect on the anode performance due to the structural distortion and stress accumulated

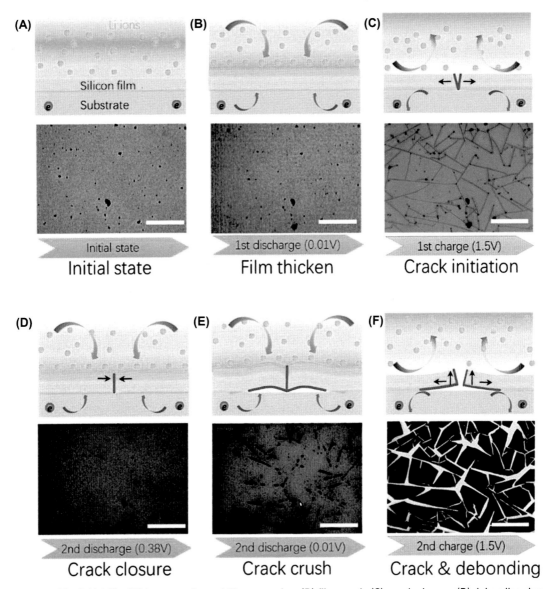

Figure 3.4 The initial film (A) has experienced film expansion, (B) film crack, (C) crack closure, (D) debonding due to crack face crushing, (E) and mode I debonding during lithium desertion (F). The scale bar is 20 mm. Reprinted with permission from Yang, L., Chen, H.S., Jiang, H., Wei, Y.J., Song, W.L., Fang, D.N. (2018). Failure mechanisms of 2D silicon film anodes: in situ observations and simulations on crack evolution. *Chemical Communications (Cambridge), 54*, 3997–4000. Copyright 2018, Royal Society of Chemistry.

during alloying with lithium and the crystalline phase transition (Huang et al., 2018). Thus a proper Si/C anode should be designed after fully considering the above factors. The silicon part should be nanosized to naturally alleviate the lithiation-induced volume change, while the introduction of carbon can facilitate the conductivity as a current collector, and act as a support that can restrict the volume change or provide additional space for it (Feng et al., 2020).

Tracking the volume changes and structural deformations of Si/C anodes after repetitive lithiation/delithiation has shown that in a core–shell structured Si/C anode, the amorphous carbon shell could shatter to expose a silicon core, though many pieces are still attached onto silicon nanoparticle surface (Wu et al., 2012). On the other hand, when graphene cages were grown over silicon particles, the elasticity promised the shell integrity while still allowed ionic permeability through shell defects (Li et al., 2016). Besides, the porous structure and yolk–shell structure might provide extra space to tolerate the volume change during lithiation (Si et al., 2009; Yin et al., 2011).

To achieve both high electrical conductivity and mechanical stability, combining various formats of carbon to construct a more complicated Si/C structure is a straightforward strategy. Hierarchical Si/C porous structures built on 3D graphene network or cross-linked CNT/Gr network with silicon nanoparticles embedded, have been proven to have good electrochemical performance and mechanical stability (Feng et al., 2020; Ji et al., 2013). Also, organic nanostructure can be introduced to provide an extraordinary mechanical property to realize an ultrahigh mass loading of silicon, that is capacity. A Si/CNTs with cellulose hybrid can have a silicon content of 92%, while the conductivity can be kept at ~ 200 S cm^{-1} (Yang et al., 2018).

3.3.2 Influence of chemical components

The chemical states of elements or molecular structures, formation of solid electrolyte interlayer (SEI), dendrite growth, as well as electrolyte distribution near SEI have been found responsible for decreasing energy storage or stability (Pekarek et al., 2020; Ruther et al., 2018; Shi, Ross, Somorjai, & Komvopoulos, 2017; Yang, Kraytsberg, & Ein-Eli, 2015). The chemical evolution is as important as morphological evolution to the performance of Si/C anodes, and implicates the charge-transfer mechanisms (Finegan et al., 2019; Tian et al., 2018). It has been also found that during the battery resting electrolyte reduction was possibly happening and a reactive phase, formed upon deep discharge, relaxed back to a more stable phase, that is a self-discharge process. Pecher, Carretero-González, Griffith, and Grey (2016).

Organic functional groups anchored over silicon nanoparticles captured lithium ions and transformed them into a stable artificial SEI layer, hence increasing the electrode performance. The difference between surface functional groups and carbon coating has been investigated via Fourier-transform infrared spectroscopy (FTIR) and nuclear magnetic resonance spectroscopy (NMR), proposing that carbon shells actually lacked the advantage of reacting with electrolyte and lithium ions (Ren et al., 2019) (Fig. 3.5) (Ren et al., 2019).

Electrochemical impedance spectrum and galvanostatic intermittent titration technique gives a general idea of the internal charge transfer resistance and the ionic diffusion coefficient within the active material, find that compositing silicon with carbon helps increase the electric or ionic transport. Apparently different formats of carbon have different thermal and mechanical properties, affecting the electrical conductivity for Si/C hybrid anodes in different ways. On the other hand, thermal analysis and X-ray photoelectron spectroscopy has found that carbon in a Si/C anode can also contribute to part of the capacity, especially in a multitype carbon-embedded Si/C hybrid nanostructure (Feng et al., 2020).

3.3.3 In situ characterization on of Si/C hybrids' properties

To fully understand how and why each component responds to lithiation and delithiation, in situ (or *operando*) instrumentation is indispensable to track real-time, real-location, real-environment dynamic evolutions of the morphology, structure, and chemical composition during electrochemical reactions at different charging stages. More importantly, in situ characterization is greatly helpful to figure out the mechanism of how active material failure comes (Liu et al., 2019; Wu & Liu, 2018).

With high-resolution in situ SEM or TEM, an anisotropic volume expansion of silicon nanowires and amorphous carbon shell fracturing could be clearly observed during lithiation/delithiation. The results suggested that the silicon lithiation front was strongly related to its crystallographic orientation, while the graphitic carbon shell could preserve its structural wholeness thanks to its mechanical flexibility. Aligning the void inside carbon shells with the silicon core expansion direction could therefore potentially keep shells free of dramatic fractures, mitigating the composite's energy storage degradation (Li et al., 2016; Liu et al., 2011). In situ XRD and

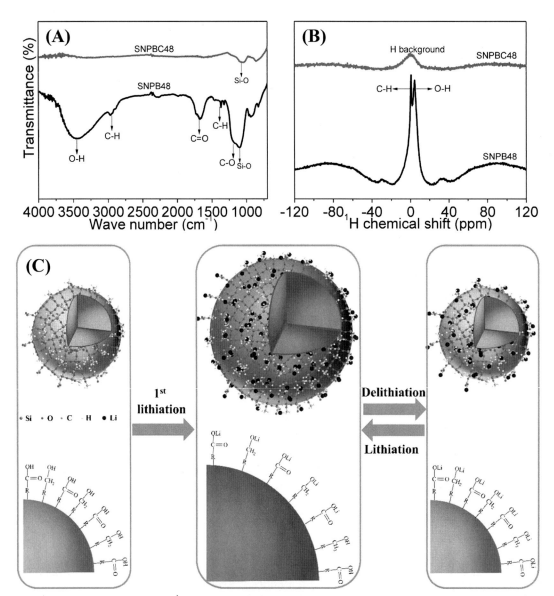

Figure 3.5 (A) FTIR spectra and (B) ¹H solid state NMR spectra of ball milled Si nanoparticles, and (C) the schematic visualization of ball milled Si nanoparticles during the first lithiation and the followed delithiation/lithiation processes. Reprinted with permission from Ren, W., Wang, Y., Tan, Q., Yu, J., Etim, U.J., Zhong, Z., Su, F. (2019). Nanosized Si particles with rich surface organic functional groups as high-performance Li-battery anodes. *Electrochimica Acta*, *320*, 134625. Copyright 2019, Elsevier.

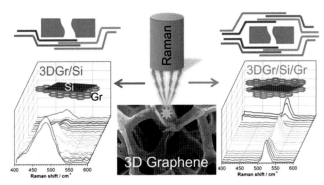

Figure 3.6 Protection mechanisms of graphene-encapsulated silicon anodes with operando Raman spectroscopy. Reprinted with permission from Huang, C., Feng, Z., Pei, F., Fu, A., Qu, B., Chen, X., ... Cui, J. (2020). Understanding protection mechanisms of graphene-encapsulated silicon anodes with operando Raman spectroscopy. *ACS Applied Materials & Interfaces, 12*, 35532−35541. Copyright 2020, American Chemical Society.

Raman characterization have revealed changes in crystal phases, chemical compositions, and structural deformations. It has been found that these changes cannot be uniform and simultaneous, and will lead to local strains in and between particles, eventually resulting in electrode cracking and delamination (Finegan et al., 2019; Tardif et al., 2017). Silicon and carbon both have distinguishable Raman signals. The changes of crystal states, chemical compositions, molecular structures, and mechanical strains in a Si/C composite could lead to position shifts or intensity variations of the peaks. With a confocal Raman microscope and free-standing Si/graphene foam electrodes assembled in a flexible, optically transparent battery, the protection mechanism of graphene over silicon has been clearly specified (Fig. 3.6) (Huang et al., 2020).

The rich experimental data have notably advanced our understanding of silicon lithiation and carbon performance and will facilitate the optimization of morphologies, architecture designs, and chemical compositions for Si/C hybrid nanoparticles.

3.4 Optical properties of ultrasmall Si/C nanoparticles

Bulk Si is an indirect bandgap semiconductor, which hinders its wide application in photodetectors, phototransistors, biosensors, and other optical applications (Herzinger, Johs, McGahan, Woollam, & Paulson, 1998; Monroy, Omnès, & Calle, 2003; Pavesi,

Negro, Mazzoleni, Franzo, & Priolo, 2000). For slicon nanoparticles with the size below 5 nm (Bohr radius), namely silicon QDs, the quantum confinement effect enlarges the bandgap to exhibit different optical properties, which has made silicon QDs very promising in the application of fluorescent probes, photodetectors, solar cells, etc. The bandgap of silicon QDs, influenced by variation in size, shape, and surface (Pavesi et al., 2000; Valenta, Juhasz, & Linnros, 2002; Wang & Zunger, 1994), can be described by

$$E_g = 0.45 + 3.33/D$$

where E_g is the bandgap in eV, D is the diameter in nm (Fig. 3.7) (Niaz & Zdetsis, 2016; Zhang, De Sarkar, Niehaus, & Frauenheim, 2012).

3.4.1 Fluorescence of Si/C quantum dots

The SiQDs are widely used for the fluorescence probes due to their quantum effect-induced fluorescent properties as well as their biocompatibility, cell permeability, and nontoxicity (Hu et al., 2019; Licciardello et al., 2018). Meanwhile, carbon QDs (CQDs) with the diameters of 2–8 nm can exhibit tunable optical properties from ultraviolet to near infrared via varying sizes and surface functionalization (Liu et al., 2020; Wolfbeis, 2015). It is found that Si/C hybrids could exhibit a stronger fluorescence, better biocompatibility, and therapeutic effects (Hu et al., 2019), showing their high potential in nanomedical applications. For example, the microwave-assisted approach-derived multifunctional Si/C nanocomposites were employed in high bacteria inhibition and expediting wound healing by grafting a functional group of Notoginseng (Fig. 3.8A) (Hu et al., 2019).

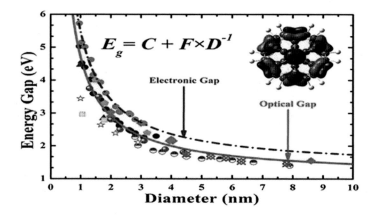

Figure 3.7 Plot corresponding to the energy gap energy dependence on the diameter of SiQDs. Reprinted with permission from Niaz, S., & Zdetsis, A.D. (2016). Comprehensive Ab initio study of electronic, optical, and cohesive properties of silicon quantum dots of various morphologies and sizes up to infinity. *The Journal of Physical Chemistry C, 120*, 11288–11298. Copyright 2016, American Chemical Society.

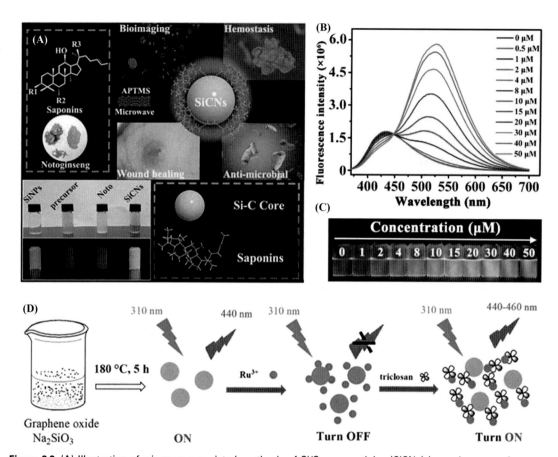

Figure 3.8 (A) Illustration of microwave-assisted synthesis of Si/C nanoparticles (SiCNs) (saponins are major functional groups within notoginseng), schematic description of versatile biofunctions of the SiCNs, picture of real products, and basic characterizations of the SiCNs. (B) Fluorescence changes of this probe upon addition of hydroquinone (0–50 μM). (C) Fluorescence images of Si/C QDs-hydroquinone systems under a 365 nm UV lamp. (D) Schematic representation of the detection mechanism of the Si/C QDs/Ru3 + as a "'turn-on" fluorescent probe for the detection of triclosan. (A) Reprinted with permission from Hu, G., Song, B., Jiang, A., Chu, B., Shen, X., Tang, J., ..., He, Y. (2019). Multifunctional silicon-carbon nanohybrids simultaneously featuring bright fluorescence, high antibacterial and wound healing activity. *Small*, 15, e1803200. Copyright 2019, Wiley-VCH. (B and C) Reprinted with permission from Liu, Y., Cao, Y., Bu, T., Sun, X., Zhe, T., Huang, C., ... Wang, L. (2019). Silicon-doped carbon quantum dots with blue and green emission are a viable ratiometric fluorescent probe for hydroquinone. *Microchimica Acta*, 186, 399. Copyright 2019, Springer Nature. (D) Reprinted with permission from Du, F., Zeng, Q., Lai, Z., Cheng, Z., & Ruan, G. (2019). Silicon doped graphene quantum dots combined with ruthenium(iii) ions as a fluorescent probe for turn-on detection of triclosan. *New Journal of Chemistry*, 43, 12907–12915. Copyright 2019, Royal Society of Chemistry.

It should be noted that silicon nanoparticles with a diameter larger than 5 nm, which had similar band structure to bulk silicon, could show a promising class of nanoplatform specially for its porous nanostructures (Godin et al., 2012; Gu et al., 2013;

Park et al., 2009; Zhang et al., 2017). Furthermore, the porous silicon nanoparticles could be also applied to combine with CQDs for the purpose of drug carriers. With the help of the near infrared (NIR) absorption of CQDs which is a harmless optical signal, and the high area of silicon porous structure which has an effective electrostatic interaction with drug molecules, the CQDs decorated silicon porous structure used in drug delivery has exhibited a high loading of drug for targeted therapy and in vitro/in vivo bioimaging for a real-time monitoring (Zhang et al., 2017).

Additionally, the Si/C nanoparticles have tunable fluorescent properties that could be quenched and recovered, which could be used in the detection of hazardous contaminants, such as hydroquinone, mercuric ions, ciprofloxacin, etc. (Guo, Liu, Li, Zhang, & Wang, 2018; Gui, Bu, He, & Jin, 2018; Liu et al., 2019). Fig. 3.8B and C shows that the blue fluorescence of Si/C nanoparticles will turn green with the presence of hydroquinone, while the detection limit can reach 0.077 μM. Silicon and CQDs work synergistically for the detection. When the pollutant contains Hg^{2+}, the blue-emission of silicon QDs will be quenched with the increase of Hg^{2+} concentration while the red emission of CQDs remains, displaying different discernable colors with a detection limit of 7.63 nM. It should be also mentioned that this quenching effect is recoverable and thus can be used in an on−off−on detection for specific molecules (Du, Zeng, Lai, Cheng, & Ruan, 2019; Gao, Jiang, Jia, Yang, & Wu, 2018). A series of certain ions can combine with the Si/C nanoparticles to quench the fluorescence, while the target molecules have higher binding energy with those ions, wresting those ions and releasing their bonding with Si/C nanoparticles. As a result, the fluorescence of Si/C nanoparticles would be recovered. Fig. 3.8D shows that Ru^{3+} combined Si−graphene QDs for the detection of triclosan in toothpaste and water samples can obtain a very low detection limit of 173 ng L^{-1}.

3.4.2 Light absorption and photocarrier behavior of silicon quantum dots/graphene

A big advantage of SiQDs/graphene-based optoelectronic devices is that the hybrid structure combines the quantum effect-induced optical properties of SiQDs and the high mobility of graphene, leading to enhanced light absorption and photocarrier separation, and thus optimized device performance such as photodetectors and solar cells (Arefinia & Asgari, 2018; Kim et al., 2018; Liu et al., 2017; Ni et al., 2017; Shin et al., 2015;

Shin et al., 2017; Shin, Jang, Kim, & Choi, 2018; Wang et al., 2019; Yu et al., 2016). Compared to the narrow and indirect bandgap of bulk silicon (Shin et al., 2015), SiQDs with varying sizes and chemical doping would significantly enlarge the window of optical response via inducing the transition from ground state to different excited states to generate electron−hole pairs when absorbing light with relevant wavelength, as Fig. 3.9A indicates (Ni et al., 2017). Studies on SiQDs/graphene have also shown that the presence of SiQDs would obviously lower the Dirac point from ∼39 V to ∼9 V in a SiQDs/graphene heterostructure compared to that of bare graphene, indicating the carrier transport from SiQDs to graphene, as shown in Fig. 3.9B (Ni et al., 2017). Thus, in such a heterostructure,

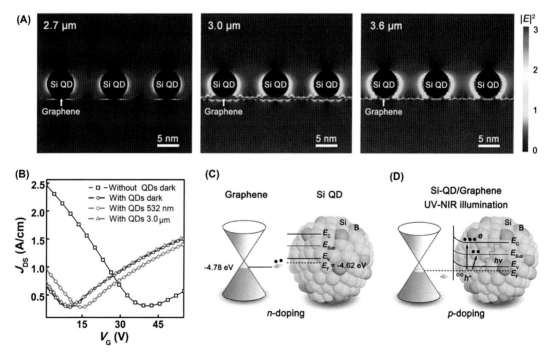

Figure 3.9 (A) Cross-section distribution of the square of electric field ($|E|^2$) at B-doped SiQDs and graphene under the illumination at the wavelength of 2.7, 3.0, and 3.6 μm. (B) Drain-source current density (J_{DS}) as a function of V_G at 3 μm and 532 nm with irradiances of 375 mW cm^{-2} and 0.2 μW cm^{-2}, respectively. The MIR measurements were conducted at 77 K. Those for SiQD/graphene and graphene-only devices in the dark are also shown. (C) Schematic diagram of the band structures of graphene and SiQDs and the tendency of charge transfer between them. (D) Schematic diagram of the band structure of SiQD/graphene and charge transfer after the ultraviolet-to-NIR illumination. *Reprinted with permission from Ni, Z., Ma, L., Du, S., Xu, Y., Yuan, M., Fang, H., ..., Yang, D. (2017). Plasmonic silicon quantum dots enabled high-sensitivity ultrabroadband photodetection of graphene-based hybrid phototransistors.* ACS Nano, *11*, 9854−9862. Copyright 2017, American Chemical Society.

graphene can enhance the transport photogenerated electrons, while the interface induced built-in electric field facilitates the separation of electron–hole pairs, which would be positive to the performance of photodetectors and solar cells (Arefinia & Asgari, 2018; Kim et al., 2018; Ni et al., 2017). In that case, SiQD/graphene/Si photodetector has been developed to have sensitive and fast response in a short wavelength window (Yu et al., 2016). Moreover, the properties of SiQD/graphene/Si photodetector can be further tunable by the changing layers of graphene, size of SiQDs, as well as dopants on either graphene or SiQDs (Ni et al., 2017; Shin et al., 2017). And a more complicated structure of Ag-nanowires-doped graphene/SiQDs/Si photodetector has demonstrated that the Ag nanowires act as an n-type impurity to provide extra electrons (Shin et al., 2018). Fig. 3.9C and D shows with the demand for the middle IR (MIR) region detection, the heavily n/p-doped SiQDs in SiQDs/graphene/Si phototransistors can act as the absorber with dopant-dependent shift of Fermi level, resulting in an ultranarrow bandgap to match the MIR detection (Ni et al., 2017). Therefore the properties of SiQDs/graphene-based optoelectronic devices can be tuned by changing SiQDs sizes, graphene layers, doping, and so on, endowing further application throughout ultrabroad UV-to-MIR range.

3.5 Conclusion

This chapter summarizes the design and electrical and optical properties of Si/C hybrid nanoparticles, discussing their application in energy storage and optoeletronic devices and medicine based on the structure–activity relationship.

References

Arefinia, Z., & Asgari, A. (2018). Optimization study of a novel few-layer graphene/silicon quantum dots/Silicon heterojunction solar cell through opto-electrical modeling. *IEEE Journal of Quantum Electronics, 54*, 1–6.

Baughman, R. H., Zakhidov, A. A., & Heer, W. A. D. (2002). Carbon Nanotubes—the route toward applications. *Science, 297*, 787–792.

Chen, Y., Nie, M., Lucht, B. L., Saha, A., Guduru, P. R., & Bose, A. (2014). High capacity, stable silicon/carbon anodes for lithium-ion batteries prepared using emulsion-templated directed assembly. *ACS Applied Materials and Interfaces, 6*, 4678–4683.

Dimov, N., Fukuda, K., Umeno, T., Kugino, S., & Yoshio, M. (2003). Characterization of carbon-coated silicon. *Journal of Power Sources, 114*, 88–95.

Durkop, T., & Getty, S. A. (2003). Enrique Cobas, M.S. Fuhrer, Extraordinary mobility in semiconducting carbon nanotubes. *Nano Letters, 4*, 35–39.

Du, F., Zeng, Q., Lai, Z., Cheng, Z., & Ruan, G. (2019). Silicon doped graphene quantum dots combined with ruthenium(iii) ions as a fluorescent probe for turn-on detection of triclosan. *New Journal of Chemistry, 43*, 12907–12915.

Evanoff, K., Magasinski, A., Yang, J., & Yushin, G. (2011). Nanosilicon-coated graphene granules as anodes for Li-ion batteries. *Advanced Energy Materials, 1*, 495–498.

Feng, Z., Huang, C., Fu, A., Chen, L., Pei, F., He, Y., ... Cui, J. (2020). A three-dimensional network of graphene/silicon/graphene sandwich sheets as anode for Li-ion battery. *Thin Solid Films, 693*, 137702.

Finegan, D. P., Vamvakeros, A., Cao, L., Tan, C., Heenan, T. M. M., Daemi, S. R., ... Ban, C. (2019). Spatially resolving lithiation in silicon-graphite composite electrodes via in situ high-energy X-ray diffraction computed tomography. *Nano Letters, 19*, 3811–3820.

Gao, G., Jiang, Y.-W., Jia, H.-R., Yang, J., & Wu, F.-G. (2018). On-off-on fluorescent nanosensor for Fe^{3+} detection and cancer/normal cell differentiation via silicon-doped carbon quantum dots. *Carbon, 134*, 232–243.

Geim, A. K. (2009). Graphene: Status and prospects. *Science, 324*, 1530–1534.

Godin, B., Chiappini, C., Srinivasan, S., Alexander, J. F., Yokoi, K., Ferrari, M., ... Liu, X. (2012). Discoidal porous silicon particles: fabrication and biodistribution in breast cancer bearing mice. *Advanced Functional Materials, 22*, 4225–4235.

Gui, R., Bu, X., He, W., & Jin, H. (2018). Ratiometric fluorescence, solution-phase and filter-paper visualization detection of ciprofloxacin based on dual-emitting carbon dot/silicon dot hybrids. *New Journal of Chemistry, 42*, 16217–16225.

Guo, X., Liu, C., Li, N., Zhang, S., & Wang, Z. (2018). Ratiometric fluorescent test paper based on silicon nanocrystals and carbon dots for sensitive determination of mercuric ions. *Royal Society Open Science, 5*, 171922.

Guo, H., Long, D., Zheng, Z., Chen, X., Ng, A. M. C., & Lu, M. (2017). Defect-enhanced performance of a 3D graphene anode in a lithium-ion battery. *Nanotechnology, 28*, 505402.

Gu, L., Hall, D. J., Qin, Z., Anglin, E., Joo, J., Mooney, D. J., ... Sailor, M. J. (2013). In vivo time-gated fluorescence imaging with biodegradable luminescent porous silicon nanoparticles. *Nature Communications, 4*, 2326.

Han, Z., & Fina, A. (2011). Thermal conductivity of carbon nanotubes and their polymer nanocomposites: A review. *Progress in Polymer Science, 36*, 914–944.

Herzinger, C. M., Johs, B., McGahan, W. A., Woollam, J. A., & Paulson, W. (1998). Ellipsometric determination of optical constants for silicon and thermally grown silicon dioxide via a multi-sample, multi-wavelength, multi-angle investigation. *Journal of Applied Physics, 83*, 3323–3336.

He, Y., Xu, G., Wang, C., Xu, L., & Zhang, K. (2018). Horsetail-derived Si@N-doped carbon as low-cost and long cycle life anode for Li-ion half/full cells. *Electrochimica Acta, 264*, 173–182.

Huang, R., Fan, X., Shen, W., & Zhu, J. (2009). Carbon-coated silicon nanowire array films for high-performance lithium-ion battery anodes. *Applied Physics Letters, 95*, 133119.

Huang, C., Feng, Z., Pei, F., Fu, A., Qu, B., Chen, X., ... Cui, J. (2020). Understanding protection mechanisms of graphene-encapsulated silicon anodes with operando Raman spectroscopy. *ACS Applied Materials & Interfaces, 12*, 35532–35541.

Huang, X. D., Zhang, F., Gan, X. F., Huang, Q. A., Yang, J. Z., Lai, P. T., & Tang, W. M. (2018). Electrochemical characteristics of amorphous silicon carbide film as a lithium-ion battery anode. *RSC Advances, 8*, 5189–5196.

Hu, Y. S., Demir-Cakan, R., Titirici, M. M., Muller, J. O., Schlogl, R., Antonietti, M., & Maier, J. (2008). Superior storage performance of a Si@SiOx/C nanocomposite as anode material for lithium-ion batteries. *Angewandte Chemie International Edition, 47*, 1645–1649.

Hu, G., Song, B., Jiang, A., Chu, B., Shen, X., Tang, J., ... He, Y. (2019). Multifunctional silicon-carbon nanohybrids simultaneously featuring bright fluorescence, high antibacterial and wound healing activity. *Small, 15*, e1803200.

Ji, J., Ji, H., Zhang, L. L., Zhao, X., Bai, X., Fan, X., ... Ruoff, R. S. (2013). Graphene-encapsulated Si on ultrathin-graphite foam as anode for high capacity lithium-ion batteries. *Advanced Materials, 25*, 4673–4677.

Kim, J. M., Kim, S., Shin, D. H., Seo, S. W., Lee, H. S., Kim, J. H., ... Lee, H. (2018). Si-quantum-dot heterojunction solar cells with 16.2% efficiency achieved by employing doped-graphene transparent conductive electrodes. *Nano Energy, 43*, 124–129.

Licciardello, N., Hunoldt, S., Bergmann, R., Singh, G., Mamat, C., Faramus, A., ... Stephan, H. (2018). Biodistribution studies of ultrasmall silicon nanoparticles and carbon dots in experimental rats and tumor mice. *Nanoscale, 10*, 9880–9891.

Liu, Y., Cao, Y., Bu, T., Sun, X., Zhe, T., Huang, C., ... Wang, L. (2019). Silicon-doped carbon quantum dots with blue and green emission are a viable ratiometric fluorescent probe for hydroquinone. *Microchimica Acta, 186*, 399.

Liu, Y., Guo, X., Li, J., Lv, Q., Ma, T., Zhu, W., & Qiu, X. (2013). Improving coulombic efficiency by confinement of solid electrolyte interphase film in pores of silicon/carbon composite. *Journal of Materials Chemistry A, 1*, 14075.

Liu, D., Shadike, Z., Lin, R., Qian, K., Li, H., Li, K., ... Li, B. (2019). Review of recent development of in situ/operando characterization techniques for lithium battery research. *Advanced Materials, 31*, e1806620.

Liu, Y., Wen, Z. Y., Wang, X. Y., Hirano, A., Imanishi, N., & Takeda, Y. (2009). Electrochemical behaviors of Si/C composite synthesized from F-containing precursors. *Journal of Power Sources, 189*, 733–737.

Liu, H., Xu, A., Feng, Z., Long, D., Chen, X., & Lu, M. (2020). pH-dependent fluorescent quenching of graphene oxide quantum dots: Towards hydroxyl. *Materials Science and Engineering B, 260*.

Liu, J., Yin, Y., Yu, L., Shi, Y., Liang, D., & Dai, D. (2017). Silicon-graphene conductive photodetector with ultra-high responsivity. *Scientific Reports, 7*, 40904.

Liu, X. H., Zheng, H., Zhong, L., Huang, S., Karki, K., Zhang, L. Q., ... Huang, J. Y. (2011). Anisotropic swelling and fracture of silicon nanowires during lithiation. *Nano Letters, 11*, 3312–3318.

Li, Y., Yan, K., Lee, H.-W., Lu, Z., Liu, N., & Cui, Y. (2016). Growth of conformal graphene cages on micrometre-sized silicon particles as stable battery anodes. *Nature Energy, 1*, 15029.

Monroy, E., Omnès, F., & Calle, F. (2003). Wide-bandgap semiconductor ultraviolet photodetectors. *Semiconductor Science and Technology, 18*, R33–R51.

Niaz, S., & Zdetsis, A. D. (2016). Comprehensive Ab initio study of electronic, optical, and cohesive properties of silicon quantum dots of various morphologies and sizes up to infinity. *The Journal of Physical Chemistry C, 120*, 11288–11298.

Ni, Z., Ma, L., Du, S., Xu, Y., Yuan, M., Fang, H., ... Yang, D. (2017). Plasmonic silicon quantum dots enabled high-sensitivity ultrabroadband photodetection of graphene-based hybrid phototransistors. *ACS Nano, 11*, 9854–9862.

Novoselov, K. S., Geim, A. K., Morozov, S. V., Jiang, D., Zhang, Y., Dubonos, S. V., ... Firsov, A. A. (2004). Electric field effect in atomically thin carbon films. *Science, 306*, 666–669.

Park, J. H., Gu, L., von Maltzahn, G., Ruoslahti, E., Bhatia, S. N., & Sailor, M. J. (2009). Biodegradable luminescent porous silicon nanoparticles for in vivo applications. *Nature Materials, 8*, 331–336.

Pavesi, L., Negro, L. D., Mazzoleni, C., Franzo, G., & Priolo, F. (2000). Optical gain in silicon nanocrystals. *Nature, 408*, 440–445.

Pecher, O., Carretero-González, J., Griffith, K. J., & Grey, C. P. (2016). Materials' methods: NMR in battery research. *Chemistry of Materials, 29*, 213–242.

Pekarek, R. T., Affolter, A., Baranowski, L. L., Coyle, J., Hou, T., Sivonxay, E., . . . Neale, N. R. (2020). Intrinsic chemical reactivity of solid-electrolyte interphase components in silicon-lithium alloy anode batteries probed by FTIR spectroscopy. *Journal of Materials Chemistry A, 8*, 7897–7906.

Ren, W., Wang, Y., Tan, Q., Yu, J., Etim, U. J., Zhong, Z., & Su, F. (2019). Nanosized Si particles with rich surface organic functional groups as high-performance Li-battery anodes. *Electrochimica Acta, 320*, 134625.

Ruther, R. E., Hays, K. A., An, S. J., Li, J., Wood, D. L., & Nanda, J. (2018). Chemical evolution in silicon-graphite composite anodes investigated by vibrational spectroscopy. *ACS Applied Materials & Interfaces, 10*, 18641–18649.

Shin, D. H., Jang, C. W., Kim, J. M., & Choi, S.-H. (2018). Self-powered Ag-nanowires-doped graphene/Si quantum dots/Si heterojunction photodetectors. *Journal of Alloys and Compounds, 758*, 32–37.

Shin, D. H., Jang, C. W., Kim, J. H., Kim, J. M., Lee, H. S., Seo, S. W., . . . Choi, S.-H. (2017). Enhancement of efficiency and long-term stability in graphene/Si-quantum-dot heterojunction photodetectors by employing bis(trifluoromethanesulfonyl)-amide as a dopant for graphene. *Journal of Materials Chemistry C, 5*, 12737–12743.

Shin, D. H., Kim, S., Kim, J. M., Jang, C. W., Kim, J. H., Lee, K. W., . . . Kim, K. J. (2015). Graphene/Si-quantum-dot heterojunction diodes showing high photosensitivity compatible with quantum confinement effect. *Advanced Materials, 27*, 2614–2620.

Shi, F., Ross, P. N., Somorjai, G. A., & Komvopoulos, K. (2017). The chemistry of electrolyte reduction on silicon electrodes revealed by in situ ATR-FTIR spectroscopy. *Journal of Physical Chemistry C, 121*, 14476–14483.

Si, Q., Hanai, K., Imanishi, N., Kubo, M., Hirano, A., Takeda, Y., & Yamamoto, O. (2009). Highly reversible carbon–nano-silicon composite anodes for lithium rechargeable batteries. *Journal of Power Sources, 189*, 761–765.

Tardif, S., Pavlenko, E., Quazuguel, L., Boniface, M., Marechal, M., Micha, J. S., . . . Lyonnard, S. (2017). Operando Raman spectroscopy and synchrotron x-ray diffraction of lithiation/delithiation in silicon nanoparticle anodes. *ACS Nano, 11*, 11306–11316.

Terranova, M. L., Orlanducci, S., Tamburri, E., Guglielmotti, V., & Rossi, M. (2014). Si/C hybrid nanostructures for Li-ion anodes: An overview. *Journal of Power Sources, 246*, 167–177.

Tian, H., Tan, X., Xin, F., Wang, C., & Han, W. (2015). Micro-sized nano-porous Si/C anodes for lithium ion batteries. *Nano Energy, 11*, 490–499.

Tian, S., Zhu, G., Tang, Y., Xie, X., Wang, Q., Ma, Y., . . . Xie, X. (2018). Three-dimensional cross-linking composite of graphene, carbon nanotubes and Si nanoparticles for lithium ion battery anode. *Nanotechnology, 29*, 125603.

Valenta, J., Juhasz, R., & Linnros, J. (2002). Photoluminescence spectroscopy of single silicon quantum dots. *Applied Physics Letters, 80*, 1070–1072.

Wang, F., Hu, Z., Mao, L., & Mao, J. (2020). Nano-silicon @ soft carbon embedded in graphene scaffold: High-performance 3D free-standing anode for lithium-ion batteries. *Journal of Power Sources, 450*, 227692.

Wang, W., & Kumta, P. N. (2010). Nanostructured hybrid Silicon/Carbon nanotube heterostructures: Reversible high-capacity lithium-ion anodes. *ACS Nano, 4*, 2233–2224.

Wang, P.-F., Liu, Y., Yin, J., Ma, W., Dong, Z., Zhang, W., ... Sun, J.-L. (2019). A tunable positive and negative photoconductive photodetector based on a gold/graphene/p-type silicon heterojunction. *Journal of Materials Chemistry C, 7*, 887–896.

Wang, Y., & Wang, C.-Y. (2005). Simulation of flow and transport phenomena in a polymer electrolyte fuel cell under low-humidity operation. *Journal of Power Sources, 147*, 148–161.

Wang, L. W., & Zunger, A. (1994). Solving Schrödinger's equation around a desired energy: Application to silicon quantum dots. *Journal of Chemical Physics, 100*, 2394–2397.

Wolfbeis, O. S. (2015). An overview of nanoparticles commonly used in fluorescent bioimaging. *Chemical Society Reviews, 44*, 4743–4768.

Woo, S.-H., Park, J.-H., Hwang, S. W., & Whang, D. (2012). Silicon embedded nanoporous carbon composite for the anode of Li ion batteries. *Journal of the Electrochemical Society, 159*, A1273–A1277.

Wu, Y., & Liu, N. (2018). Visualizing battery reactions and processes by using in situ and in operando microscopies. *Chem, 4*, 438–465.

Wu, H., Zheng, G., Liu, N., Carney, T. J., Yang, Y., & Cui, Y. (2012). Engineering empty space between Si nanoparticles for lithium-ion battery anodes. *Nano Letters, 12*, 904–909.

Xie, J., Tong, L., Su, L., Xu, Y., Wang, L., & Wang, Y. (2017). Core-shell yolk-shell Si@C@Void@C nanohybrids as advanced lithium ion battery anodes with good electronic conductivity and corrosion resistance. *Journal of Power Sources, 342*, 529–536.

Yang, L., Chen, H. S., Jiang, H., Wei, Y. J., Song, W. L., & Fang, D. N. (2018). Failure mechanisms of 2D silicon film anodes: in situ observations and simulations on crack evolution. *Chemical Communications (Cambridge), 54*, 3997–4000.

Yang, J., Kraytsberg, A., & Ein-Eli, Y. (2015). In-situ Raman spectroscopy mapping of Si based anode material lithiation. *Journal of Power Sources, 282*, 294–298.

Yin, Y.-X., Xin, S., Wan, L.-J., Li, C.-J., & Guo, Y.-G. (2011). Electrospray synthesis of Silicon/Carbon nanoporous microspheres as improved anode materials for lithium-ion batteries. *Journal of Physical Chemistry C, 115*, 14148–14154.

Yun, Q., Qin, X., He, Y.-B., Lv, W., Kaneti, Y. V., Li, B., ... Kang, F. (2016). Micron-sized spherical Si/C hybrids assembled via water/oil system for high-performance lithium ion battery. *Electrochimica Acta, 211*, 982–988.

Yu, T., Wang, F., Xu, Y., Ma, L., Pi, X., & Yang, D. (2016). Graphene coupled with silicon quantum dots for high-performance bulk-silicon-based schottky-junction photodetectors. *Advanced Materials, 28*, 4912–4919.

Zhang, R.-Q., De Sarkar, A., Niehaus, T. A., & Frauenheim, T. (2012). Excited state properties of Si quantum dots. *Physica Status Solidi (b), 249*, 401–412.

Zhang, J., Zhang, J., Li, W., Chen, R., Zhang, Z., Zhang, W., ... Lee, C. S. (2017). Degradable hollow mesoporous silicon/carbon nanoparticles for photoacoustic imaging-guided highly effective chemo-thermal tumor therapy in vitro and in vivo. *Theranostics, 7*, 3007–3020.

Zhou, X., Yin, Y. X., Cao, A. M., Wan, L. J., & Guo, Y. G. (2012). Efficient 3D conducting networks built by graphene sheets and carbon nanoparticles for high-performance silicon anode. *ACS Applied Materials & Interfaces, 4*, 2824–2828.

Zhu, Y., Murali, S., Cai, W., Li, X., Suk, J. W., Potts, J. R., & Ruoff, R. S. (2010). Graphene and graphene oxide: synthesis, properties, and applications. *Advanced Materials, 22*, 3906–3924.

Zuo, P., Yin, G., & Ma, Y. (2007). Electrochemical stability of silicon/carbon composite anode for lithium ion batteries. *Electrochimica Acta, 52*, 4878–4883.

4

Properties of silicon−ZnO hybrid nanoparticles

Adem Kocyigit[1,2]

[1]*Department of Electrical and Electronics Engineering, Faculty of Engineering, Igdir University, Igdir, Turkey* [2]*Department of Electronics and Automation, Vocational High School, Bilecik Şeyh Edebali University, Bilecik, Turkey*

4.1 Introduction

ZnO has a high direct bandgap (3.37 eV) for optoelectronic applications and a large exciton binding energy (60 meV) for long carrier lifetime (Yilmaz, & Grilli, 2016). It is also a low-cost, nontoxic material, and can be synthesized at low deposition temperatures. These properties make it great material for various applications, such as chemical sensors, solar cells, varistors, light-emitting diodes, photodetectors, and photodiodes (Yıldırım & Kocyigit, 2018). ZnO is employed in many applications in its nanostructure forms, such as nanowires, nanoplates, nanorods, and nanoparticles. Furthermore, the silicon is the most important and suitable material for electronics. However, it is not suitable for all optoelectronic applications (Belyakov, et al., 2008). Unlike bulk silicon, nanoforms of silicon such as quantum dots, nanowires, and nanoparticles exhibit good properties, such as adjustable bandgap energy and luminescence, depending on size and electronic behaviors (Belomoin et al., 2002). Both the silicon and ZnO can be combined in nanostructured forms to obtain more suitable properties for optoelectronic applications.

This chapter discusses the synthesis and properties of the ZnO nanoparticles, and fabrication of the hybrid ZnO and silicon structures. Furthermore, the structural, electrical, and optical properties and applications of the hybrid ZnO and silicon are reviewed.

4.2 Zinc oxide nanoparticles: synthesis and properties

Zinc oxide is a great material and can be used for many applications, such as solar cells, gas sensors, photodetectors,

UV-lasers, and catalysis. It can be synthesized as rods, wires, belts in one dimension (1D); plates and sheets in 2D; and flowers, snowflakes etc. as 3D materials (Kolodziejczak-Radzimska & Jesionowski, 2014). The synthesis method is so important to obtain the various forms of nanoparticles. In a zero-dimension, they are usually called nanoparticles. The ZnO are obtained in nanoparticle form using various methods, which generally can be divided into two methods: bottom-up and top-down. While the bottom-up methods show synthesis from molecules to nanoparticles, the top-down methods obtain nanoparticles from bulk, as shown in Fig. 4.1.

The synthesis of the ZnO nanoparticles is generally conducted physically, chemically, and biologically (Mazitova et al., 2019). The physical methods are laser ablation, spray pyrolysis, and magnetron sputtering, and the chemical methods are microemulsion, sol−gel, hydrothermal, and chemical deposition.

Various researchers have investigated the properties of the ZnO nanoparticles. The structural properties revealed that the ZnO nanoparticles exhibited hexagonal wurtzite phase by space group of $P6_3mc$ (Kaddes et al., 2018). The TEM images confirm the spherical shape of the ZnO particles in various solutions (Kumar Jangir et al., 2017). The morphology of the ZnO nanoparticles were studied by SEM and proved the agglomeration of the particles (Geetha et al., 2016; Khalil et al., 2014). The optical properties were obtained by UV−Vis spectrometry of ZnO nanoparticles, and exhibited an absorption edge at around the 300−350 nm (Tachikawa et al., 2011). The bandgap values of the ZnO nanoparticles change by the dimension of the nanoparticles in the range of 3.30−3.43 eV for 12−3.5 nm particle sizes as quantum dots (Lin et al., 2005). They exhibit good transparency in the visible region as well as photoluminescence peaks in blue and green regions. The

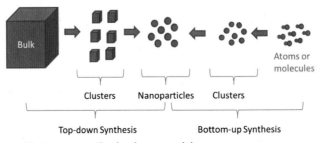

Figure 4.1 The top-down and bottom-up synthesis of nanoparticles.

electrical behaviors of the ZnO particles reveal that they have high dielectric permittivity and n-type conductivity due to having oxygen vacancies (Cao et al., 2017; Lanje et al., 2013).

There many applications of ZnO nanoparticles in electronics, the textile industry, medicine, and agriculture due to having various physical and chemical properties, such as wide bandgap, catalytic and antibacterial activity, and semiconducting behavior with anisotropic crystal structure (Mazitova et al., 2019). In electronics, ZnO nanoparticles can be used in photodetectors, solar cells, lasers etc. In the textile industry, they can be employed as UV absorbers. In the case of medicine, the ZnO nanoparticles can be used in bioimaging, anticancer activity, diebatic therapy, etc. due to their biocompatibility (Barui et al., 2018). They can be employed to enhance crops in agricultural applications. Fig. 4.2 displays various usage areas of the ZnO nanoparticles.

4.3 Preparation of silicon–zinc oxide hybrid structures

The silicon nanoparticles can be prepared chemically or physically in colloidal form in a solvent. However, there is not so much study of the core–shell or composite nanoparticles of silicon with ZnO nanoparticles directly. Instead, usually ZnO

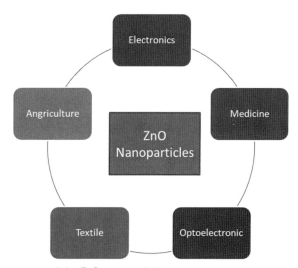

Figure 4.2 Some applications areas of the ZnO nanoparticles.

nanoparticles have been deposited on the porous silicon to obtain composite or hybrid structure. Due to that, this chapter focuses on the porous silicon and ZnO nanoparticles hybrid structures. Most studies have used anodization or the chemical etching method of silicon wafers to obtain porous silicon. A discussion of both anodization and chemical etching of the silicon wafers follows.

4.3.1 Anodization of silicon wafers

Silicon wafer can be n-type or p-type for anodization. Firstly, silicon wafers are cleaned by various alcohol solvents in an ultrasonic cleaner to obtain pure porous silicon. The used solvents are usually acetone, isopropanol, and deionized water. The researchers usually want to remove the native oxide layer before the anodization of silicon. The diluted HF can be used to remove a native oxide layer that naturally forms on the surface at room temperature after manufacturing. After this cleaning procedure, the wafers are transferred into anodization cells for the anodization process. The anodization cells have an electrolyte, counterelectrode, and silicon electrode or silicon wafers, wires, and DC sources. A typically anodization cell for obtaining porous silicon is illustrated in Fig. 4.3. Here, the DC source provides the required voltage to transfer ions between electrodes in an electrolyte solution. The time taken for the porous silicon on the silicon wafer surface to be obtained depends on the current that passes between the electrodes.

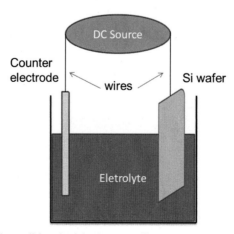

Figure 4.3 The typically anodization cell for obtaining porous silicon.

4.3.2 Metal assisted chemically etching of silicon wafers

Pt or Au atoms can chemically etch silicon wafer and porous silicon can be obtained after this treatment, according to Nayfeh's Group (Smith et al., 2005). Firstly, the solution of the Pt or Au metals is coated on the cleaned silicon by dumping into solution with hydrogen peroxide to cave the silicon surface; metal ions connect to the silicon surface and make caves on the surface of silicon. At this stage, the SiO_2 also composes. The diluted HF solution is used to remove this oxide layer as well as metal ions from the surfaces. At the end, the porous silicon can be obtained. The process of obtaining porous silicon with the metal-assisted chemical etching technique is shown schematically in Fig. 4.4A. The SEM image of the platinated Si wafer surface is illustrated in Fig. 4.4B. The size of the platinum particles changes between 0.2 microns and about 1.0 microns with a bright color. The SEM image of porous silicon after platinum and H_2O_2 treatment has been exhibited in Fig. 4.4C. The surface of the porous silicon is obtained homogenously after the metal-assisted chemical etching procedure. The obtained porous silicon usually emits red light, which originates around 650 nm, that is in the visible region, when illuminated under UV light.

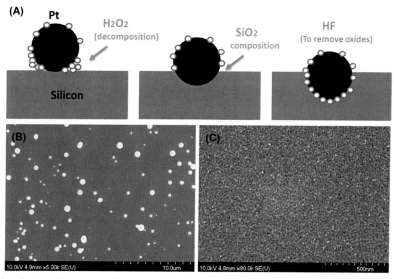

Figure 4.4 (A) Chemical etching of the Si surface for obtaining porous silicon. (B) SEM image of the platinum particles on Si surface. (C) SEM images of the porous silicon after metal assisted etching (Nayfeh, 2018). Permission of the Nayfeh's Group from the University of Illinois at Urbana-Champaign.

4.3.3 Synthesis of silicon–zinc oxide hybrid structures

The synthesis of the hybrid of silicon and ZnO hybrid structure usually contains deposition of the ZnO nanoparticle on the porous silicon. Some studies have been based on the silicon nanowires and ZnO particle film on the silicon. Researchers prepared a colloidal solution of the ZnO nanoparticles/thin films, and directly spin coated or sprayed them on the porous silicon to obtain ZnO and porous silicon hybrid structure for wide-range photoluminescence response and better optoelectronic behaviors (Khashan, 2011; Singh et al., 2007). Both the spin-coating and spray pyrolysis are cheap and easy techniques to obtain various types of nanomaterials from a prepared solution of the desired material. In the case of the spin-coating technique, the solution is dropped on the substrate and rotates around the vertical axis. The rotation of the substrate provides the spreading and evaporating of the solution when the solution layer becomes thinner. The layers usually are dried in an oven at around 150°C–200°C and, if it is necessary, the other layers are deposited after the drying process. When the layer deposition process is finished, the sample is usually annealed above 300°C for inorganic materials. The schematic illustration of the spin-coating technique is shown in Fig. 4.5A. In the case of the spray pyrolysis technique, the solution of the materials is sprayed onto the preheated substrates by a spray gun, as shown in Fig. 4.5B. There is no kind of second heating or annealing process in this method. The ZnO nanoparticles can be spin-coated or sprayed on the porous silicon by these techniques.

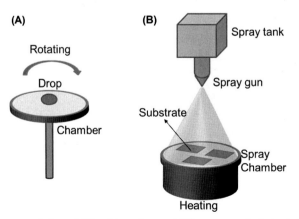

Figure 4.5 Schematically representation of (A) spin-coating and (B) spray pyrolysis technique.

Chan et al. prepared flower-like ZnO/Si nanostructures, both of them nanowires, via self-catalytic chemical vapor transport growth of ZnO nanowires and hydrothermal etching for the nanoporous Si pillar array (NSPA) (Chan et al., 2012). They deposited Au nanoparticles onto the ZnO/Si by the hydrothermal method for SERS (surface-enhanced Raman scattering) application. The schematically geometric structure and SEM image are illustrated in Figs. 4.6A and B, respectively.

Ramadan et al. studied the composite structure of the ZnO and porous silicon for light-emitting diode applications (Ramadan et al., 2020). They used UV lithography to obtain a hexagonal micropattern of a photoresist and ZnO composition on silicon. The ZnO thin film first was deposited on the silicon substrate by an e-beam evaporation technique. Then, Ariston 20 series negative photoresist was coated on the ZnO by a spin-casting method. The UV photolithography method was employed to obtain hexagonal micropatterned ZnO thin films, as shown in Fig. 4.7. Before removing the photoresist, the porous silicon was fabricated on the UV exposed silicon areas by an electrochemical etching method. At the end, the photoresist layer was removed, and the sample was rinsed and dried. The morphology and photoluminescence behavior of the micropatterned ZnO/porous silicon sample was investigated for solid-state lighting applications.

Pavlenko et al. fabricated porous silicon and zinc oxide nanocomposite by metal-assisted chemical etching and atomic layer deposition technique, respectively, for optical/PL biosensing of mycotoxins (Aflatoxin B1) (Pavlenko et al., 2020). They used Ag silver nanoparticles to obtain n-type and p-type porous

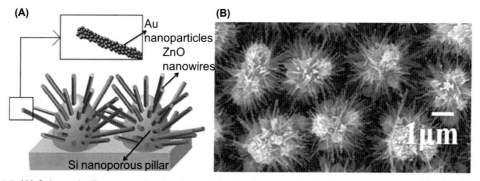

Figure 4.6 (A) Schematically geometric structure of the Au decorated ZnO silicon nanoporous pillar. (B) SEM image of the flower-like Au decorated ZnO and Si nanoporous pillars. Permission of American Institute of Physics (Chan, Y.F., Xu, H.J., Cao, L., Tang, Y., Li, D.Y., Sun, X.M. (2012). ZnO/Si arrays decorated by Au nanoparticles for surface-enhanced Raman scattering study. *Journal of Applied Physics, 111*, 033104. https://doi.org/10.1063/1.3682462).

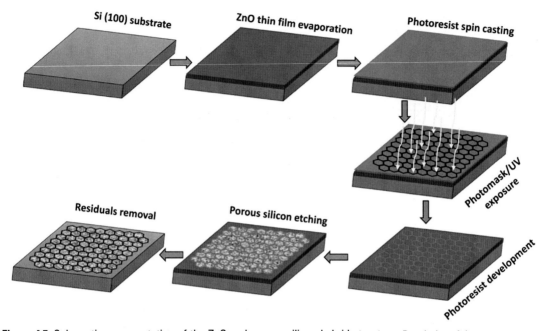

Figure 4.7 Schematic representation of the ZnO and porous silicon hybrid structure. Permission of the Multidisciplinary Digital Publishing Institute (MDPI) (Ramadan, R., Torres-Costa, V., Martín-Palma, R.J. (2020). Fabrication of zinc oxide and nanostructured porous silicon composite micropatterns on silicon. *Coatings, 10,* 529. https://doi.org/10.3390/COATINGS10060529).

silicon by a metal-assisted chemical etching method. ALD technique was employed for coating of the porous n-type and p-type silicon substrates. The results revealed that the obtained hybrid ZnO and porous silicon structure gave good results for detection of Aflatoxin B1.

Zhang et al. fabricated a composite of the ZnO nanowires and silicon nanocrystals for photodgradation of methylene blue performance under various illumination intensities (Zhang et al., 2020). Firstly, the zinc oxide nanowires were fabricated by a hydrothermal method on glass and silicon substrates. The silicon nanocrystals were synthesized in a nonthermal reactor from the gas phase of silane (SiH_4). The obtained Si nanocrystals were deposited on the zinc oxide nanowires by a spin-coating technique. Thus ZnO nanowire and silicon nanocrystal hybrid structures were obtained, and their photocatalytic behaviors were investigated under various illumination conditions.

As can be seen from the above studies, the silicon and ZnO hybrid materials can be synthesized by various methods for different aims and can be employed in various applications.

4.4 Structural and morphological properties of silicon–zinc oxide hybrid nanoparticles

Zinc oxide and silicon have been composed in the literature and characterized by various structural characterization techniques such as X-ray diffraction (XRD), Fourier-transform infrared spectroscopy (FTIR), and Raman spectroscopy. The morphology of the zinc oxide and silicon hybrid structures have been studied by scanning electron microscopy (SEM), transmission electron microscopy (TEM), and atomic force microscopy (AFM). Furthermore some researchers have also studied the structural amount of the composition by energy dispersive spectroscopy (EDS). Zinc oxide and silicon hybrid structures have good structural and morphological behaviors. In this part, both the structural and morphological characterization will be discussed according to instrument type.

4.4.1 X-ray diffraction (XRD) patterns of the silicon–zinc oxide hybrid structures

Generally materials can be classified as crystalline or amorphous structures. While the crystalline structure shows good order according to its atomic planes, the amorphous structure represents irregular arrangement. Most materials including zinc oxide and silicon have a crystalline structure in nature. While the silicon usually is crystallized in a diamond cubic crystal structure, zinc oxide has a hexagonal wurtzite crystal structure. The XRD is used to detect the crystalline structures of the materials by the collection of scattering beams from the crystalline planes. If a material has a crystalline structure, the XRD pattern reveals some peaks which belong to its characteristic crystal behavior.

The XRD peaks of the ZnO nanoparticles belongs to hexagonal wurtzite structure of the ZnO and usually contains (100), (002), (101), (102), (110), (103), (200), (112), (201), and (202) crystal planes (Muhammad et al., 2019). The average crystalline size is obtained 20–50 nm from the Debye–Scherrer method (Chikkanna et al., 2019; Kalpana et al., 2018). The typical XRD pattern of the ZnO nanoparticles has been shown in Fig. 4.8A for a wide range of 2θ degrees. When the peak intensity and sharpness increases, the particle size usually increases due to increasing the signal from the nanoparticles. The most striking peak or preferred orientation of the materials in XRD peaks is usually changeable, but in the case of ZnO nanoparticles, the (101) peak is the most striking or preferred orientation.

In the case of the XRD pattern of the silicon, especially in the particle form, the (111), (220), (311), (400), (331), (422), and (511) crystallographic plane can be identified (Card No.027−1402) (shown in Fig. 4.8B) (Hossain et al., 2018). Usually, the (111) plane is the preferred orientation according to the literature. Again, the peak position changes and peak width increases with a decreasing of the silicon size in both the porous or nanoparticle forms.

In the case of zinc oxide and silicon hybrid structures, both the silicon peaks and zinc oxide peaks are usually detected in the XRD patterns. In other words, the hybrid structure generally indicates the characteristic structure of the silicon and zinc oxide. Chan et al. fabricated Au-decorated zinc oxide silicon hybrid structure and obtained the broadened (111) peak of silicon as well as hexagonal structure peaks of the zinc oxide in the XRD pattern (Chan et al., 2012). Singh et al. synthesized a hybrid film structure of zinc oxide and porous silicon nanocomposite for broadband photoluminescence material, and they obtained both zinc oxide and weak porous silicon peaks in the XRD patterns (Singh et al., 2007). Sometimes, the compound of

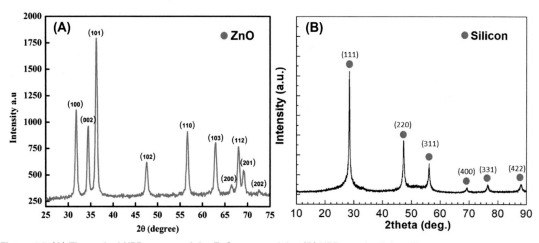

Figure 4.8 (A) The typical XRD pattern of the ZnO nanoparticles (B) XRD graph of the silicon. (A) Permission of RCS advances in Muhammad, W., Ullah, N., Haroon, M., Abbasi, B.H. (2019). Optical, morphological and biological analysis of zinc oxide nanoparticles (ZnO NPs) using: Papaver somniferum L. *RSC Advances, 9*, 29541−29548. https://doi.org/10.1039/c9ra04424h. (B) Permission of the Multidisciplinary Digital Publishing Institute (MDPI) (Hossain, S.T., Johra, F.T., Jung, W.G. (2018). Fabrication of silicon carbide from recycled silicon wafer cutting sludge and its purification. *Applied Sciences, 8*, 1841. https://doi.org/10.3390/app8101841).

the zinc oxide and silicon is obtained when they are composed or hybridized. Verma et al. studied zinc oxide nanoparticles embedded porous silicon for improving efficiency of solar cells with the increasing of the photoluminescence of the porous silicon (Verma et al., 2014). The XRD patterns showed the silicon, zinc oxide, and zinc silicate (Zn_2SiO_4) after annealing at 600°C temperatures. When the silicon was deposited on zinc oxide nanowire structures, especially low ratio deposition, only ZnO peaks are observed. Zhang et al. fabricated hydrothermally grown ZnO nanowires and gas-phase produced silicon nanocrystals as hybrid structures for photodegradation (Zhang et al., 2020). The XRD patterns of the ZnO/silicon nanocrystals exhibited only characteristic peaks of the ZnO nanowires due to the low level of the silicon nanocrystals in the hybrid structures.

4.4.2 Fourier-transform infrared spectroscopy or Raman graph of the silicon—zinc oxide hybrid structures

Fourier-transform infrared spectroscopy (FTIR) and Raman spectroscopy techniques are used to determine usual bonding vibration of the materials in the infrared region when the materials are stimulated. Thus materials can be classified according to their bond vibrations. Actually, FTIR and Raman spectroscopies reveal the same results, but the Raman is more sensitive than FTIR for fingerprint regions. Moreover, FTIR is mostly used to obtain the vibration of the heteronuclear functional groups. However, the Raman spectroscopy is employed mostly to detect the molecular bonds of the homonuclear vibrations. The zinc oxide and silicon hybrid structure can be investigated by FTIR or Raman spectroscopy to understand the bonds between them. Furthermore, the sharpness of the FTIR or Raman peaks exhibits the degree of materials crystallinity (Londoño-Restrepo et al., 2019).

FTIR spectra of the ZnO particles reveals characteristic stretching mode of Zn-O in the range of 350 cm^{-1} and 550 cm^{-1} wavenumbers in the fingerprint region (Nagaraju et al., 2017). There are other peaks in the FTIR spectra of the ZnO particles changes depending on the solvent compositions which can belong to hydroxyl group, —OH, and C=O bond vibrations. Raman spectra of the ZnO particles display peaks showing the characteristic band of the wurtzite ZnO at around 300—600 cm^{-1} wavenumbers (Handore et al., 2014). These peaks can be various modes such as A, B, and E of the ZnO phonons (Sankara Reddy & Venkatramana Reddy, 2013).

In the case of silicon as porous or nanoparticles forms, the FTIR spectra shows Si-O-Si vibrational modes at around $1000\,\text{cm}^{-1}$ and $1200\,\text{cm}^{-1}$, and vibrational modes Si-H (Si-H, Si-H$_2$, and Si-H$_3$ stretches) in between $2000\,\text{cm}^{-1}$ and $2200\,\text{cm}^{-1}$ wavenumbers (Amonkosolpan et al., 2012). The Si-H stretch mode, Si-H$_3$ symmetric deformation mode, Si-H$_2$ scissors mode can be seen in-between $600\,\text{cm}^{-1}$ and $1000\,\text{cm}^{-1}$ wavenumbers in FTIR spectra of nanostructured silicon (Macias-Montero et al., 2016). Furthermore, since the nanostructured silicon can be easily oxidized, the Si-O-Si bond vibration can be easily detected in the FTIR spectrum of the nanostructured silicon (Mansour et al., 2012). The other bond vibrations of −OH, C = O, C−H, or Si-C can be detected in the FTIR spectrum of the silicon due to solution compositions but usually there is no Si−Si vibration in the FTIR spectrum of the nanostructured silicon (Tan et al., 2011). Raman spectrum of the nanostructured silicon gives a peak at around $500-550\,\text{cm}^{-1}$ wavenumbers. Normally, bulk silicon gives the same Raman spectrum, but the nanostructured silicon displays broadening and shifts of the Raman peak and its position due to the quantum confinement effect of the phonon vibration (Shirahata et al., 2009; Tan et al., 2011).

When the ZnO and silicon hybrid structure are thought, the FTIR and Raman spectra have both ZnO and silicon vibration bonds and modes. Salem et al. fabricated a thin-layer of ZnO thin film coated on porous silicon for various molarities of the ZnO (Salem et al., 2015). They obtained FTIR spectra of the films for various molarities and confirmed both the ZnO and silicon bond vibrations, but silicon bonded to both ZnO and Zn according to the FTIR spectra. ZnO bond vibration stayed in its specific peak region, and Zn-Si bond vibration $1615\,\text{cm}^{-1}$ wavenumbers were obtained due to hybridization of silicon and ZnO. The FTIR spectra of the ZnO thin film on the porous silicon have been shown in Fig. 4.9 for various ZnO molarities. According to FTIR spectra, the Si-H bond vibration disappeared and Si-O-Si or Si-O vibration bonds appeared or sharpened with increasing molarity of the ZnO. The same results have been obtained by Verma et al. (2014). The depositing of the porous silicon with ZnO nanoparticles and heat treatment caused a decrease in peak intensity of the Si-H vibration and an increase in the Si-O-Si and Si-O peak intensity.

The Raman spectrum of the ZnO and silicon hybrid structure reveals again combined Raman peaks of ZnO and silicon.

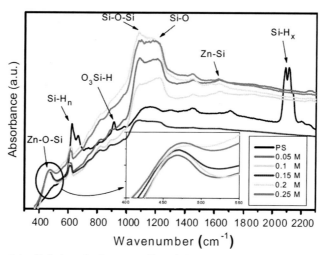

Figure 4.9 FTIR spectra of the ZnO deposited porous silicon hybrid structure for various molarities of the ZnO. Permission of the Springer (Salem, M., Alami, Z.Y., Bessais, B., Chahboun, A., Gaidi, M. (2015). Structural and optical properties of ZnO nanoparticles deposited on porous silicon for mc-Si passivation. *Journal of Nanoparticle Research, 17*, 1–9. https://doi.org/10.1007/s11051-015-2944-2).

Zhang et al. studied Si nanocrystals/ZnO nanowires hybrid structures for various amount loading of the silicon nanocrystals (ZS1, ZS2, and ZS3 refers to 5 μL, 10 μL, and 15 μL silicon nanocrystals) for photodegrading applications (Zhang et al., 2020). The Raman spectrum showed that the loading amount of the silicon nanocrystals caused an increase in the Raman peaks of the silicon at around 515 cm^{-1} as shown in Fig. 4.10. The peaks sharpness of the ZnO at around 485 cm^{-1} wavenumbers decreased with increasing silicon loading amount according to Zhang et al. Some of the bulk silicon (c-Si) peaks at around 305 cm^{-1} wavenumber have become clear with an increase of the silicon amount (Nayef et al., 2014).

4.4.3 Morphological characterization of the silicon–zinc oxide hybrid structures

The morphological structures as well as obtained patterns the fabricated materials can be determined by various imaging techniques, such as SEM and AFM. In the case of the SEM technique, the electrons are scattered from the samples and collected by a detector. Thus a real image of the sample is obtained on a screen. In the AFM technique, the image of the sample is obtained by a tip without touching the surface.

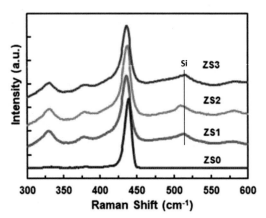

Figure 4.10 Raman spectrum of the ZnO and silicon hybrid structure for various molarities of the silicon. Permission of MDPI (Zhang, Y., Mandal, R, Ratchford, D.C., Anthony, R., Yeom, J. (2020). Si nanocrystals/ZnO nanowires hybrid structures as immobilized photocatalysts for photodegradation, *Nanomaterials*, *10*, 491. https://doi.org/10.3390/nano10030491).

The piezoelectric effect is employed to collect surface morphological data in the AFM technique. While the SEM technique can damage the surface of the samples, the AFM technique is harmless.

SEM and AFM techniques have been performed to determine the surface morphology of the zinc oxide and silicon hybrid structures. Singh et al. fabricated ZnO thin film on the porous silicon and investigated surface morphology by AFM technique for a wide range of surface areas (Singh et al., 2007). The surface of the obtained films was not homogenous, and the RMS values increased with increasing annealing temperature. Eswar et al. studied the morphology of the ZnO deposited porous silicon by SEM and AFM methods for various annealing temperatures (Eswar et al., 2014). The particle sizes of the ZnO on the porous silicon changed by annealing temperature due to thermal expansion of the ZnO. Verma et al. fabricated the ZnO nanoparticle deposited porous silicon hybrid structure and employed SEM to understand the surface morphology of the hybrid structures (Verma et al., 2014). They also annealed the ZnO deposited porous silicon hybrid structure and obtained more compact morphology. The SEM images of the porous silicon and ZnO deposited porous silicon surfaces have been illustrated in Fig. 4.11A and B, respectively. If the ZnO nanoparticles are fine enough, they can fill the pores of the porous silicon and reduce the surface roughness (Eswar et al., 2014). Salem et al. also investigated the effect of the various molarities amount of the ZnO on porous silicon and revealed that ZnO grain sizes

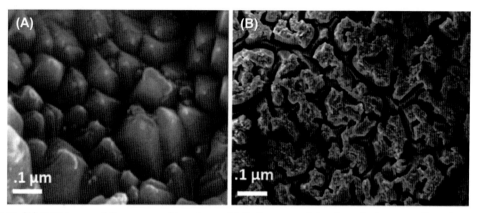

Figure 4.11 (A) SEM image of the porous silicon layer. (B) ZnO nanoparticles deposited porous silicon layer.
From: Reproduced from D. Verma, A. Kharkwal, S.N. Singh, P.K. Singh, S.N. Sharma, S.S. Mehdi, M. Husain, Application of ZnO nanoparticles to enhance photoluminescence in porous silicon and its possible utilization for improving the short wavelength quantum efficiency of silicon solar cell, Solid State Sci. 37: 13–17. Copyright 2014 Elsevier Masson SAS. All rights reserved.

increased with increasing molar ratio of the ZnO according to AFM images (Salem et al., 2015). Pavlenko et al. fabricated porous silicon–zinc oxide nanocomposites by atomic layer deposition for biosensing of the Aflatoxin B1 and studied the morphological behaviors of the ZnO–porous silicon nanocomposites by SEM (Pavlenko et al., 2020). The results showed that the surfaces had uniform pore distribution.

4.5 Electrical properties of silicon–zinc oxide hybrid nanoparticles

The electrical behavior of materials usually are related to the conductivity behavior or conduction of electrical current. There many parameters such as resistivity and dielectric constant of the materials to determine electrical behaviors. Furthermore, researchers study also electrical conductivity types such as n-type and p-type if the investigated material is a semiconductor. In addition, carrier concentration and mobility of the materials can be studied as electrical behavior.

Salem et al. studied electrical properties of the ZnO deposited porous silicon by measuring the effective minority lifetimes and solar cell parameters by $I-V$ measurements (Salem et al., 2015). The effective carrier lifetime slightly decreased with porous silicon according to multicrystalline silicon due to the increasing surface area and structural defects. However, the effective carrier lifetime increased from 1 μs to 90 μs at 3×10^{13} cm^{-3} excess carrier densities with increasing ZnO molarity which deposited on the porous silicon. This case was attributed to decreased surface recombination velocity by the lowering of the dangling bond density. They also tested the ZnO deposited porous silicon as solar cell by $I-V$ measurements using Au front and Al back contacts under illumination intensity of 100 mW/cm^2 (AM 1.5). The results revealed that both the short circuit current and open circuit voltage increased with increasing ZnO molarity. Thus, the power conversion efficiency increased up to 12% with 0.25 M ZnO and porous silicon composition. Khawla fabricated the ZnO nanoparticles by chemical method and sprayed them onto porous silicon by an electrochemical etching technique. The $I-V$ measurements under various illumination powers were performed on the ZnO/porous silicon structure with the contacting of Al by the thermal evaporation technique. The results revealed that light was absorbed in the depletion region and caused to form electron–hole pairs and photocurrent at reverse biases. Furthermore, the responsivity of the ZnO/porous silicon was studied and found to be 0.59 A/W for 400 nm wavelength. Porous silicon has good photoluminescence behavior, but its electrical conductivity is low. Both the electrical conductivity and photoluminescence spectral areas can be improved by the combining of the ZnO and porous silicon (Ramadan et al., 2020).

4.6 Optical properties of silicon–zinc oxide hybrid nanoparticles

Optical properties of a material can be illuminated by the determining of the bandgap, optical transmission, reflectance, and photoluminescence properties. While the bandgap, optical transmittance, and reflectance values are obtained by UV–Vis spectrometer, the photoluminescence behavior can be determined by a photoluminescence spectroscopy technique. Both the UV–Vis and photoluminescence spectroscopy techniques are obtained in UV, visible, and near-infrared regions. In other words, both the UV–Vis and photoluminescence spectroscopy measurements usually are obtained from 200 nm to 1100 nm wavelengths. In the case of UV–Vis spectrometry techniques, the electrons are excited to higher energy levels and return directly to a lower energy state, as shown in Fig. 4.12A. Thus the absorption or transmission as well as reflectance of the light

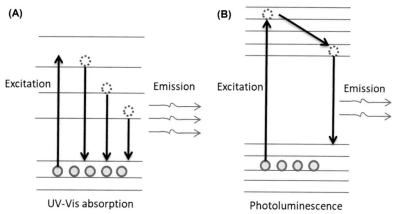

Figure 4.12 Working principles of (A) UV–Vis and (B) photoluminescence spectrometers.

from matter can be measured in given wavelength ranges. The photoluminescence occurs when light energy or photons stimulate the electrons of matter and electrons emit a photon from matter at lower energy states, as shown in Fig. 4.12B. There is a nonradiative or thermal radiation region in conduction bands that electrons pass through. Both UV–Vis and photoluminescence spectroscopy techniques are nondestructive and noncontact for probing of materials.

Both the UV–Vis spectroscopy results and photoluminescence spectrums will be discussed for silicon-ZnO hybrid structures in below.

4.6.1 UV–Vis spectroscopy results of silicon–zinc oxide hybrid structures

The ZnO nanoparticles exhibits absorbance peak at around 350–400 nm wavelengths and transmit the light in thin film form as 80%–90% UV–Vis range (Pudukudy & Yaakob, 2015; Yilmaz et al., 2015). In the case of nanostructured silicon, the absorption peaks can be seen at around the 250–500 nm depending on the size of silicon (Shiohara et al., 2011; Wang et al., 2011; Zhang & Yu, 2014). In the literature, ZnO nanoparticles usually are combined with porous silicon on the silicon substrate and the diffuse reflection spectra are usually used to characterize the ZnO and nanostructured silicon composites for that reason. Sampath et al. studied ZnO nanoparticle deposited porous silicon for photodegradation and photocatalytic activity by atomic layer deposition technique (Sampath et al., 2016).

The absorption spectra of the plane and porous silicon, which obtained increasing etching time, have been illustrated in Fig. 4.13A according to Sampath et al. The absorbance of the plane silicon is low, but the porous silicon has high and absorbance values increase with increasing wavelength. It is possible to see some peaks of the nanostructured silicon in the absorption spectra. Fig. 4.13B exhibits the absorption spectra of the ZnO/glass and ZnO/porous silicon. When the peaks at the 365 nm comes from the ZnO nanoparticles, the strong peaks at the 526 nm attributes to ZnO/porous silicon structure. The peak intensity of the ZnO/porous silicon increased with increasing porous silicon etching time. The absorbance of the obtained ZnO/porous silicon is useful for solar spectrum. The researchers uses the diffuse reflectance spectra of the ZnO/porous silicon to calculate the bandgap (Pavlenko et al., 2020). Zhang et al. used to diffuse reflectance spectra to determine absorption and bandgap behaviors of the silicon nanocrystal loaded ZnO nanowires (Zhang et al., 2020). The absorption of the silicon nanocrystal loaded ZnO increased with increasing ZnO loading. The bandgap values were extracted from diffuse reflectance spectra, and their values decreased from 3.22 eV (nonloaded ZnO nanowires) to 3.07 eV (15 μL silicon nanocrystal loaded).

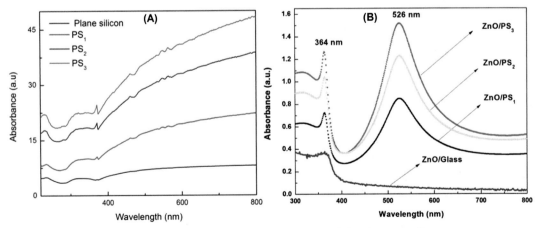

Figure 4.13 Absorption spectra of (A) the plane and porous silicon and (B) ZnO/porous silicon. Permission of Royal Society of Chemistry (Sampath, S., Shestakova, M., Maydannik, P., Ivanova, T., Homola, T., Bryukvin, A., ... Alagan, V. (2016). Photoelectrocatalytic activity of ZnO coated nano-porous silicon by atomic layer deposition, *RSC Advances*, *6*, 25173–25178. https://doi.org/10.1039/c6ra01655c).

4.6.2 Photoluminescence spectroscopy results of silicon-zinc oxide hybrid structures

The ZnO nanoparticles give photoluminescence peaks at around 400–600 nm due to band-edge excitonic transition, oxygen vacancies, and interstitials (Kumar et al., 2012). The nanostructured silicon has changing photoluminescence spectrum from blue to red emission or 400–750 nm wavelengths depending on the size of the nanostructure (Gupta et al., 2009). However, the porous silicon usually gives red luminescence due to having oxygen bonding. The ZnO nanoparticles are combined with silicon to obtain wide luminescence peak. Many researchers have deposited ZnO nanoparticles on the porous silicon to widen the photoluminescence peaks to whole UV and visible regions. Singh et al. investigated the photoluminescence of the ZnO deposited porous silicon structure and obtained a broadband luminescence graph across most of the visible region and obtained stable photoluminescence by annealing (Singh et al., 2007). Verma et al. studied to enhance the photoluminescence spectra of the ZnO/porous silicon to improve quantum efficiency of silicon solar cell (Verma et al., 2014). The obtained photoluminescence graphs (shown in Fig. 4.14) revealed that the photoluminescence peaks shifted with increasing temperature from red to blue region due to composition of the ZnO and porous silicon. There are many studies on the photoluminescence of ZnO and porous silicon hybrid structures to obtain wide range or white light luminescence spectrum (Marin et al., 2015; Zhang & Jia, 2017). Thus ZnO nanoparticles can be employed with nanostructured silicon to obtain wide range luminescence behavior.

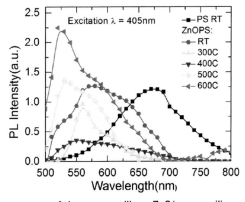

Figure 4.14 Photoluminescence spectrum of the porous silicon ZnO/porous silicon obtained after annealed various temperature. From: Reproduced from D. Verma, A. Kharkwal, S.N. Singh, P.K. Singh, S.N. Sharma, S.S. Mehdi, M. Husain, Application of ZnO nanoparticles to enhance photoluminescence in porous silicon and its possible utilization for improving the short wavelength quantum efficiency of silicon solar cell, Solid State Sci. 37: 13–17. Copyright 2014 Elsevier Masson SAS. All rights reserved.

4.7 Applications of the silicon−zinc oxide hybrid nanoparticles

The zinc oxide and silicon hybrid structures can be employed in light-emitting diodes, solar cells, optical biosensors, or photonic and photocatalysis, or photodegradation applications. Most studies use porous silicon to combine ZnO nanoparticles with nanostructured silicon. The porous silicon has strong absorption behavior, high surface area, and can hold ZnO nanoparticles for various hybrid applications. Thus, the red emission of the porous silicon can be complete with the blue−green emission of the ZnO nanoparticles, and they can be used in screens and white light-emitting diodes (Algün & Akçay, 2019). Verma et al. obtained porous silicon from silicon wafer and filled the pores with ZnO nanoparticles to fabricate high quantum efficiency silicon solar cells (Verma et al., 2014). This design caused an increase in the power conversion efficiency above a certain critical value. Martinez et al. fabricated a memristive device by combining ZnO and mesoporous silicon substrate by spin-coating of the ZnO thin film on porous silicon (Martínez et al., 2014). The device showed the current−voltage (I-V) curve characteristic of memristive systems with zero-crossing pinched hysteresis curve. Pavlenko et al. fabricated porous silicon and ZnO composite structure by metal-assisted chemical etching and atomic layer deposition techniques for optically sensing aflatoxin B1 (Pavlenko et al., 2020). The aflatoxin B1 was detected optically from photoluminescence changes with increasing aflatoxin B1 amount. According to the literature, the application area of the ZnO and silicon hybrid structure can be improved when the silicon and ZnO nanoparticles combined in nanodimensions.

References

Algün, G., & Akçay, N. (2019). The role of etching current density and porous structure on enhanced photovoltaic performance of ZnO/PS heterojunction solar cells. *Applied Physics A: Materials Science and Processing, 125*, 568. Available from https://doi.org/10.1007/s00339-019-2867-3.

Amonkosolpan, J., Wolverson, D., Goller, B., Polisski, S., Kovalev, D., Rollings, M., ... Birks, T. A. (2012). Porous silicon nanocrystals in a silica aerogel matrix. *Nanoscale Research Letters, 7*, 397. Available from https://doi.org/10.1186/1556-276X-7-397.

Barui, A. K., Kotcherlakota, R., & Patra, C. R. (2018). Biomedical applications of zinc oxide nanoparticles. *Inorganic Frameworks as Smart Nanomedicines*, 239−278. Available from https://doi.org/10.1016/B978-0-12-813661-4.00006-7, William Andrew.

Belomoin, G., Therrien, J., Smith, A., Rao, S., Twesten, R., Chaieb, S., ... Mitas, L. (2002). Observation of a magic discrete family of ultrabright Si nanoparticles. *Applied Physics Letters, 80*, 841–843. Available from https://doi.org/10.1063/1.1435802.

Belyakov, V. A., Burdov, V. A., Lockwood, R., & Meldrum, A. (2008). Silicon nanocrystals: Fundamental theory and implications for stimulated emission. *Advanced Optical Technologies*. Available from https://doi.org/10.1155/2008/279502.

Cao, Y., Saygili, Y., Ummadisingu, A., Teuscher, J., Luo, J., Pellet, N., ... Grätzel, M. (2017). 11% efficiency solid-state dye-sensitized solar cells with copper(II/I) hole transport materials. *Nature Communications, 8*, 15390. Available from https://doi.org/10.1038/ncomms15390.

Chan, Y. F., Xu, H. J., Cao, L., Tang, Y., Li, D. Y., & Sun, X. M. (2012). ZnO/Si arrays decorated by Au nanoparticles for surface-enhanced Raman scattering study. *Journal of Applied Physics, 111*, 033104. Available from https://doi.org/10.1063/1.3682462.

Chikkanna, M. M., Neelagund, S. E., & Rajashekarappa, K. K. (2019). Green synthesis of Zinc oxide nanoparticles (ZnO NPs) and their biological activity. *SN Applied Sciences, 1*, 1–10. Available from https://doi.org/10.1007/s42452-018-0095-7.

Eswar, K. A., Husairi, F. S., Ab Aziz, A., Rusop, M., & Abdullah, S. (2014). Photoluminescence spectra of ZnO thin film composed nanoparticles on silicon and porous silicon. *Advanced Materials Research, Trans Tech Publications Ltd*, 843–847. Available from https://doi.org/10.4028/http://www.scientific.net/AMR.8320.843.

Eswar, K. A., Rouhi, J., Husairi, H. F., Rusop, M., & Abdullah, S. (2014). Annealing heat treatment of ZnO nanoparticles grown on porous si substrate using spin-coating method. *Advances in Materials Science and Engineering*. Available from https://doi.org/10.1155/2014/796759.

Geetha, M. S., Nagabhushana, H., & Shivananjaiah, H. N. (2016). Green mediated synthesis and characterization of ZnO nanoparticles using Euphorbia Jatropa latex as reducing agent. *Journal of Science: Advanced Materials and Devices, 1*, 301–310. Available from https://doi.org/10.1016/j.jsamd.2016.060.015.

Gupta, A., Swihart, M. T., & Wiggers, H. (2009). Luminescent colloidal dispersion of silicon quantum dots from microwave plasma synthesis: Exploring the photoluminescence behavior across the visible spectrum. *Advanced Functional Materials, 19*, 696–703. Available from https://doi.org/10.1002/adfm.200801548.

Handore, K., Bhavsar, S., Horne, A., Chhattise, P., Mohite, K., Ambekar, J., ... Chabukswar, V. (2014). Novel green route of synthesis of ZnO nanoparticles by using natural biodegradable polymer and its application as a catalyst for oxidation of aldehydes. *Journal of Macromolecular Science, Part A: Pure and Applied Chemistry, 51*, 941–947. Available from https://doi.org/10.1080/10601325.2014.967078.

Hossain, S. T., Johra, F. T., & Jung, W. G. (2018). Fabrication of silicon carbide from recycled silicon wafer cutting sludge and its purification. *Applied Sciences, 8*, 1841. Available from https://doi.org/10.3390/app8101841.

Kaddes, M., Omri, K., Kouaydi, N., & Zemzemi, M. (2018). Structural, electrical and optical properties of ZnO nanoparticle: combined experimental and theoretical study. *Applied Physics A: Materials Science and Processing, 124*, 518. Available from https://doi.org/10.1007/s00339-018-1921-x.

Kalpana, V. N., Kataru, B. A. S., Sravani, N., Vigneshwari, T., Panneerselvam, A., & Devi Rajeswari, V. (2018). Biosynthesis of zinc oxide nanoparticles using culture filtrates of Aspergillus niger: Antimicrobial textiles and dye degradation studies. *OpenNano, 3*, 48–55. Available from https://doi.org/10.1016/j.onano.2018.06.001.

Khalil, M. I., Al-Qunaibit, M. M., Al-zahem, A. M., & Labis, J. P. (2014). Synthesis and characterization of ZnO nanoparticles by thermal decomposition of a curcumin zinc complex. *Arabian Journal of Chemistry, 7*, 1178–1184. Available from https://doi.org/10.1016/j.arabjc.2013.100.025.

Khashan, K. S. (2011). Optoelectronic properties of ZnO nanoparticles deposition on porous silicon. *International Journal of Modern Physics B, 25*, 277–282. Available from https://doi.org/10.1142/S0217979211054744.

Kolodziejczak-Radzimska, A., & Jesionowski, T. (2014). Zinc oxide-from synthesis to application: A review. *Materials (Basel) Switzerland, 7*, 2833–2881. Available from https://doi.org/10.3390/ma7042833.

Kumar Jangir, L., Kumari, Y., Kumar, A., Kumar, M., & Awasthi, K. (2017). Investigation of luminescence and structural properties of ZnO nanoparticles, synthesized with different precursors. *Materials Chemistry Frontiers, 1*, 1413–1421. Available from https://doi.org/10.1039/c7qm00058h.

Kumar, Y., Herrera-Zaldivar, M., Olive-Méndez, S. F., Singh, F., Mathew, X., & Agarwal, V. (2012). Modification of optical and electrical properties of zinc oxide-coated porous silicon nanostructures induced by swift heavy ion. *Nanoscale Research Letters, 7*, 2–17. Available from https://doi.org/10.1186/1556-276X-7-366.

Lanje, A. S., Sharma, S. J., Ningthoujam, R. S., Ahn, J. S., & Pode, R. B. (2013). Low temperature dielectric studies of zinc oxide (ZnO) nanoparticles prepared by precipitation method. *Advanced Powder Technology, 24*, 331–335. Available from https://doi.org/10.1016/j.apt.2012.080.005.

Lin, K. F., Cheng, H. M., Hsu, H. C., Lin, L. J., & Hsieh, W. F. (2005). Band gap variation of size-controlled ZnO quantum dots synthesized by sol-gel method. *Chemical Physics Letters, 409*, 208–211. Available from https://doi.org/10.1016/j.cplett.2005.050.027.

Londoño-Restrepo, S. M., Zubieta-Otero, L. F., Jeronimo-Cruz, R., Mondragon, M. A., & Rodriguez-García, M. E. (2019). Effect of the crystal size on the infrared and Raman spectra of bio hydroxyapatite of human, bovine, and porcine bones. *Journal of Raman Spectroscopy, 50*, 1120–1129. Available from https://doi.org/10.1002/jrs.5614.

Macias-Montero, M., Askari, S., Mitra, S., Rocks, C., Ni, C., Svrcek, V., ... Mariotti, D. (2016). Energy band diagram of device-grade silicon nanocrystals. *Nanoscale, 8*, 6623–6628. Available from https://doi.org/10.1039/c5nr07705b.

Mansour, N., Momeni, A., Karimzadeh, R., & Amini, M. (2012). Blue-green luminescent silicon nanocrystals fabricated by nanosecond pulsed laser ablation in dimethyl sulfoxide. *Optical Materials Express, 2*, 740. Available from https://doi.org/10.1364/ome.2.000740.

Marin, O., Grinblat, G., María Gennaro, A., Tirado, M., Koropecki, R. R., & Comedi, D. (2015). On the origin of white photoluminescence from ZnO nanocones/porous silicon heterostructures at room temperature. *Superlattices and Microstructures, 79*, 29–37. Available from https://doi.org/10.1016/j.spmi.2014.120.016.

Martínez, L., Ocampo, O., Kumar, Y., & Agarwal, V. (2014). ZnO-porous silicon nanocomposite for possible memristive device fabrication. *Nanoscale Research Letters, 9*, 437. Available from https://doi.org/10.1186/1556-276X-9-437.

Mazitova, G. T., Kienskaya, K. I., Ivanova, D. A., Belova, I. A., Butorova, I. A., & Sardushkin, M. V. (2019). Synthesis and properties of zinc oxide nanoparticles: Advances and prospects. *Journal of Chemical Reviews, 9*, 127–152. Available from https://doi.org/10.1134/s207997801902002x.

Muhammad, W., Ullah, N., Haroon, M., & Abbasi, B. H. (2019). Optical, morphological and biological analysis of zinc oxide nanoparticles (ZnO NPs) using: Papaver somniferum L. *RSC Advances, 9*, 29541–29548. Available from https://doi.org/10.1039/c9ra04424h.

Nagaraju, G., Udayabhanu., Shivaraj., Prashanth, S. A., Shastri, M., Yathish, K. V., ... Rangappa, D. (2017). Electrochemical heavy metal detection, photocatalytic, photoluminescence, biodiesel production and antibacterial activities of Ag–ZnO nanomaterial. *Materials Research Bulletin, 94*, 54–63. Available from https://doi.org/10.1016/j.materresbull.2017.05.043.

Nayef, U. M., Muayad, M. W., & Khalaf, H. A. (2014). Effect of ZnO layers on porous silicon properties. *International Journal of Electrochemical Science, 9*, 2278–2284. Available from http://www.electrochemsci.org, accessed November 28, 2020.

Nayfeh, M., H. (2018). Fundamentals and applications of nano silicon in plasmonics and fullerines, Nanosilicon, Elsevier. Chapter 9.

Pavlenko, M., Myndrul, V., Gottardi, G., Coy, E., Jancelewicz, M., & Iatsunskyi, I. (2020). Porous silicon-zinc oxide nanocomposites prepared by atomic layer deposition for biophotonic applications. *Materials (Basel) Switzerland, 13*, 1987. Available from https://doi.org/10.3390/MA13081987.

Pavlenko, M., Myndrul, V., Gottardi, G., Coy, E., Jancelewicz, M., & Iatsunskyi, I. (2020). Porous silicon-zinc oxide nanocomposites prepared by atomic layer deposition for biophotonic applications. *Materials (13, p. 1987). Switzerland: Basel.* Available from https://doi.org/10.3390/MA13081987.

Pudukudy, M., & Yaakob, Z. (2015). Facile synthesis of quasi spherical ZnO nanoparticles with excellent photocatalytic activity. *Journal of Cluster Science, 26*, 1187–1201. Available from https://doi.org/10.1007/s10876-014-0806-1.

Ramadan, R., Torres-Costa, V., & Martín-Palma, R. J. (2020). Fabrication of zinc oxide and nanostructured porous silicon composite micropatterns on silicon. *Coatings, 10*, 529. Available from https://doi.org/10.3390/COATINGS10060529.

Salem, M., Alami, Z. Y., Bessais, B., Chahboun, A., & Gaidi, M. (2015). Structural and optical properties of ZnO nanoparticles deposited on porous silicon for mc-Si passivation. *Journal of Nanoparticle Research, 17*, 1–9. Available from https://doi.org/10.1007/s11051-015-2944-2.

Sampath, S., Shestakova, M., Maydannik, P., Ivanova, T., Homola, T., Bryukvin, A., ... Alagan, V. (2016). Photoelectrocatalytic activity of ZnO coated nanoporous silicon by atomic layer deposition. *RSC Advances, 6*, 25173–25178. Available from https://doi.org/10.1039/c6ra01655c.

Sankara Reddy, B., & Venkatramana Reddy, S. (2013). N. Koteeswara Reddy, Physical and magnetic properties of (Co, Ag) doped ZnO nanoparticles. *Journal of Materials Science: Materials in Electronics, 24*, 5204–5210. Available from https://doi.org/10.1007/s10854-013-1545-z.

Shiohara, A., Prabakar, S., Faramus, A., Hsu, C. Y., Lai, P. S., Northcote, P. T., & Tilley, R. D. (2011). Sized controlled synthesis, purification, and cell studies with silicon quantum dots. *Nanoscale, 3*, 3364–3370. Available from https://doi.org/10.1039/c1nr10458f.

Shirahata, N., Linford, M. R., Furumi, S., Pei, L., Sakka, Y., Gates, R. J., & Asplund, M. C. (2009). Laser-derived one-pot synthesis of silicon nanocrystals terminated with organic monolayers. *Chemical Communications*, 4684–4686. Available from https://doi.org/10.1039/b905777c.

Singh, R. G., Singh, F., Agarwal, V., & Mehra, R. M. (2007). Photoluminescence studies of ZnO/porous silicon nanocomposites. *Journal of Physics D: Applied Physics, 40*, 3090–3093. Available from https://doi.org/10.1088/0022-3727/40/10/012.

Smith, A., Yamani, Z. H., Roberts, N., Turner, J., Habbal, S. R., Granick, S., & Nayfeh, M. H. (2005). Observation of strong direct-like oscillator strength in the photoluminescence of Si nanoparticles. *Physical Review B: Condensed Matter and Materials Physics, 72*, 205307. Available from https://doi.org/10.1103/PhysRevB.72.205307.

Tachikawa, S., Noguchi, A., Tsuge, T., Hara, M., Odawara, O., & Wada, H. (2011). Optical properties of zno nanoparticles capped with polymers. *Materials* (4, pp. 1132-1143). *Switzerland: Basel*. Available from https://doi.org/10.3390/ma4061132

Tan, D., Ma, Z., Xu, B., Dai, Y., Ma, G., He, M., ... Qiu, J. (2011). Surface passivated silicon nanocrystals with stable luminescence synthesized by femtosecond laser ablation in solution. *Physical Chemistry Chemical Physics, 13*, 20255–20261. Available from https://doi.org/10.1039/c1cp21366k.

Verma, D., Kharkwal, A., Singh, S. N., Singh, P. K., Sharma, S. N., Mehdi, S. S., & Husain, M. (2014). Application of ZnO nanoparticles to enhance photoluminescence in porous silicon and its possible utilization for improving the short wavelength quantum efficiency of silicon solar cell. *Solid State Sciences, 37*, 13–17. Available from https://doi.org/10.1016/j.solidstatesciences.2014.080.008.

Wang, J., Sun, S., Peng, F., Cao, L., & Sun, L. (2011). Efficient one-pot synthesis of highly photoluminescent alkyl-functionalised silicon nanocrystals. *Chemical Communications, 47*, 4941–4943. Available from https://doi.org/10.1039/c1cc10573f.

Yıldırım, M., & Kocyigit, A. (2018). Characterization of Al/In:ZnO/p-Si photodiodes for various In doped level to ZnO interfacial layers. *Journal of Alloys and Compounds.* Available from https://doi.org/10.1016/j.jallcom.2018.070.295.

Yilmaz, M., Caldiran, Z., Deniz, A. R., Aydogan, S., Gunturkun, R., & Turut, A. (2015). Preparation and characterization of sol–gel-derived n-ZnO thin film for Schottky diode application. *Applied Physics A: Materials Science and Processing, 119*, 547–552. Available from https://doi.org/10.1007/s00339-015-8987-5.

Yilmaz, M., & Grilli M., L. (2016). The modification of the characteristics of nanocrystalline ZnO thin films by variation of Ta doping content, Philosophical Magazine, 96, 2125–2142. Available from https://doi.org/10.1080/14786435.2016.1195023.

Zhang, H., & Jia, Z. (2017). Application of porous silicon microcavity to enhance photoluminescence of ZnO/PS nanocomposites in UV light emission. *Optik (Stuttg), 130*, 1183–1190. Available from https://doi.org/10.1016/j.ijleo.2016.110.131.

Zhang, Y., Mandal, R., Ratchford, D. C., Anthony, R., & Yeom, J. (2020). Si nanocrystals/ZnO nanowires hybrid structures as immobilized photocatalysts for photodegradation. *Nanomaterials, 10*, 491. Available from https://doi.org/10.3390/nano10030491.

Zhang, J., & Yu, S. H. (2014). Highly photoluminescent silicon nanocrystals for rapid, label-free and recyclable detection of mercuric ions. *Nanoscale, 6*, 4096–4101. Available from https://doi.org/10.1039/c3nr05896d.

Assembly and electroluminescence of sheet-like zinc oxide/silicon light-emitting diode by a radio frequency magnetron sputtering technique

L. Castañeda

Sección de Estudios de Posgrado e Investigación de la Escuela Superior de Medicina, Instituto Politécnico Nacional, Mexico City, México

5.1 Introduction

A light-emitting diode (LED) is a semiconductor light source that emits light when current flows through it. Electrons in the semiconductor recombine with electron holes, releasing energy in the form of photons. The color of the light (corresponding to the energy of the photons) is determined by the energy required for electrons to cross the bandgap of the semiconductor. White light is obtained by using multiple semiconductors or a layer of light-emitting phosphor on the semiconductor device. Early LEDs were often used as indicator lamps, replacing small incandescent bulbs, and in seven-segment displays. Recent developments have produced high-output white light LEDs suitable for room and outdoor area lighting. LEDs have led to new displays and sensors, while their high switching rates are useful in advanced communications technology.

LEDs have many advantages over incandescent light sources, including lower energy consumption, longer lifetime, improved physical robustness, smaller size, and faster switching. LEDs are used in applications as diverse as aviation lighting, automotive headlamps, advertising, general lighting, traffic signals, camera flashes, lighted wallpaper, and plant growing light and medical devices.

Unlike a laser, the light emitted from a LED is neither spectrally coherent nor even highly monochromatic. However, its spectrum is sufficiently narrow that it appears to the human eye as a pure (saturated) color. Nor, unlike most lasers, is its radiation spatially coherent, so that it cannot approach the very high brightnesses characteristic of lasers (Zheludev, 2007).

Zinc oxide (ZnO) is crystallized in two main forms, hexagonal wurtzite, and cubic zinc blende. The wurtzite structure is more stable under standard conditions and is therefore more common. The zinc blende form can be stabilized by forming ZnO in substrates with latticed cubic structure. In both cases, the center of zinc and oxide are tetrahedral, the most characteristic geometry of Zn(II). The hexagonal polymorphs and zinc blende do not have inversion symmetry (the reflection of a crystal relative to any point, does not transform it into itself). This and other properties of lattice symmetries result in the piezoelectricity of the hexagonal shape and the zinc blende of ZnO, and the piezoelectricity of the hexagonal form of ZnO. Zinc oxide [ZnO] is one of the most important group II–VI semiconductor materials. It is a wide-bandgap oxide semiconductor with a direct energy gap of about 3.37 eV. ZnO has high chemical and mechanical stability, together with being nontoxic in nature and highly abundant. ZnO is a member of the hexagonal wurtzite class; it is a semiconducting, piezoelectric, and optical waveguide material. Recently, transparent conducting oxides (TCOs) have been widely studied. ZnO thin films are one of the most prominent transparent conducting oxides for the fabrication of the next generation of advanced applications such as a window layer in heterojunction solar cells (Ariyanto, Abdullah, Shaari, Yuliarto, & Junaidi, 2009), multilayer photothermal conversion systems (Castañeda & Avendaño-Alejo, 2014), heat mirrors (Castaneda et al., 2012), piezoelectric devices (Castañeda et al., 2006; Castañeda, 2018), and solid state gas sensors (Castañeda, 2018).

The thin solid films of the ZnO have been prepared by using several deposition techniques, which include radio frequency magnetron sputtering (Castañeda, 2018b), thermal evaporation (Girtan, Rusu, Dabos-Seignon, & Rusu, 2008; Lee, Saif Islam, & Kim, 2007) sol–gel technique (Castañeda, Avendaño-Alejo, Gómez, Olvera, & Maldonado, 2013) chemical vapor deposition (Castañeda, 2009), chemical bath deposition (Ku & Wu, 2007), and spray pyrolysis (Castañeda, 2013; Castañeda, Torres-Torres et al., 2013). Among these methods, ultrasonic spray pyrolysis is useful for large area applications. There are a number of different sputter techniques available, including ion beam, reactive, diode, radio frequency and magnetron sputtering. They have a long history of use in the electrical industry but have more recently been used to produce bioactive coatings. Generally, such processes are slow but versatile as they can be used to produce coatings on a wide range of materials—polymers, ceramics, and metals. The coatings are dense, homogeneous, and adhere well to the substrate which is normally flat. Again, the coatings are very thin in nature (1.0–10.0 μm), with a deposition rate in the order of 1.0–1.5 μm/h. The preparation of thin solid films using sputtering is a method in which accelerated high-energy inert gas ions are incident on the target material to eject, or sputter, atoms or small clusters from the surface, which are subsequently deposited on the substrate in high pressure. In sputtering, two electrodes are used, one electrode is the target material and the other is the substrate, and in-between are inert gas ions (see Fig. 5.1).

Due to being environment-friendly, easily prepared, and showing excellent stability, as well as its high exciton binding energy (≈60.0 meV), ZnO has been studied and applied in the field of optoelectronic devices, solar cells, humidity/gas sensors, and flat panel displays (Ding et al., 2018; Luo, Wang, & Zhang, 2018; Park, 2019). In particular, nanoscale ZnO has attracted many scientists since the nanomaterial preparation has been greatly developed (Aziz et al., 2018; Chen, Ma, & Yang, 2007; Jeem et al., 2017; Zhai et al., 2011; Zhu, Zhang, Li, & Yang, 2013;

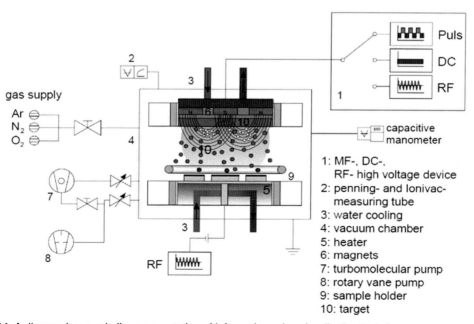

Figure 5.1 A diagram is a symbolic representation of information using visualization techniques.

Powder Diffraction File, 1967). Chemical bath deposition, vapor–liquid–solid, magnetron sputtering, and metal–organic chemical vapor deposition, and so on, have been used to fabricate nanoscale ZnO (Hong & Lee, 2011; Lupan, Pauporté, & Viana, 2010). Various morphologies of ZnO have been obtained and investigated. In the field of light-emitting diodes (LED), many research efforts have focused on the application of ZnO to generate a blue or violet electroluminescence (EL) with emissions from 350 nm to 420 nm (Li, Song, Ji, & Zhou, 2017; Wook Kim et al., 2015). Some researchers have applied the blue or violet emission from the bandgap of ZnO and green emission from the defects of ZnO to construct white light. This method does not need to consider the different attenuation of three basic color LEDs, and the matching between different emission layers and self-absorption from the color mixing and color conversion (Cassette et al., 2012; Dellis et al., 2017). However, this direct white light generation method is very sensitive to the preparation conditions (Song et al., 2016).

In the present research work, sheet-like ZnO/Si LED has been fabricated through depositing ZnO on the single crystal silicon by using a radio frequency (RF) magnetron sputtering method. The effect of sputtering time and power on the structures of ZnO has been investigated. UV, green, and orange emissions from EL spectra of ZnO/Si LED have been discussed in detail.

5.2 Experimental details

5.2.1 Fabrication of the samples

Sheet-like ZnO in Si LED has been fabricated through depositing ZnO on the p-type crystal silicon by using the RF magnetron sputtering in a NSC-4000 with autoload/unload sputter-up configuration (NANO-MASTER, USA).

The ZnO target had purity of 99.999% (from Sigma-Aldrich Química, S.L. Toluca, México) and the single crystal silicon had (111) orientation and a resistivity of 0.007~0.008Ω cm (from Sigma-Aldrich Química, S.L. Toluca, México). It is very important that the substrate is free of all organic residues and surface particles are removed. Cleaning is accomplished by a thorough "P" solution rinse followed by baking at temperatures of 120.0–200.0°C for up to 1200.0 s. To remove the native oxide from the silicon wafers, they were chemically etched with "P" solution (15.0 parts of hydrogen fluoride (HF) 99.999%, from Sigma-Aldrich Química, S.L. Toluca, México) 10.0 parts of nitric acid (HNO_3) 99.999%, from Sigma-

Aldrich Química, S.L. Toluca, México, and 300.0 parts of deionized water (H2O) (resistivity of water produced at ~23.0°C, >5.0 MΩ cm, ZLXL62080 Milli-Q, from Merck Milipore, México). Before the sputtering began, the sputtering chamber was pumped to less than 10^{-8} Torr. In succession, the argon gas was poured into the chamber to maintain the constant pressure of 7.5 Torr. A power of 100 W was applied to the ZnO target. The depositions was carried out for 120 s, 150 s, and 180 s, respectively, which are named as Sample_1, Sample_2, and Sample_3. In order to obtain the better (002)-oriented growth ZnO, the effect of the sputtering power on orientation has been investigated. In the present work, only one fabricated ZnO with the power of 60.0 W has been shown, which was been named as Sample_4. In the process of deposition, all substrates were kept at 500.0°C. And sheet-like ZnO in Si heterojunctions have been fabricated and naturally cooled to room temperature ≈23.0°C. To investigate the EL spectrum from ZnO in Si heterojunctions, ITO and silver (Ag) (99.999%, from Sigma-Aldrich Química, S.L. Toluca, México) electrodes have been fabricated as the front transparent electrode and back electrode by the magnetron sputtering and vacuum electron evaporation methods, respectively. So, the prototypical LED with a structure of ITO/ZnO/p-Si/Ag has been prepared.

5.2.2 Analysis of the samples

The morphological properties of ZnO in Si LED were characterized by a field emission scanning electron microscopy (FESEM, JSM 6700 F), a high-resolution transmission electron microscope (HR-TEM, JEM-2100) and X-ray diffraction (XRD, Panalytical X'Pert Pro) using K_α-Cu as the X-ray source ($\lambda = 0.154056$ nm). The absorption spectra were obtained by a UV–Vis–NIR spectrophotometer (Shimadzu, UV-3150) with the integrated sphere detector. The EL spectra have been collected using a double grating spectrofluorometer (HORIBA, FL3-22).

5.3 Results and Discussions

In order to investigate the structures and growth orientation, XRD patterns of ZnO in Si LED have been shown in Fig. 5.2. From the XRD pattern of Sample_1, three diffraction peaks can be observed, which are located at 31.52°, 34.21°, and 35.98° are indexed to the diffraction from (100), (002), and (101) reflection planes of hexagonal-ZnO. With the increasing sputtering time, as can be seen from Fig. 5.2, the intensity of (100) and (101)

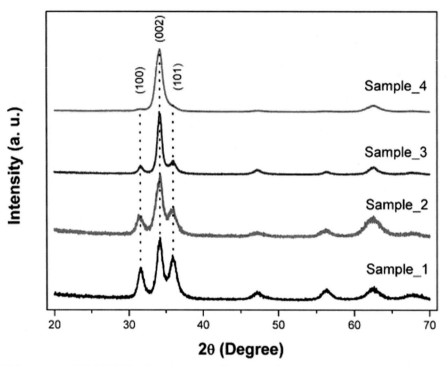

Figure 5.2 XRD patterns of ZnO/Si LEDs. Sample_1: 100.0 W, 120.0 s; Sample_2: 100.0 W, 150.0 s; Sample_3: 100.0 W, 180.0 s; Sample_4: 60.0 W, 180.0 s.

diffraction peaks has gradually decreased and the intensity of (002) diffraction peak has gradually increased. The crystallite sizes were calculated from (100), (101), and (002) peaks, using the Debye–Scherer formula (Warren, 1990):

$$D = \frac{0.9\lambda}{B\cos\theta} \quad (5.1)$$

where D is the crystallite size in nanometers, λ is the wavelength value of the Cu $K_{\alpha 1}$ line ($\lambda = 0.154056$ nm), θ is the Bragg diffraction angle, and B is the FWHM of the diffraction peak measured in radians. Average crystallite sizes with an accuracy of 10.0%.

According to the Debye–Scherer formula (Warren, 1990), the average sizes of ZnO are calculated to be ~16.3 nm, ~17.8 nm, and ~20.9 nm, respectively. The reflection intensities from each of the X-ray diffraction patterns contain information related to the preferential or random growth of polycrystalline thin films using the following expression (Barret & Massalki, 1980):

$$T_{c(hkl)} = \frac{\left(\frac{I_{(hkl)}}{I_{r(hkl)}}\right)}{\frac{1}{n}\sum\left(\frac{I_{(hkl)}}{I_{r(hkl)}}\right)} \qquad (5.2)$$

where $T_{c(hkl)}$ corresponds to the texture coefficient, $I_{(hkl)}$ indicates the X-ray diffraction intensities obtained from the films, and n is the number of diffraction peaks considered. $I_{r(hkl)}$ is the intensity of the reference diffraction pattern (which has been obtained from the ASTM ZnO card). Thus from the texture coefficient calculations, the first three peaks of the ZnO characteristic series, namely (100), (002), and (101), prevail over the rest, and the corresponding T_c values approximate to 2. Therefore, and since these values are higher than 1, it is possible to conclude an abundance of grains growing in a preferential (002) direction.

To study the effect of the sputtering power on the structures, ZnO with sputtering power of 60.0 W and sputtering time of 180.0 s has been shown to compare with Sample_3. From XRD pattern of Sample_4, only (002) diffraction peak can be observed. According to the Debye–Scherer formula, the average sizes of ZnO are calculated to be ≈14.9 nm. It is indicated that the sputtering power is an important factor for the structure of ZnO.

The production of zinc oxide nanocrystallites by radio frequency magnetron sputtering technique is a process in which atoms or ions sputtered from the target react with oxidizing atmosphere (e.g., oxygen) to form ZnO. The oxygen atoms occupy the positions of some zinc atoms or disperse in the intergranular region of the samples. Furthermore, high oxygen concentration is conducive for the nanocrystallites of ZnO on the substrate. Zinc oxide nanocrystallites also benefit from heating substrate properly, adding radio frequency bias, activating the reactants, and providing the required energy for the reaction.

In the process of producing the samples by radio frequency magnetron sputtering, the movement of charged particles in the plasma directly affects the growth of the samples, while the charged particles are controlled by sputtering parameters. The intensity of (002) peak is low at small radio frequency powers. It increases rapidly with increasing power, reaches a maximum, and starts decreasing with a further increase in radio frequency power. The energy of sputtered atoms arriving at the substrate as well as the effect of high energy electron bombardment on the growing film increases with increasing radio frequency power. These two factors provide thermal energy for the deposited atoms, thus enhancing their mobility on the substrate and consequently improving the crystallinity of the deposited samples. This is evidenced by the increasing intensity of (002) peak up.

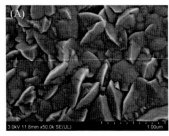 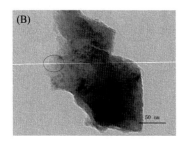 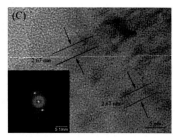

Figure 5.3 (A) SEM, (B) TEM, and (C) HR-TEM images of sheet-like ZnO/Si LEDs (Sample_3: 100.0 W, 180.0 s). Inset: the corresponding FFT pattern.

SEM image of Sample_3 is shown in Fig. 5.3(A). From Fig. 5.3(A), the sheet-like ZnO can be observed. In order to illuminate the fine structures, ZnO have been carefully cleaved to the ethanol solution. After ultrasonic for 300.0 s, the sample has been moved to the copper net. A piece of ZnO thin film has been shown in Fig. 5.3(B), in which the marked area by a circle is studied by the HR-TEM. Fig. 5.3(C) shows the HR-TEM image and the corresponding fast Fourier-transform (FFT) pattern. As shown in the HR-TEM image, a set of lattice fringe can clearly be observed. The lattice distance is confirmed to be ≈0.267 nm, which is obtained by measuring the distance of 10 fringes. The lattice distance of ∼0.267 nm has been corresponded to the (002) lattice family planes of hexagonal ZnO. From the analysis of FFT pattern, the growth of ZnO in this work is along (002) orientation, which is consistent with the result of XRD data (Powder Diffraction File, 1967).

Fig. 5.4(A) presents the reflectance from the four samples. For all samples, a peak located at ∼369 nm can be observed. It is indicated that the bandgap from the sheet-like ZnO is ∼3.36 eV (∼369 nm) (Chen, Ma, & Yang, 2007; Lupan, Pauporté, & Viana, 2010). At room temperature the EL spectra of ZnO/Si LED have been recorded at the forward bias of 10.0 V and shown in Fig. 5.4(B). In order to investigate the origins of emissions, the EL spectra have been decomposed into three bands, that is UV, green, and orange emissions. The positions and full width at half maximum (FWHM) from those emission peaks are shown in Table 5.1. The peaks of ∼369 nm (∼3.36 eV) are consistent with the analysis from the reflectance, which are confirmed as the bandgap emissions of ZnO. From the results of the present works, the values of bandgap are independent on the sputtering time and power of RF sputtering method. However, the FWHM of bandgap emissions have varied with the sputtering time and sputtering power, which are generally improved with increasing the sputtering time and

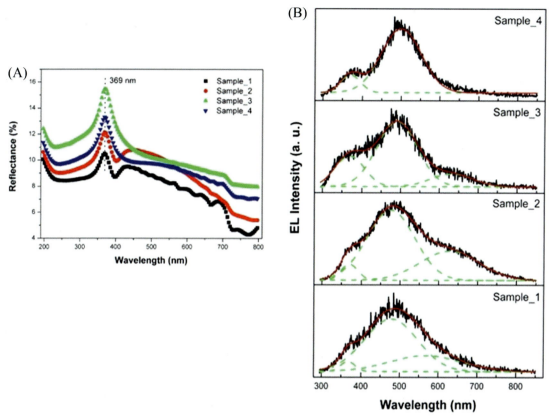

Figure 5.4 (A) Reflectance and (B) EL spectrum of ZnO/Si LEDs at 10.0 V forward bias. Sample_1: 100.0 W, 120.0 s; Sample_2: 100.0 W, 150.0 s; Sample_3: 100.0 W, 180.0 s; Sample_4: 60.0 W, 180.0 s.

Table 5.1 Positions and FWHM of UV, green, and orange emissions from EL spectra.

	A: Position	A: FWHM	B: Position	B: FWHM	C: Position	C: FWHM
Sample_1	369 nm	36 nm	476 nm	138 nm	568 nm	185 nm
Sample_2	369 nm	43 nm	477 nm	123 nm	628 nm	153 nm
Sample_3	369 nm	75 nm	490 nm	104 nm	632 nm	108 nm
Sample_4	369 nm	49 nm	499 nm	104 nm		

A, UV emission; B, Green emission; C, Orange emission.

sputtering power. It is likely to be due to the broad size distribution and more defects with the increasing sputtering time. The lower sputtering power is likely conducive to growth of ZnO nanocrystallites. Green emissions maybe originate from the zinc interstitials (I_{Zn}) or oxygen vacancies (V_O) (Hong & Lee, 2011; Wook Kim et al., 2015). There is no apparent trend of change with the increasing sputtering time. But the green emission shows redshift when reducing the sputtering power. The FWHM of green emissions have been decreasing with the sputtering time. The positions of orange emissions show redshift with the increasing sputtering time, and the FWHM of orange emissions have slowed. The orange emission is believed to origin from the shallow levels in the interface of ZnO/Si heterojunctions (Li, Song, Ji, & Zhou, 2017; Wook Kim et al., 2015). For Sample_4, there is no orange emission. It is indicated that the orange emission can be removed by optimizing the preparation conditions. This occurs due to filling of energy states near the conduction band minimum by excess carriers donated by impurities. Since states near the conduction band minimum are already full, a carrier transition from valence band must happen to the higher energy states in the conduction band, thus requiring more energy than the actual bandgap of the material. Also, the bandgap of the samples are significantly higher than that of pure undoped ZnO (3.3 eV), which is again a consequence of bandgap widening resulting from significantly higher carrier concentration in the samples compared to that in undoped ZnO. Defects are often electrically active and introduce levels in the bandgap of the semiconductor, which involve transitions between different charge states of the same defect. These transition levels are observable quantities that can be derived directly from the calculated formation energies. The transition levels are not to be confused with the Kohn–Sham states that result from band-structure calculations. So, how to optimize and tune the preparation conditions to obtain the purposeful ZnO/Si will be the focus in the future works.

5.4 Conclusions

In summary, sheet-like ZnO/Si LED has been fabricated through depositing ZnO on Si by using a RF magnetron sputtering method. The effects of the sputtering time and power on the structures and EL spectra from ZnO/Si LED have investigated in detail. The structures and characterizations from EL spectra have been sensitive to the fabrication conditions. It is believed that understanding of fabrication conditions will guide us to purposely apply the sheet-like ZnO by RF magnetron sputtering method.

Conflicts of Interest

The authors declare no conflict of interest.

Acknowledgments

The authors gratefully acknowledge the financial support from the Escuela Superior de Medicina del Instituto Politécnico Nacional, through Project no. 20210385.

References

Ariyanto, N. P., Abdullah, H., Shaari, S., Yuliarto, B., & Junaidi, S. (2009). Preparation and characterisation of porous nanosheets zinc oxide films: Based on chemical bath deposition. *World Applied Sciences Journal, 6*, 764−768.

Aziz, A., Tiwale, N., Hodge, S. A., Attwood, S. J., Divitini, G., & Welland, M. E. (2018). Core−shell electrospun polycrystalline ZnO nanofibers for ultrasensitive NO_2 gas sensing. *ACS Applied Materials & Interfaces, 10*(50), 43817−43823.

Barret, C., & Massalki, T. B. (1980). *Structure of Metals* (p. 73) Oxford: Pergamon Press.

Cassette, E., Mahler, B., Guigner, J.-M., Patriarche, G., Dubertret, B., & Pons, T. (2012). Colloidal CdSe/CdS dot-in-plate nanocrystals with 2D-polarized emission. *ACS Nano, 6*, 6741−6750.

Castañeda, L. (2009). Synthesis and characterization of ZnO micro-and nanocages. *Acta Materialia, 57*, 1385−1391.

Castañeda, L. (2013). Photoluminescence and morphological characterization of silver-doped zinc oxide novel nanostructures obtained by ultrasonic spray pyrolysis. *Journal of Nanoelectronics and Optoelectronics, 8*(4), 373−377.

Castañeda, L. (2018a). Evaluating propane sensing properties of ultrasonic spray deposited zinc oxide thin solid films. *Journal of Nanoelectronics and Optoelectronics, 13*(6), 819−824.

Castañeda, L. (2018b). The effects of warmness plasma, and high frequency electrical field on beam-1 plasma interaction in plasma waveguide. *Journal of Computational and Theoretical Nanoscience, 15*(6−7), 1830−1834.

Castañeda, L., & Avendaño-Alejo, M. (2014). Physical characterization of nickel-doped zinc oxide thin solid films deposited by the ultrasonic chemically sprayed technique: Gas-sensing performance in a propane atmosphere. *Journal of Nanoelectronics and Optoelectronics, 9*(3), 419−426.

Castañeda, L., Avendaño-Alejo, M., Gómez, H., Olvera, M. L., & Maldonado, A. (2013). Physical characterization of ruthenium-doped zinc oxide thin solid films deposited by the sol−gel technique: Gas-sensing performance in a propane atmosphere. *Sensor Letters, 11*(2), 286−293.

Castañeda, L., García-Valenzuela, A., Zironi, E. P., Cañetas-Ortega, J., Terrones, M., & Maldonado, A. (2006). Formation of indium-doped zinc oxide thin films using chemical spray techniques: The importance of acetic acid content in the aerosol solution and the substrate temperature for enhancing electrical transport. *Thin Solid Films, 503*(1−2), 212−218.

Castaneda, L., Maldonado, A., Rodríguez-Baez, J., Cheang-Wong, J. C., López-Fuentes, M., & Olvera, M. L. (2012). Chemical spray pyrolysis deposited fluorine-doped zinc oxide thin films: Effect of acetic acid content in the starting solution on the physical properties. *Materials Science in Semiconductor Processing, 15*(13), 232–239.

Castañeda, L., Torres-Torres, C., Trejo-Valdez, M., Castro-Chacón, M. J., Graciano-Armenta, G. A., & Khomenko, A. V. (2013). Optical and photoconductive properties exhibited by silver doped zinc oxide thin films. *Journal of Nanoelectronics and Optoelectronics, 8*(3), 267–272.

Chen, P., Ma, X., & Yang, D. (2007). Ultraviolet electroluminescence from ZnO/p-Si heterojunctions. *Journal of Applied Physics, 101*, 053103.

Dellis, S., Kalfagiannis, N., Kassavetis, S., Bazioti, C., Dimitrakopulos, G. P., Koutsogeorgis, D. C., & Patsalas, P. (2017). Photoluminescence enhancement of ZnO via coupling with surface plasmons on Al thin films. *Journal of Applied Physics, 121*, 103104.

Ding, J., Zhou, Y., Dong, G., Liu, M., Yu, D., & Liu, F. (2018). Solution-processed ZnO as the efficient passivation and electron selective layer of silicon solar cells. *Progress in Photovoltaics: Research and Applications, 26*, 974–980.

Girtan, M., Rusu, G. G., Dabos-Seignon, S., & Rusu, M. (2008). Structural and electrical properties of zinc oxides thin films prepared by thermal oxidation. *Applied Surface Science, 254*(13), 4179–4185.

Hong, K., & Lee, J.-L. (2011). Recent developments in light extraction technologies of organic light emitting diodes. *Electronic Materials Letters, 7*, 77–91.

Hussain, S., Liu, T., Aslam, N., Kashif, M., Cao, S., Rashad, M., … Javed, M. S. (2015). Polymer-assisted co-axial multi-layered circular ZnO nanodisks. *Materials Letters, 152*, 260–263.

Hussain, S., Liu, T., Kashif, M., Cao, S., Zeng, W., Xu, S., … Hashim, U. (2014). A simple preparation of ZnO nanocones and exposure to formaldehyde. *Materials Letters, 128*, 35–38.

Jeem, M., Zhang, L., Ishioka, J., Shibayama, T., Iwasaki, T., Kato, T., & Watanabe, S. (2017). Tuning optoelectrical properties of ZnO nanorods with excitonic defects via submerged illumination. *Nano Letters, 17*, 2088–2093.

Kim, J. W., Yoo, S., Kang, J. S., Lee, S. E., Kim, Y. K., Yu, H. H., Turak, A., & Kim, W. Y. (2015). Color stable white phosphorescent organic light emitting diodes with red emissive electron transport layer. *Journal of Applied Physics, 117*, 245503.

Ku, C. H., & Wu, J. J. (2007). Chemical bath deposition of ZnO nanowire–nanoparticle composite electrodes for use in dye-sensitized solar cells. *Nanotechnology, 18*(50), 505706.

Lee, J. S., Saif Islam, M., & Kim, S. (2007). Photoresponses of ZnO nanobridge devices fabricated using a single-step thermal evaporation method. *Sensors and Actuators B: Chemical, 126*(1), 73–77.

Li, Y., Song, Y. L., Ji, P. F., & Zhou, F. Q. (2017). White light emission with tuneable colour temperature and high colour rendering index from CdS/Si multi-interface nanoheterojunctions. *Nanoscale, 9*, 5922–5926.

Luo, J., Wang, Y., & Zhang, Q. (2018). Progress in perovskite solar cells based on ZnO nanostructures. *Solar Energy, 163*, 289–306.

Lupan, O., Pauporté, T., & Viana, B. (2010). Low-voltage UV-electroluminescence from ZnO-Nanowire array/p-GaN light-emitting diodes. *Advanced Materials, 22*, 3298–3302.

Park, S. (2019). Enhancement of hydrogen sensing response of ZnO nanowires for the decoration of WO3 nanoparticles. *Materials Letters, 234*, 315–318.

Powder Diffraction File, Joint committee on powder diffraction standards, *ASTM International*, Philadelphia, PA, 1967, Card 36–1451.

Song, Y. L., Zhang, T. J., Jie Du, H., Ji, P. F., Li, Y., & Zhou, F. Q. (2016). Synthesis, structures and temperature-dependent photoluminescence from ZnO nano/micro-rods on Zn foil. *Materials Letters, 176*, 139–142.

Suchea, M., Christoulakis, S., Moschovis, K., Katsarakis, N., & Kiriakidis, G. (2006). ZnO transparent thin films for gas sensor applications. *Thin Solid Films, 515*(2), 551–554.

Warren, B. E. (1990). *X-Ray Diffraction* (p. 253) New York: Dover.

Zhai, T., Li, L., Ma, Y., Liao, M., Wang, X., Fang, X., ... Golberg, D. (2011). One-dimensional inorganic nanostructures: synthesis, field-emission and photodetection. *Chemical Society Reviews, 40*, 2986–3004.

Zheludev, N. (2007). The life and times of the LED – a 100-year history. *Nature Photonics, 1*, 189–192.

Zhu, R., Zhang, W., Li, C., & Yang, R. (2013). Uniform zinc oxide nanowire arrays grown on nonepitaxial surface with general orientation control. *Nano Letters, 13*, 5171–5176.

Silicon-based nanomaterials for energy storage

Shumaila Ibraheem[1,2], Ghulam Yasin[1,2], Rashid Iqbal[1,2], Adil Saleem[1,2], Tuan Anh Nguyen[3] and Sehrish Ibrahim[4]

[1]Institute for Advanced Study, Shenzhen University, Shenzhen, P.R. China
[2]College of Physics and Optoelectronic Engineering, Shenzhen University, Shenzhen, P.R. China [3]Institute for Tropical Technology, Vietnam Academy of Science and Technology, Hanoi, Viet Nam [4]College of Life Science and Technology, Beijing University of Chemical Technology, Beijing, China

6.1 Introduction

Presently, the energy crisis is a critically elevated profound societal problem, which eventually impedes the economic development of the globe (Goodenough, 2014; Mehtab et al., 2019). The efficacious development and advancement of green, clean, safe, and viable energy conversion and storage systems have, therefore, been considered as the hot field of research nowadays (Ullah et al., 2020; Wang et al., 2020; Yasin, Arif, Mehtab, Lu, et al., 2020; Yasin, Arif, Mehtab, Shakeel, et al., 2020; Yu, Kumar, Nguyen, Nazir, & Yasin, 2020). Consequently, the improvements in designing innocuous, cost-effective, and renewable energy storage and/or conversion technologies, including batteries and supercapacitors (Ibraheem, Chen, Li, Li et al., 2019; Ibraheem, Chen, Li, Wang, & Wei, 2019; Ibraheem et al., 2020; Mehtab et al., 2019; Muhammad et al., 2020; Yasin et al., 2019; Yu et al., 2020), as well as fuel cells have paid widespread consideration by researchers (Arif et al., 2019; Kumar et al., 2020; Kumar et al., 2021; Nadeem et al., 2018, 2020, 2021). Among various energy technologies, ESTs have involved copious interest because of their high power and/or energy efficiencies and ecologically benign nature (Lin et al., 2015; Mehtab et al., 2019). Secondary batteries and electrochemical supercapacitors are regarded as the most broadly explored energy systems for next-generation storage devices (Dubal, Ayyad, Ruiz, & Gómez-Romero, 2015). LIBs with considerable large energy density and low weight are largely employed in various electronic gadgets (mobile phones, computers, and portable electronics), as well as now in environmentally friendly electric automobiles (Goodenough & Park, 2013; Su et al., 2014). Similarly, supercapacitors have drawn noteworthy appeal and are being employed in automobiles and aerospace systems due to their modest price, high power density, as well as satisfactory cycle life (Iqbal, Ahmad, et al., 2020; Iqbal, Badshah, Ma, & Zhi, 2020; Yan, Wang, Wei, & Fan, 2014). Hence, the potential for worthwhile solutions to the challenges of future energy storage systems entails the novel and unique materials for high-performance energy storage to be constructed from low-priced sources for large-scale electricity storage applications (Dunn, Kamath, & Tarascon, 2011).

Silicon (Si) is an enormously imperative preliminary material with various conventional applications, such as metallurgical (Garibaldi, Ashcroft, Simonelli, & Hague, 2016), semiconductor industries (Reece et al., 2011), as well as solar energy (Liu, Ma, Long, Gao, & Xiong, 2017). Owing to its exceptional chemical and physical properties, nanostructured silicon (nano Si) displays a diversity of significant uses in optics (Zhong et al., 2013), biocatalysts (Kan, Lewis, Chen, & Arnold, 2016), nanoelectronics (Tian et al., 2007), sensors (Lin, Motesharei, Dancil, Sailor, & Ghadiri, 1997), supercapacitors (Oakes et al., 2013), and LIBs (Zhu, Luo, Wang, Jiang, & Yang, 2019). The consistently mounting propulsion behind ESTs demands cutting-edge anode materials to alternative for graphite that are the only commercially available anodes (Liu et al., 2019). Amid the copious new anode materials, Si is viewed as the most favorable potential candidate material for high-performance anodes owing to its large abundance (27.7% of the Earth's crust), relatively low potential plateau and large identified theoretical storage capacity (Yang et al., 2015; Zhu, Gu, Wang, Qu, & Zheng, 2019).

Besides, Si-based nanomaterials are explored less for supercapacitors because of their excessive reactivity with the electrolytes. Nevertheless, doped Si nanomaterials enable excellent conductivity, boast a low mass density, and a controllably etched nanoporous structure, leading to appealing as a promising option for a variety of next-generation energy storage frameworks. Various strategies were adopted to develop Si nanostructures to enhance the specific surface area and modify the pores as well. Several reports also of this field directed on the construction of silicon nanowires (Si NWs) by employing the unique growth mechanism for various morphologies, doping categories, as well as doping level that enables the high performance micro-supercapacitors (Bencheikh et al., 2019).

In this chapter, the brief progress of the fundamentals, fabrication strategies for Si-based nanomaterials, and their utilization in high-performance ESTs, such as LIBs and supercapacitors are summarized. Finally, we propose the existing challenges and possible outlook for future directions toward commercial applications of Si-based nanomaterials for energy storage applications.

6.2 General background and progress

Si was first studied by a Swedish chemist named Jöns Jacob Berzelius in 1824 by heating potassium chips in a silica vessel and then removed the residual by-products by washing. Si is the seventh most abundantly available element on the earth (Sommers, 2007; Wisniak & Jöns Jacob Berzelius, 2000). Owing to the high abundance of this essential element, Si has a widespread range of chemical compounds extending from organics to inorganic materials. Until now, various Si compounds have been discovered as monomers to yield nano Si-based and/or Si nanomaterials. Due to the adjustable environment of chemical bonding, diverse precursors possess variation in stabilities which considerably interrupt the reduction activity. It has been well established that the salt permanency alters typically on the decomposition and on the kinetics bond dissociation enthalpy (De Marco et al., 2018).

Previously, an extensive range of approaches has been established for formulating nano Si. The commonly used methods can be usually categories into silicon halides reduction (Boettcher et al., 2010), plasma or heat aided decomposition of silane (Shi, Tuzer, Fenollosa, & Meseguer, 2012), metal silicides oxidation (Atkins et al., 2013), metallothermic reduction (Nguyen, Hamad, & MacLachlan, 2016), molten salt reduction (Zhu, Yang, Yu, Chen, & Pan, 2017), chemical etching (Zhu, Wu, et al., 2019), electrochemical etching (Li et al., 2014) or reduction (Zhou et al., 2019), and bulk Si from ball-milling (Zong et al., 2016). With the green energy demand of the LIB field (Zheng, Wang, Feng, & He, 2018), emerging cost-effective strategies and ecologically benign for Si or Nano Si applied to anode materials is extremely required (Bao, Huang, Lan, Chen, & Duh, 2015). Incidentally, all substances involved in the reaction should be affordable, inoffensive and practical. In addition, all the reaction process circumstances should be feasible, easy, and, evading high temperature/pressure. When linking these necessities with the accessible approaches, there are only a few of them that can supply this mandate.

6.2.1 Structure—property relationship

Since numerous Si precursors morphologies, tunable structures can be synthesized by means of diverse reducing agents based on different mechanisms. In order to understand the structure-property relationship could be advantageous to evolve impactful reduction methods. Conventionally, the structure-property relationship typically mentions the association amongst the intrinsic and the structural properties of nanomaterials (Shang et al., 2017). Enhancing the electrochemical performance of electrode materials involves a profound understanding of the impact of the structural skeleton on the stability, coulombic efficiency, rate capability, etc. In this heading, we will elucidate Si-based anode materials dependency for structure-property relationship from crystallinity, size, dispersity, and porosity.

6.2.1.1 Size

The size of Si at the nanometer scale has been well authenticated that alleviates the problem of pulverization for the bulk Si anode (Kim, Seo, Park, & Cho, 2010). Verbrugge and his team cautiously examined theoretical diffusion-induced stresses in the relationship for spherical nanoparticles properties of surface modulus and surface tension (Cheng & Verbrugge, 2008). The considerably reduced tensile stress in extent with reducing particle radius may be accountable for the pliability for damage of nanoparticles. Nevertheless, elementary queries leftovers unclear: whether there is an acute size below which the strain on surface persuaded through lithiation in a Si electrode. Liu's group revealed a robust size is a key factor for damage of Si nanoparticles throughout the lithiation process by in situ transmission electron microscopy (TEM) (Liu et al., 2012). Their nanoparticle size should be $\sim$150 nm, below which claps would be subdued, and above this size, surface fractures would be introduced after several cycles. Consequently, incredible works have been dedicated to revealing diverse structured based on Si anode nanomaterials, and productive accomplishments have been fashioned. In spite of those striking developments in Si anode nanomaterials, the unavoidable disadvantages, for instance, expensive, sluggish CE, and low-slung tapped density have hindered its extensive applications (Li et al., 2019).

6.2.1.2 Porosity

The participation of a structure with pores enhanced the electrochemical activity mostly revealed in two characteristics (Li et al., 2014; Shen et al., 2016). Initially, the pores originally available in the nanomaterial can contain the huge volume variations throughout the cycling performance test, can assist to stabilize the entire network of material, conserving cycling stability (An et al., 2019). Secondly, the huge surface area can advance the availability of the electrolyte, reducing lithium-ion diffusion distances and therefore enhancing the obtainable capacity at higher current rates. Hypothetically, unified nanopores permit inner volume growth deprived of noticeable nanoparticle-level external enlargement that advances porous structures and makes them appealing candidates for LIBs (An et al., 2019). Nevertheless, several drawbacks are still remaining for future application, containing low electronic conductivity, cumulative side reactions, and a huge-sized solid electrolyte interphase (SEI) since amplified interaction interface amongst the electrode and electrolyte (Zhu, Luo, et al., 2019).

6.2.1.3 Crystallinity

Besides, the size property of Si affecting the electrochemical performance, numerous scientists studied the consequence of Si crystallinity on the performance. In prior investigations, amorphous Si possesses a higher cycling performance as compared to Si in the crystalline form (Jung, Park, Yoon, Kim, & Joo, 2003). The following report exposed that Si crystalline alloys with lithium at the somewhat lower potential of 120 mV than 220 mV potential for Si in amorphous form, which skillful to the knowledge that by means of a core (Si crystalline)-shell (Si amorphous) composite as the anode material concluded devotedly modifying the preventive charging potential (Cui, Ruffo, Chan, Peng, & Cui, 2009). Kim's group established a fast-charging Si anode material by means of embedding a Si amorphous few layers at the edges of activated graphite, where the enhanced lithium diffusion rate was accredited to the Si amorphous few-layered material (Kim, Chae, Ma, Ko, & Cho, 2017). It should be noted that the Si in the crystalline form will renovate into amorphous after the initial lithiation procedure and experience a similar lithiation in the following cycling process. Though Si amorphous in state has higher kinetic behavior as compared to Si in crystalline form, synthesis of Si-amorphous constantly includes violent circumstances, for instance, bulky flows, huge pressure/temperature based on a CVD (Zhu, Luo, et al., 2019).

6.2.1.4 Dispersity

Monodispersion is largely considered as a serious feature that impacts the consistency of performance. Nevertheless, Si monodisperse particle synthesis is an enormous method built on existing methods (Kim et al., 2010). Therefore, there exists insufficient research associating the electrochemical performance with dispersity. Until now, tendencies can be briefed by a vigilant review of the reports. Zuo's group invented a macro-/mesoporous Si three-dimensional (3D) ordered structure from Stöber silica particles, which are zero-dimensional (0D) with a huge density of mass loading (Zuo et al., 2017). A self-templating mechanism was projected to clarify the construction of a 3D network. This accumulating tendency is perhaps produced by the upright Stöber silica particles monodispersity. Inversely, Xu's group constructed Si/C watermelon-like microspheres by means of ball milling along with spray drying (Zhang et al., 2017). The enhanced size dispersal of the pristine Si/C microspheres subsidizes to attain effective empty space and consolidation the supportive outcome by satisfying the interspace, which results in a huge demanding density and an enhanced areal capacity. Consequently, dispersity chiefly alters the areal capacity of a prepared anode by means of loading patterns or assembling.

6.3 Si-based nanomaterials for lithium storage

The Si nanoparticles are the utmost superior applicants for LIB electrodes for the subsequent motives. Primarily, silicon possesses a huge theoretical capacity of 4200 mAh g^{-1} by creating Li$_{4.4}$Si and additionally, the second most plentiful element in the earth-crust (Martin et al., 2009). This suggestion its countless scope to made-up LIB anodes at a comparatively low price. Second, preparation approaches for Si nanoparticles are somewhat established, and industrially accessible (Hwang, Lee, Kong, Seo, & Choi, 2012). Third, Si nanoparticles are well-suited with the present industrial procedure for LIB electrodes (Liu et al., 2011). Lastly, decreasing the dimensions of Si particles will assist to decrease the strain and avoid the rapid damage of Si throughout the Li insertion, which will knowingly advance the cycling activity of the anodes (Hwang et al., 2012; Oumellal et al., 2011).

Owing to the enormous volume alteration of Si nanoparticles throughout the discharge and charge procedure of the LIBs, still there are few challenges for the application of Si nanoparticles in LIBs. As shown in Fig. 6.1, Si nanoparticles will be unwrapped

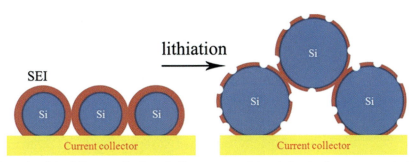

Figure 6.1 Schematic of the failure mechanism of silicon nanoparticles during cycling. Reproduced with permission from Su X, Wu Q, Li J, Xiao X, Lott A, Lu W, et al. (2014). silicon-based nanomaterials for lithium-ion batteries: a review. *Advanced Energy Materials, 4,* 1300882, Wiley-VCH.

from current collectors because of the volume alteration throughout their discharging/charging cycling (Bridel, Azaïs, Morcrette, Tarascon, & Larcher, 2010; Su et al., 2014). In the meantime, the huge volume alteration of Si nanoparticles throughout discharging/charging will result in the disruption of the SEI layer and produce a newly exposed surface to the electrolyte. Altogether of these pieces of evidence will source an enhanced capacity declining of the Si nanoparticles anode. To overwhelmed these difficulties, porous core-shell Si nanoparticles have been established; moreover, numerous nanomaterials such as nanotubes (Wang & Kumta, 2010), graphene (Zhou, Yin, Wan, & Guo, 2012), and mesoporous carbon (Park, Kim, Nam, Park, & Yi, 2012), have been acquainting with as conductive conditions for Si nanoparticles (Magasinski et al., 2010).

6.3.1 Si/carbon-based nanomaterials for Li storage

Carbon nanomaterials is the most capable composited with Si nanoparticles for LIBs since it is obviously ductile and conductive. Numerous carbon architectures have been fused with Si nanoparticles to achieve a sophisticated battery performance. Initially, carbon consumption for Si nanoparticles LIB anodes, the Si nano are typically distributed in the carbonous medium. Due to the enhanced conductivity and shielded volume alteration, the cycling stability of the anode can be enhanced to some degree. Nevertheless, the difficulties of constant accumulation of the Si nano and determined fracturing/renaissance of the SEI films are still tough to resolve, and are probable to source abrupt performance decline (Yang et al., 2020).

Structural engineering of Si nanoparticles to advance their cycling performance is also an appreciated way. Meanwhile, Si has a huge volume variation throughout cycling, spongy/porous Si nanoparticles are capable applicants as LIB anodes by giving adequate free room for enlargement/expansion throughout the lithiation process. Kim's group constructed 3D porous bulky Si as LIB anodes. This Si porous in nature was prepared by engraving SiO_2 from the fusion of carbon/SiO_2 layered with silicon. The capacity remained sustained at 0.2 C, 2780 mAh g^{-1} after 100 cycles (Kim, Han, Choo, & Cho, 2008). Lately, Ge's group established a large-scale production of porous Si nanoparticles. The porous Si nanoparticles were attained by a two-step technique containing (1) doping of boron; and (2) etching with electroless. The Si porous nanoparticles remained enfolded by graphene for LIBs as the anode. The capacity of these Si porous nanoparticles-based anode material can be preserved at 1000 mAh g^{-1} at 0.5 C in 200 cycles (Ge et al., 2013). As stated above, producing structure of Si porous nanoparticles can progress their cycling stability. Though, the volumetric capacity of the anode is extremely low. These results will meaningfully decrease the whole storage capacity of the LIB anode.

Carbon covering was then projected as an operative way out and has been rigorously stated (Sourice et al., 2016). The carbon layers covering Si nanoparticles (Si-NPs) are helpful for increasing the electrical conductivity of the electrode by supplying sufficient electrical connection points and decreasing the electron transportation paths, consequently improving electrical isolation and increasing the electrode kinetics throughout cycles (Liu et al., 2018). The carbon covering can also disperse the Si-NPs and electrolyte. This assists in conquering the development of the SEI and dropping the usage of active nanomaterials, therefore enhancing the ICE and cycling stability of the anodes. In addition, the carbon covering can confine and cushion the large volume variation of Si-NPs and control their accumulation throughout repetitive delithiation/lithiation procedures. This is promising to reserve the physical reliability and cumulative the cyclic performance of the Si-NPs based anodes. It is extraordinary that covering is essential since the uncoated portions of the active nanomaterials are still open to the electrolyte, which might additional worsen the capacity deterioration of the anodes afterward long-lasting cycles (Yang et al., 2020). The extensiveness of the carbon covering also has extreme significance on battery stability. The Si-NPs with varied sizes of the carbon covering yield different electrochemical activities. Luo's group considered the properties of the carbon covering and the coating width on the Li storage competence of Si-NPs (Fig. 6.2)

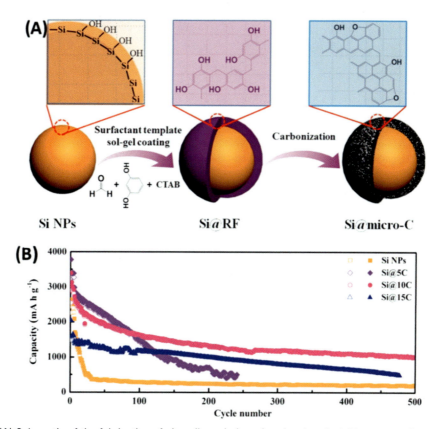

Figure 6.2 (A) Schematic of the fabrication of phenolic resin-based carbon interfacial layer coated commercial silicon nanoparticles through the surfactant template sol-gel approach. (B) Charge-discharge cycling performance (based on the total weight of the electrode) and (C) capacity retention of Si NPs, Si@5 C, Si@10 C, and Si@15 C electrodes at a current density of 500 mA g^{-1}. Reproduced with permission from Luo W, Wang Y, Chou S, Xu Y, Li W, Kong B, et al. (2016). Critical thickness of phenolic resin-based carbon interfacial layer for improving long cycling stability of silicon nanoparticle anodes. *Nano Energy*, 27, 255–264, Elsevier. Copyright @ Elsevier B.V. (2016) (Luo et al., 2016).

(Luo et al., 2016). This validates that the carbon covering cannot efficiently preserve the physical veracity of the anodes throughout long-lasting cyclic stability when a tremendously minor width is cast-off. In divergence, by means of an extremely thick covering might limit the diffusion of Li$^+$ into Si-NPs.

Abundant exertion on emerging improved Si-NPs based anode for LIBs has been completed on perusing fused anodes with Si-NPs. Graphene is an emergent 2D network of carbon atoms, has sp^2-hybridization that contains an extensive variety of properties together with astonishing superb mechanical properties, electrical conductivity, and impermeability (Geim &

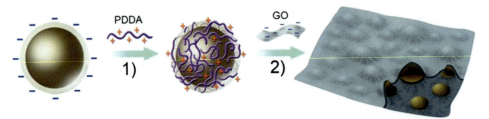

Figure 6.3 Schematic process for fabricating the Si-NP@G nanocomposite: (1) render the Si nanoparticle charged positively by self-assembly of Si nanoparticles and poly (diallyl dimethylammonium) chloride (PDDA); and (2) combine the silicon nanoparticles and graphene oxide by self-assembly of positively charged Si-PDDA nanoparticles and negatively charged GO, which is followed by freeze-drying, thermal reduction, and HF treatment. Reproduced with permission from Zhou X, Yin Y-X, Wan L-J, Guo Y-G. (2012). Self-assembled nanocomposite of silicon nanoparticles encapsulated in graphene through electrostatic attraction for lithium-ion batteries. *Advanced Energy Materials*, 1086–1090, Wiley-VCH. Copyright @ Wiley-VCH (2012).

Novoselov, 2007; Jabbar et al., 2017; Yasin, Anjum, et al., 2020; Yasin, Arif, Nizam, et al., 2018; Yasin, Arif, Shakeel, et al., 2018; Yasin, Khan, et al., 2018). Owing to the mechanical flexibility, huge surface area, and virtuous chemical stability of graphene, Si-NPs covered with graphene have converted a capable applicant as LIB anode (Zhao, Hayner, Kung, & Kung, 2011; Zhou, Cao, Wan, & Guo, 2012). Researchers have industrialized numerous approaches to construct Si-NPs and composite of graphene for LIB lately. As revealed in Fig. 6.3, Zhou's group used electrostatic powers to cover Si-NPs homogeneously in graphene by means of a self-assembly method. An extremely steady cyclic stability of the composite has been experientially observed. The capacity leftovers $1205\ mAh\ g^{-1}$ until and beyond 150 cycles (Zhou, Yin, et al., 2012). Particularly, the Si-NPs/graphene oxide suspension can also be straight spin-coated on a substrate of copper to acquire the LIBs anode (Zhou, Cao, et al., 2012).

6.3.2 Si/metals-based nanomaterials for Li storage

Metal-based nanomaterials, for instance., Cu, Ag, Ni, Co, Fe, with equally high mechanical strength and electrical conductivity as compared to carbon nanomaterials, have also been composited with Si-NPs for LIBs (Yang et al., 2020; Yoo, Lee, Ko, & Park, 2013). Amongst these metals, Ag and Cu are the records examined metals, whereas others appear to appeal less attention. The outcome of these metals on the electrochemical activity of the Si-NP/metal composites is normally the similar. Comparable to the operation of carbon nanomaterials, the Si-NPs are also distributed in

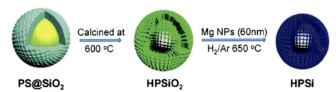

Figure 6.4 Preparation of hollow porous Si (HPSi) nanoparticles. Reproduced with permission from Chen D, Mei X, Ji G, Lu M, Xie J, Lu J, et al. (2012). Reversible lithium-ion storage in silver-treated nanoscale hollow porous silicon particles. *Angewandte Chemie International Edition*, 51, 2409–2413. Copyright @ Wiley-VCH (2012), Wiley-VCH.

the metal atmosphere or covered with metallic nanoparticles. To disclose the behavior of metals, Ag and Cu are held as illustrations in this heading (Yang et al., 2020).

Such as, the electrochemical activity of the Ag-coated Si-NPs can be supplementary enhanced by generating a muffled network built on the Si-NPs. Chen's group. fashioned echoing permeable Ag-coated Si-NPs for LIBs, as shown in Fig. 6.4 (Chen et al., 2012). The pristine porous Si-NPs (HPSi) are covered with Ag nanoparticles, wherein the mass ratio of Si ranges 92.3%. The permitted capacity in the hollow nanoparticle internal and the porous structure in the outer surface (shell) efficiently mitigated the dimension variations in Li-Si alloying/de-alloying reactions, and better-quality convenience of the Si host to Li-ions. The well-ordered structure and the nanoparticle source of the shell subsidized to tumble the charge-transfer reactions by diffusion resistance. The Ag NPs were important for stabilizing the electrical conductivity of the Si-NPs. Consequently, underneath the joined properties of the Ag-NPs and porous hollow structure, the composite displays a capacity of approximately 3000 mAh g^{-1} after 100 cycles at 0.5 Ag^{-1}. Moreover, at a higher current density of 4 Ag^{-1}, a discharge capacity of 2000 mA h g^{-1} was still attained (Chen et al., 2012).

6.3.3 Si/other materials-based nanostructures for Li storage

6.3.3.1 Si/polymers-based LIB anodes

Numerous polymers materials have also been joined with Si-NPs for enhanced electrochemical activity. Amongst these polymers, the focally examined nanomaterial is conductive polymers (CPs), for instance, polyaniline (PANI) (Wu et al., 2013), polypyrrole (PPy) (Luo et al., 2015) and polyethylenedioxythiophene

(PEDOT) (McGraw et al., 2016), since their decent electrical conductivity and mechanical elasticity. Perhaps, an extraordinary improvement of the electrochemical activity of the Si-NPs based anode was understood by inserting the Si-NPs inside a PANI 3D structure (Wu et al., 2013). This composite material can be enclosed onto current collectors of copper deprived of any addition of binders and electrical conductors. In the composited nanomaterial, the Si-NPs are also conformally covered with a thin layer of PANI. The Si-NPs are certain with the PANI 3D structure over whichever an electrostatic interaction or crosslinker hydrogen-bonded by phytic acid through the polymer, which is charged positively. The pores of a polymer matrix deliver abundant annulled room which can take the immense volume variation of Si-NPs throughout lithiation. The 3D PANI structure with the PANI covering provides extremely continuous and conductive paths for equally electrons and ions, instantaneously safeguarding worthy electrical linking with the Si-NPs. Furthermore, even if the huge Si-NPs are probably crash throughout the cyclic performance, they are limited in the consistent small pores of the PANI structure, illuminating the outstanding electrical veracity of the composite. This is verified by the circumstance that the Si-NPs are even reserved inside the PANI network afterward 2000 cycles deprived of detachment electrical connection (Wu et al., 2013).

6.3.3.2 Si/transition metal nitrides/carbides-based LIB anodes

Transition metal carbides or nitrides also own outstanding mechanical properties and electrical conduction. While nearly these nanomaterials display Li storage competence, their performances are not analogous to that of Si. Henceforth, these transition metal carbides or nitrides incline to be used to advance the conduction and physical structure veracity of Si-NPs based anodes. Such as, TiN-coated Si-NPs were examined by Tang's group (Tang et al., 2014). The Si content in the composite is around 67.4%. The TiN coating can enable electron transport and assistance decrease the volume variation of Si-NPs throughout lithiation, subsequent to a stabilized SEI and faster kinetics. This composited anode material displays a primary discharge capacity of 2560 mA h g^{-1} at 0.4 A g^{-1} with an ICE of 76%, which is somewhat inferior to that (79%) of pure Si materials at the similar investigation circumstances. This is owing to the creation of likely SiO$_2$ contamination in the construction procedure of the TiN-coated Si-NPs upon tempering at 1000°C. Though, the composite preserves a high adjustable discharge capacity of over

1900 mA h g^{-1} after 100 cycles at 0.4 A g^{-1} with a capacity retention of more than 75% (Tang et al., 2014).

Particularly, MXenes are a big group of transition metal nitrides and carbides with a 2D network. MXenes have a joint formulation of $M_{n+1}X_nT_x$, where M signifies an initial transition metal, X is C and/or N, and T_x is abbreviated for the exterior functional groups such as $-$OH, $-$F, and $-$O (Yan et al., 2017). Owed to their outstanding mechanical flexibility, electrical conduction, and hydrophilicity, these materials are extensively functioned to numerous electrochemical energy storage devices (Yang et al., 2020). A SiNP/MXene ($Ti_3C_2T_x$) composite was studied by Tian's group (Tian, An, & Feng, 2019). In the composited material, SiNPs are covalently bonded on the exterior of MXene and arbitrarily distributed among equivalent MXene nanosheets. The MXene nanosheets not only avoid the accumulation of Si-NPs and cushion their dimensions variation throughout the cyclic performance, nonetheless also deliver fast ionic and electronic diffusion paths. Furthermore, Si-NPs concurrently avert the restacking of MXene nanosheets. This freestanding and flexible composited nanomaterial can be right away applied as the anode. Consequently, the composite nanomaterial displays equally excellent rate and cycling performances. It provides discharge capacities of 2118 mA h g^{-1} after 100 cycles at 100 mA g^{-1} and 1672 mA h g^{-1} after 200 cycles at 1000 mA g^{-1} (Tian et al., 2019).

6.4 Si-based nanomaterials for supercapacitors

The Si-based nanomaterials attracted low supercapacitors owing to large Si reactivity with electrolytes. Nevertheless, doped Si nanomaterials claim a low-slung mass density, exceptional electrical conduction, a controllably engraved nanoporous network and earth-abundance and scientific existence motivates for varied energy storage structures. Pint's group (Oakes et al., 2013) established a widespread direction to alter porous Si (P-Si) into steady anodes for electrochemical energy storage devices by the development of few-layered, conformal graphene covering on the P-Si exterior (Fig. 6.5). This graphene covering concurrently passivates surface control corners and delivers a perfect electrode/electrolyte electrochemical boundary. This results in up to 40 times energy density development and two times wider electrochemical window associated with similar structured non-passivated P-Si. Their work validates a method generalized to nanoporous and mesoporous nanomaterials that dissociates the manufacturing of anode

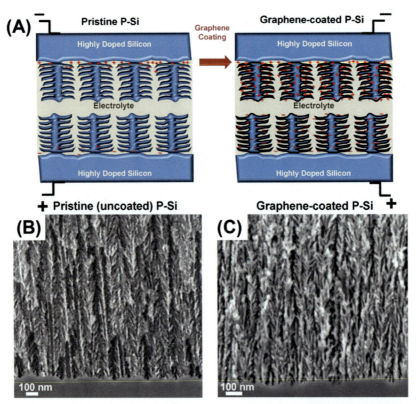

Figure 6.5 (A) Scheme of the effect of coating P-Si on the capacitive charge storage properties. SEM cross-sectional images of porous silicon showing the interface between the etched porous silicon and the silicon wafer for the case of (B) uncoated, pristine porous silicon and (C) graphene-coated porous silicon. Reproduced with permission from Oakes L, Westover A, Mares JW, Chatterjee S, Erwin WR, Bardhan R, et al. (2013). Surface engineered porous silicon for stable, high performance electrochemical supercapacitors. *Scientific Reports*, 3, 3020, Nature. Copyright @ Springer Nature.

network and electrochemical stable surface to tune performance in electrochemical surroundings. Explicitly, they established P-Si as a capable new stage for industrial scale and unified electrochemical energy storage devices.

Furthermore, the continual surge of minor electronic devices stresses for slight energy storage parts, usually recognized as microsupercapacitors, that contain the independent process of these strategies. Bencheikh's group (Bencheikh et al., 2019) strategically plan a fruitful technique to achieve enormous power and energy densities of a Si-built micro-supercapacitor, covering Si nanowires decorated with NPs of ruthenium (Ru/Si NWs). The Si NWs are attained over the vapor-liquid-solid (VLS) development mechanism, although a modest electroless procedure is exploited to credit Ru NPs. Although

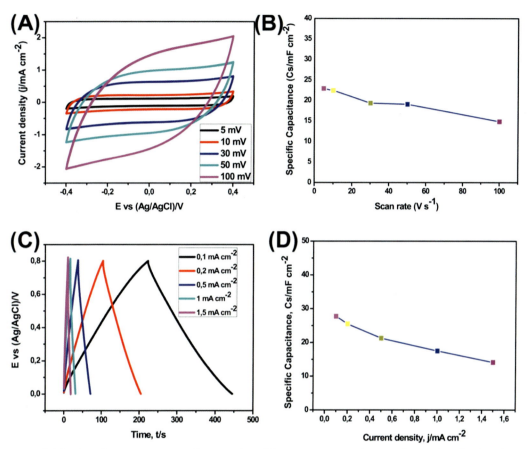

Figure 6.6 (A) Cyclic voltammetry curves of symmetric solid state supercapacitor, consisting of Ru/Si NWs//PVA-H_2SO_4//Ru/Si NWs, recorded at a different scan rates (B) Effect of scan rate on specific capacitance, (C) Galvanostatic charge-discharge curves at different current densities, (D) Effect of current density on specific capacitance. Reproduced with permission from Bencheikh Y, Harnois M, Jijie R, Addad A, Roussel P, Szunerits S, et al. (2019). Elsevier. High performance silicon nanowires/ruthenium nanoparticles micro-supercapacitors. *Electrochimica Acta*, 311, 150–159. Copyright @ Elsevier.

Si nanostructure permits to upsurge the surface area, covering through Ru NPs presents a pseudocapacitance essential to reach huge power and energy densities. The Ru/Si NWs micro-supercapacitor displays a current density of $1\,mA\,cm^{-2}$ with a capacitance of $36.25\,mF\,cm^{-2}$ in a Na_2SO_4 electrolyte and charge-discharge at $1\,mA\,cm^{-2}$ an outstanding steadiness over 25 000 cycles. A supercapacitor fabricated based on solid-state technique (Fig. 6.6) with symmetric electrodes parted by a sulfuric acid/polyvinyl alcohol electrolyte. The device exhibits at a current density of

1 mA cm^{-2} a specific capacity of ~18 mF cm^{-2} along with a specific power density of 0.5 mW cm^{-2}.

6.5 Conclusion

In this chapter, the fundamentals and most new developments in the utilization of Si-based nanostructured materials for LIB anodes and supercapacitors are briefly reviewed. Although Si nanoparticles have shown outstanding storage capacity for Li-ions and excellent electrochemical performance for supercapacitors, the problem of large volumetric variation of Si nanoparticle-based anodes upon repetitive cycling performance causes the substantial pulverization and eventually results in delamination from the current collectors. To overcome the aforementioned issues, various strategies, including the addition of binders, compositing Si nanoparticles with carbon materials and constructing the nanoarchitectures, have been employed to develop mechanically stable and high-performance anodes for energy storage devices. Despite this, how to efficiently combine these strategies to produce the ideal and efficacious Si-based hybrid nanostructures for energy storage applications that can ensure adequate performance remains a big challenge for the researchers of the battery community and material scientists.

References

An, W., Gao, B., Mei, S., Xiang, B., Fu, J., Wang, L., et al. (2019). Scalable synthesis of ant-nest-like bulk porous silicon for high-performance lithium-ion battery anodes. *Nature Communications, 10*, 1447.

Arif, M., Yasin, G., Shakeel, M., Mushtaq, M. A., Ye, W., Fang, X., et al. (2019). Hierarchical CoFe-layered double hydroxide and g-C$_3$N$_4$ heterostructures with enhanced bifunctional photo/electrocatalytic activity towards overall water splitting. *Materials Chemistry Frontiers, 3*, 520–531.

Atkins, T. M., Cassidy, M. C., Lee, M., Ganguly, S., Marcus, C. M., & Kauzlarich, S. M. (2013). Synthesis of long T1 silicon nanoparticles for hyperpolarized 29Si magnetic resonance imaging. *ACS Nano, 7*, 1609–1617.

Bao, Q., Huang, Y.-H., Lan, C.-K., Chen, B.-H., & Duh, J.-G. (2015). Scalable upcycling silicon from waste slicing sludge for high-performance lithium-ion battery anodes. *Electrochimica Acta, 173*, 82–90.

Bencheikh, Y., Harnois, M., Jijie, R., Addad, A., Roussel, P., Szunerits, S., et al. (2019). High performance silicon nanowires/ruthenium nanoparticles micro-supercapacitors. *Electrochimica Acta, 311*, 150–159.

Boettcher, S. W., Spurgeon, J. M., Putnam, M. C., Warren, E. L., Turner-Evans, D. B., Kelzenberg, M. D., et al. (2010). Energy-conversion properties of vapor-liquid-solid–grown silicon wire-array photocathodes. *Science, 327*, 185.

Bridel, J. S., Azaïs, T., Morcrette, M., Tarascon, J. M., & Larcher, D. (2010). Key parameters governing the reversibility of Si/carbon/CMC electrodes for li-ion batteries. *Chemistry of Materials, 22*, 1229–1241.

Chen, D., Mei, X., Ji, G., Lu, M., Xie, J., Lu, J., et al. (2012). Reversible lithium-ion storage in silver-treated nanoscale hollow porous silicon particles. *Angewandte Chemie International Edition, 51*, 2409–2413.

Cheng, Y.-T., & Verbrugge, M. W. (2008). The influence of surface mechanics on diffusion induced stresses within spherical nanoparticles. *Journal of Applied Physics, 104*, 083521.

Cui, L.-F., Ruffo, R., Chan, C. K., Peng, H., & Cui, Y. (2009). Crystalline-amorphous core−shell silicon nanowires for high capacity and high current battery electrodes. *Nano Letters, 9*, 491–495.

De Marco, M. L., Semlali, S., Korgel, B. A., Barois, P., Drisko, G. L., & Aymonier, C. (2018). Silicon-based dielectric metamaterials: focus on the current synthetic challenges. *Angewandte Chemie International Edition, 57*, 4478–4498.

Dubal, D. P., Ayyad, O., Ruiz, V., & Gómez-Romero, P. (2015). Hybrid energy storage: the merging of battery and supercapacitor chemistries. *Chemical Society Reviews, 44*, 1777–1790.

Dunn, B., Kamath, H., & Tarascon, J.-M. (2011). Electrical energy storage for the grid: a battery of choices. *Science, 334*, 928.

Garibaldi, M., Ashcroft, I., Simonelli, M., & Hague, R. (2016). Metallurgy of high-silicon steel parts produced using selective laser melting. *Acta Materialia, 110*, 207–216.

Ge, M., Rong, J., Fang, X., Zhang, A., Lu, Y., & Zhou, C. (2013). Scalable preparation of porous silicon nanoparticles and their application for lithium-ion battery anodes. *Nano Research, 6*, 174–181.

Geim, A. K., & Novoselov, K. S. (2007). The rise of graphene. *Nature Materials, 6*, 183–191.

Goodenough, J. B. (2014). Electrochemical energy storage in a sustainable modern society. *Energy & Environmental Science, 7*, 14–18.

Goodenough, J. B., & Park, K.-S. (2013). The Li-ion rechargeable battery: a perspective. *Journal of the American Chemical Society, 135*, 1167–1176.

Hwang, T. H., Lee, Y. M., Kong, B.-S., Seo, J.-S., & Choi, J. W. (2012). Electrospun core−shell fibers for robust silicon nanoparticle-based lithium ion battery anodes. *Nano Letters, 12*, 802–807.

Ibraheem, S., Chen, S., Li, J., Li, W., Gao, X., Wang, Q., et al. (2019). Three-dimensional Fe,N-decorated carbon-supported NiFeP nanoparticles as an efficient bifunctional catalyst for rechargeable zinc−O_2 batteries. *ACS Applied Materials & Interfaces, 11*, 699–705.

Ibraheem, S., Chen, S., Li, J., Wang, Q., & Wei, Z. (2019). In situ growth of vertically aligned FeCoOOH-nanosheets/nanoflowers on Fe, N co-doped 3D-porous carbon as efficient bifunctional electrocatalysts for rechargeable zinc−O_2 batteries. *Journal of Materials Chemistry A, 7*, 9497–9502.

Ibraheem, S., Chen, S., Peng, L., Li, J., Li, L., Liao, Q., et al. (2020). Strongly coupled iron selenides-nitrogen-bond as an electronic transport bridge for enhanced synergistic oxygen electrocatalysis in rechargeable zinc-O_2 batteries. *Applied Catalysis B: Environmental, 265*, 118569.

Iqbal, R., Ahmad, A., Mao, L.-J., Ghazi, Z. A., Imani, A., Lu, C.-X., et al. (2020). A high energy density self-supported and bendable organic electrode for redox supercapacitors with a wide voltage window. *Chinese Journal of Polymer Science, 38*, 522–530.

Iqbal, R., Badshah, A., Ma, Y.-J., & Zhi, L.-J. (2020). An electrochemically stable 2D covalent organic framework for high-performance organic supercapacitors. *Chinese Journal of Polymer Science, 38*, 558–564.

Jabbar, A., Yasin, G., Khan, W. Q., Anwar, M. Y., Korai, R. M., Nizam, M. N., et al. (2017). Electrochemical deposition of nickel graphene composite coatings: effect of deposition temperature on its surface morphology and corrosion resistance. *RSC Advances, 7*, 31100–31109.

Jung, H., Park, M., Yoon, Y.-G., Kim, G.-B., & Joo, S.-K. (2003). Amorphous silicon anode for lithium-ion rechargeable batteries. *Journal of Power Sources, 115*, 346–351.

Kan, S. B. J., Lewis, R. D., Chen, K., & Arnold, F. H. (2016). Directed evolution of cytochrome c for carbon–silicon bond formation: bringing silicon to life. *Science, 354*, 1048.

Kim, H., Han, B., Choo, J., & Cho, J. (2008). Three-dimensional porous silicon particles for use in high-performance lithium secondary batteries. *Angewandte Chemie International Edition, 47*, 10151–10154.

Kim, H., Seo, M., Park, M.-H., & Cho, J. (2010). A critical size of silicon nano-anodes for lithium rechargeable Batteries. *Angewandte Chemie International Edition, 49*, 2146–2149.

Kim, N., Chae, S., Ma, J., Ko, M., & Cho, J. (2017). Fast-charging high-energy lithium-ion batteries via implantation of amorphous silicon nanolayer in edge-plane activated graphite anodes. *Nature Communications, 8*, 812.

Kumar, A., Yasin, G., Korai, R. M., Slimani, Y., Ali, M. F., Tabish, M., et al. (2020). Boosting oxygen reduction reaction activity by incorporating the iron phthalocyanine nanoparticles on carbon nanotubes network. *Inorganic Chemistry Communications, 120*, 108160.

Kumar, A., Yasin, G., Vashistha, V. K., Das, D. K., Rehman, M. U., Iqbal, R., et al. (2021). Enhancing oxygen reduction reaction performance via CNTs/graphene supported iron protoporphyrin IX: a hybrid nanoarchitecture electrocatalyst. *Diamond and Related Materials, 113*, 108272.

Li, J.-Y., Li, G., Zhang, J., Yin, Y.-X., Yue, F.-S., Xu, Q., et al. (2019). Rational design of robust Si/C microspheres for high-tap-density anode materials. *ACS Applied Materials & Interfaces, 11*, 4057–4064.

Li, X., Gu, M., Hu, S., Kennard, R., Yan, P., Chen, X., et al. (2014). Mesoporous silicon sponge as an anti-pulverization structure for high-performance lithium-ion battery anodes. *Nature Communications, 5*, 4105.

Lin, M.-C., Gong, M., Lu, B., Wu, Y., Wang, D.-Y., Guan, M., et al. (2015). An ultrafast rechargeable aluminium-ion battery. *Nature, 520*, 324–328.

Lin, V. S. Y., Motesharei, K., Dancil, K.-P. S., Sailor, M. J., & Ghadiri, M. R. (1997). A porous silicon-based optical interferometric biosensor. *Science, 278*, 840.

Liu, D., Ma, J., Long, R., Gao, C., & Xiong, Y. (2017). Silicon nanostructures for solar-driven catalytic applications. *Nano Today, 17*, 96–116.

Liu, G., Xun, S., Vukmirovic, N., Song, X., Olalde-Velasco, P., Zheng, H., et al. (2011). Polymers with tailored electronic structure for high capacity lithium battery electrodes. *Advanced Materials, 23*, 4679–4683.

Liu, N., Mamat, X., Jiang, R., Tong, W., Huang, Y., Jia, D., et al. (2018). Facile high-voltage sputtering synthesis of three-dimensional hierarchical porous nitrogen-doped carbon coated Si composite for high performance lithium-ion batteries. *Chemical Engineering Journal, 343*, 78–85.

Liu, X. H., Zhong, L., Huang, S., Mao, S. X., Zhu, T., & Huang, J. Y. (2012). Size-dependent fracture of silicon nanoparticles during lithiation. *ACS Nano, 6*, 1522–1531.

Liu, Z., Yu, Q., Zhao, Y., He, R., Xu, M., Feng, S., et al. (2019). Silicon oxides: a promising family of anode materials for lithium-ion batteries. *Chemical Society Reviews, 48*, 285–309.

Luo, L., Zhao, P., Yang, H., Liu, B., Zhang, J.-G., Cui, Y., et al. (2015). Surface coating constraint induced self-discharging of silicon nanoparticles as anodes for lithium ion batteries. *Nano Letters, 15*, 7016–7022.

Luo, W., Wang, Y., Chou, S., Xu, Y., Li, W., Kong, B., et al. (2016). Critical thickness of phenolic resin-based carbon interfacial layer for improving long cycling stability of silicon nanoparticle anodes. *Nano Energy, 27*, 255–264.

Magasinski, A., Zdyrko, B., Kovalenko, I., Hertzberg, B., Burtovyy, R., Huebner, C. F., et al. (2010). Toward efficient binders for Li-ion battery Si-based anodes: polyacrylic acid. *ACS Applied Materials & Interfaces, 2*, 3004–3010.

Martin, C., Alias, M., Christien, F., Crosnier, O., Bélanger, D., & Brousse, T. (2009). Graphite-grafted silicon nanocomposite as a negative electrode for lithium-ion batteries. *Advanced Materials, 21*, 4735–4741.

McGraw, M., Kolla, P., Yao, B., Cook, R., Quiao, Q., Wu, J., et al. (2016). One-step solid-state in-situ thermal polymerization of silicon-PEDOT nanocomposites for the application in lithium-ion battery anodes. *Polymer, 99*, 488–495.

Mehtab, T., Yasin, G., Arif, M., Shakeel, M., Korai, R. M., Nadeem, M., et al. (2019). Metal-organic frameworks for energy storage devices: batteries and supercapacitors. *Journal of Energy Storage, 21*, 632–646.

Muhammad, N., Yasin, G., Li, A., Chen, Y., Saleem, H. M., Liu, R., et al. (2020). Volumetric buffering of manganese dioxide nanotubes by employing 'as is' graphene oxide: an approach towards stable metal oxide anode material in lithium-ion batteries. *Journal of Alloys and Compounds, 842*, 155803.

Nadeem, M., Yasin, G., Arif, M., Bhatti, M. H., Sayin, K., Mehmood, M., et al. (2020). Pt-Ni@PC900 hybrid derived from layered-structure Cd-MOF for fuel cell ORR activity. *ACS Omega, 5*, 2123–2132.

Nadeem, M., Yasin, G., Arif, M., Tabassum, H., Bhatti, M. H., Mehmood, M., et al. (2021). Highly active sites of Pt/Er dispersed N-doped hierarchical porous carbon for trifunctional electrocatalyst. *Chemical Engineering Journal, 409*, 128205.

Nadeem, M., Yasin, G., Bhatti, M. H., Mehmood, M., Arif, M., & Dai, L. (2018). Pt-M bimetallic nanoparticles (M = Ni, Cu, Er) supported on metal organic framework-derived N-doped nanostructured carbon for hydrogen evolution and oxygen evolution reaction. *Journal of Power Sources, 402*, 34–42.

Nguyen, T.-D., Hamad, W. Y., & MacLachlan, M. J. (2016). Hard photonic glasses and corundum nanostructured films from aluminothermic reduction of helicoidal mesoporous silicas. *Chemistry of Materials, 28*, 2581–2588.

Oakes, L., Westover, A., Mares, J. W., Chatterjee, S., Erwin, W. R., Bardhan, R., et al. (2013). Surface engineered porous silicon for stable, high performance electrochemical supercapacitors. *Scientific Reports, 3*, 3020.

Oumellal, Y., Delpuech, N., Mazouzi, D., Dupré, N., Gaubicher, J., Moreau, P., et al. (2011). The failure mechanism of nano-sized Si-based negative electrodes for lithium ion batteries. *Journal of Materials Chemistry, 21*, 6201–6208.

Park, J., Kim, G.-P., Nam, I., Park, S., & Yi, J. (2012). One-pot synthesis of silicon nanoparticles trapped in ordered mesoporous carbon for use as an anode material in lithium-ion batteries. *Nanotechnology, 24*, 025602.

Reece, S. Y., Hamel, J. A., Sung, K., Jarvi, T. D., Esswein, A. J., Pijpers, J. J. H., et al. (2011). Wireless solar water splitting using silicon-based semiconductors and earth-abundant catalysts. *Science, 334*, 645.

Shang, T., Wen, Y., Xiao, D., Gu, L., Hu, Y.-S., & Li, H. (2017). Atomic-scale monitoring of electrode materials in lithium-ion batteries using in situ transmission electron microscopy. *Advanced Energy Materials, 7*, 1700709.

Shen, C., Ge, M., Luo, L., Fang, X., Liu, Y., Zhang, A., et al. (2016). In Situ and Ex Situ TEM study of lithiation behaviours of porous silicon nanostructures. *Scientific Reports, 6*, 31334.

Shi, L., Tuzer, T. U., Fenollosa, R., & Meseguer, F. (2012). A new dielectric metamaterial building block with a strong magnetic response in the sub-1.5-micrometer region: silicon colloid nanocavities. *Advanced Materials, 24*, 5934–5938.

Sommers MA. Silicon: The Rosen Publishing Group, Inc; 2007.

Sourice, J., Bordes, A., Boulineau, A., Alper, J. P., Franger, S., Quinsac, A., et al. (2016). Core-shell amorphous silicon-carbon nanoparticles for high performance anodes in lithium ion batteries. *Journal of Power Sources, 328*, 527–535.

Su, X., Wu, Q., Li, J., Xiao, X., Lott, A., Lu, W., et al. (2014). silicon-based nanomaterials for lithium-ion batteries: a review. *Advanced Energy Materials, 4*, 1300882.

Tang, D., Yi, R., Gordin, M. L., Melnyk, M., Dai, F., Chen, S., et al. (2014). Titanium nitride coating to enhance the performance of silicon nanoparticles as a lithium-ion battery anode. *Journal of Materials Chemistry A, 2*, 10375–10378.

Tian, B., Zheng, X., Kempa, T. J., Fang, Y., Yu, N., Yu, G., et al. (2007). Coaxial silicon nanowires as solar cells and nanoelectronic power sources. *Nature, 449*, 885–889.

Tian, Y., An, Y., & Feng, J. (2019). Flexible and freestanding silicon/MXene composite papers for high-performance lithium-ion batteries. *ACS Applied Materials & Interfaces, 11*, 10004–10011.

Ullah, S., Yasin, G., Ahmad, A., Qin, L., Yuan, Q., Khan, A. U., et al. (2020). Construction of well-designed 1D selenium–tellurium nanorods anchored on graphene sheets as a high storage capacity anode material for lithium-ion batteries. *Inorganic Chemistry Frontiers, 7*, 1750–1761.

Wang, H., Sheng, L., Yasin, G., Wang, L., Xu, H., & He, X. (2020). Reviewing the current status and development of polymer electrolytes for solid-state lithium batteries. *Energy Storage Materials, 33*, 188–215.

Wang, W., & Kumta, P. N. (2010). Nanostructured hybrid silicon/carbon nanotube heterostructures: reversible high-capacity lithium-ion anodes. *ACS Nano, 4*, 2233–2241.

Wisniak, J. (2000). Jöns Jacob Berzelius: a guide to the perplexed chemist. *The Chemical Educator, 5*, 343–350.

Wu, H., Yu, G., Pan, L., Liu, N., McDowell, M. T., Bao, Z., et al. (2013). Stable Li-ion battery anodes by in-situ polymerization of conducting hydrogel to conformally coat silicon nanoparticles. *Nature Communications, 4*, 1943.

Yan, J., Ren, C. E., Maleski, K., Hatter, C. B., Anasori, B., Urbankowski, P., et al. (2017). Flexible MXene/graphene films for ultrafast supercapacitors with outstanding volumetric capacitance. *Advanced Functional Materials, 27*, 1701264.

Yan, J., Wang, Q., Wei, T., & Fan, Z. (2014). Recent advances in design and fabrication of electrochemical supercapacitors with high energy densities. *Advanced Energy Materials, 4*, 1300816.

Yang, J., Wang, Y.-X., Chou, S.-L., Zhang, R., Xu, Y., Fan, J., et al. (2015). Yolk-shell silicon-mesoporous carbon anode with compact solid electrolyte interphase film for superior lithium-ion batteries. *Nano Energy, 18*, 133–142.

Yang, Y., Yuan, W., Kang, W., Ye, Y., Yuan, Y., Qiu, Z., et al. (2020). Silicon-nanoparticle-based composites for advanced lithium-ion battery anodes. *Nanoscale, 12*, 7461–7484.

Yasin, G., Anjum, M. J., Malik, M. U., Khan, M. A., Khan, W. Q., Arif, M., et al. (2020). Revealing the erosion-corrosion performance of sphere-shaped morphology of nickel matrix nanocomposite strengthened with reduced graphene oxide nanoplatelets. *Diamond and Related Materials*, 104, 107763.

Yasin, G., Arif, M., Mehtab, T., Lu, X., Yu, D., Muhammad, N., et al. (2020). Understanding and suppression strategies toward stable Li metal anode for safe lithium batteries. *Energy Storage. Materials*, 25, 644–678.

Yasin, G., Arif, M., Mehtab, T., Shakeel, M., Mushtaq, M. A., Kumar, A., et al. (2020). A novel strategy for the synthesis of hard carbon spheres encapsulated with graphene networks as a low-cost and large-scalable anode material for fast sodium storage with an ultralong cycle life. *Inorganic Chemistry Frontiers*, 7, 402–410.

Yasin, G., Arif, M., Nizam, M. N., Shakeel, M., Khan, M. A., Khan, W. Q., et al. (2018). Effect of surfactant concentration in electrolyte on the fabrication and properties of nickel-graphene nanocomposite coating synthesized by electrochemical co-deposition. *RSC Advances*, 8, 20039–20047.

Yasin, G., Arif, M., Shakeel, M., Dun, Y., Zuo, Y., Khan, W. Q., et al. (2018). Exploring the nickel–graphene nanocomposite coatings for superior corrosion resistance: manipulating the effect of deposition current density on its morphology, mechanical properties, and erosion-corrosion performance. *Advanced Engineering Materials*, 20, 1701166.

Yasin, G., Khan, M. A., Arif, M., Shakeel, M., Hassan, T. M., Khan, W. Q., et al. (2018). Synthesis of spheres-like Ni/graphene nanocomposite as an efficient anti-corrosive coating; effect of graphene content on its morphology and mechanical properties. *Journal of Alloys and Compounds*, 755, 79–88.

Yasin, G., Khan, M. A., Khan, W. Q., Mehtab, T., Korai, R. M., Lu, X., et al. (2019). Facile and large-scalable synthesis of low cost hard carbon anode for sodium-ion batteries. *Results in Physics*, 14, 102404.

Yoo, S., Lee, J.-I., Ko, S., & Park, S. (2013). Highly dispersive and electrically conductive silver-coated Si anodes synthesized via a simple chemical reduction process. *Nano Energy*, 2, 1271–1278.

Yu, D., Kumar, A., Nguyen, T. A., Nazir, M. T., & Yasin, G. (2020). High-voltage and ultrastable aqueous zinc–iodine battery enabled by N-doped carbon materials: revealing the contributions of nitrogen configurations. *ACS Sustainable Chemistry & Engineering*, 8, 13769–13776.

Zhang, Z., Yi, Z., Wang, J., Tian, X., Xu, P., Shi, G., et al. (2017). Nitrogen-enriched polydopamine analogue-derived defect-rich porous carbon as a bifunctional metal-free electrocatalyst for highly efficient overall water splitting. *Journal of Materials Chemistry A*, 5, 17064–17072.

Zhao, X., Hayner, C. M., Kung, M. C., & Kung, H. H. (2011). In-plane vacancy-enabled high-power si–graphene composite electrode for lithium-ion batteries. *Advanced Energy Materials*, 1, 1079–1084.

Zheng, S., Wang, L., Feng, X., & He, X. (2018). Probing the heat sources during thermal runaway process by thermal analysis of different battery chemistries. *Journal of Power Sources*, 378, 527–536.

Zhong, Y., Peng, F., Bao, F., Wang, S., Ji, X., Yang, L., et al. (2013). Large-scale aqueous synthesis of fluorescent and biocompatible silicon nanoparticles and their use as highly photostable biological probes. *Journal of the American Chemical Society*, 135, 8350–8356.

Zhou, X., Cao, A.-M., Wan, L.-J., & Guo, Y.-G. (2012). Spin-coated silicon nanoparticle/graphene electrode as a binder-free anode for high-performance lithium-ion batteries. *Nano Research*, 5, 845–853.

Zhou, X., Yin, Y.-X., Wan, L.-J., & Guo, Y.-G. (2012). Self-assembled nanocomposite of silicon nanoparticles encapsulated in graphene through electrostatic attraction for lithium-ion batteries. *Advanced Energy Materials, 2*, 1086–1090.

Zhou, Z., Dong, P., Wang, D., Liu, M., Duan, J., Nayaka, G. P., et al. (2019). Silicon-titanium nanocomposite synthesized via the direct electrolysis of SiO_2/TiO_2 precursor in molten salt and their performance as the anode material for lithium ion batteries. *Journal of Alloys and Compounds, 781*, 362–370.

Zhu, B., Wu, X., Liu, W.-J., Lu, H.-L., Zhang, D. W., Fan, Z., et al. (2019). High-performance on-chip supercapacitors based on mesoporous silicon coated with ultrathin atomic layer-deposited In_2O_3 films. *ACS Applied Materials & Interfaces, 11*, 747–752.

Zhu, G., Gu, Y., Wang, Y., Qu, Q., & Zheng, H. (2019). Neuron like Si-carbon nanotubes composite as a high-rate anode of lithium ion batteries. *Journal of Alloys and Compounds, 787*, 928–934.

Zhu, G., Luo, W., Wang, L., Jiang, W., & Yang, J. (2019). Silicon: toward eco-friendly reduction techniques for lithium-ion battery applications. *Journal of Materials Chemistry A, 7*, 24715–24737.

Zhu, M., Yang, J., Yu, Z., Chen, H., & Pan, F. (2017). Novel hybrid Si nanocrystals embedded in a conductive SiOx@C matrix from one single precursor as a high performance anode material for lithium-ion batteries. *Journal of Materials Chemistry A, 5*, 7026–7034.

Zong, L., Jin, Y., Liu, C., Zhu, B., Hu, X., Lu, Z., et al. (2016). Precise perforation and scalable production of si particles from low-grade sources for high-performance lithium ion battery anodes. *Nano Letters, 16*, 7210–7215.

Zuo, X., Xia, Y., Ji, Q., Gao, X., Yin, S., Wang, M., et al. (2017). Self-templating construction of 3d hierarchical macro-/mesoporous silicon from 0d silica nanoparticles. *ACS Nano, 11*, 889–899.

Application of silicon-based composite in batteries

Runwei Mo

School of Mechanical and Power Engineering, East China University of Science and Technology, Shanghai, P.R. China

7.1 Introduction

Nowadays, the application of high-energy lithium ion battery (LIB) is not limited to small mobile devices commercialized in the early 1990s, it has been extended to electric vehicles for long-distance driving and large-scale storage of renewable energy. So the demand is increasing rapidly (Choi & Aurbach, 2016; Dunn, Kamath, & Tarascon, 2011; Mo et al., 2020). The principle based on LIB is to use the reversible electrochemical reaction of the electrode for energy storage, so the selection of high-capacity electrode materials will effectively increase the overall energy density of full battery (Lin, Liu, Ai, & Liang, 2019; Manj, Zhang, Rehman, Luo, & Yang, 2020). Compared with the commercial graphite electrode material (372 mA h g^{-1}), silicon (Si) has a higher theoretical specific capacity (4200 mA h g^{-1}), which is also the largest theoretical capacity among the anode materials reported in the current. And it also has a low discharge potential of 0.37 V (Iaboni & Obrovac, 2016; Shi et al., 2016). In addition, Si also has the advantages of abundant resources, low cost, and environmental friendliness. It is precisely because of the above advantages that Si becomes the most promising candidate material to replace commercial graphite to achieve high-energy LIB. However, Si materials also have some shortcomings, such as huge volume expansion ($\sim$400%) and poor conductivity ($<10^{-3}$ S cm^{-1} under 25°C), which affect their long-cycle performance and fast-charging ability, respectively (Kim, Seo, Park, & Cho, 2010; Wu & Cui, 2012; Yoon, Nguyen, Seo, & Lucht, 2015). Therefore it is difficult to directly use the bare Si material as the anode materials in LIB.

In order to obtain an applicable Si-based anode, it is a good strategy to construct a Si-based composite material. The combination of two or more materials can solve the problems of Si itself, thereby achieving the advantages of high capacity, high power, and long cycle life. Therefore innovative structural design and large-scale manufacturing process are the key to achieving the above research goals. This chapter focuses on the capacity failure mechanism of Si electrodes and the latest research progress of Si-based composite electrodes in LIB. In particular, it will introduce in detail the effective strategies and innovative structures in the construction of Si-based composite electrodes. This will provide a perspective for the future application of Si-based composite electrodes in commercial LIB.

7.2 Capacity failure mechanism of Si electrode in LIB

On the one hand, Si materials are considered a semiconductor material, with poor electrical conductivity (Forney, Ganter, Staub, Ridgley, & Landi, 2013). On the other hand, Si materials have huge volume expansion during the lithiation/delithiation process. These two defects are not conducive to the electrochemical performance of Si anodes in LIB, and also seriously affect the structural stability of the solid electrolyte interface membrane (SEI) (Park et al., 2009). Among them, the structural stability of SEI is very important for commercial LIB. The formation of unstable SEI will continuously deplete the limited lithium ions in the cathode material during the charging and discharging process, resulting in poor cycle stability and low coulombic efficiency. Furthermore, due to the huge volume expansion, the Si particles are prone to crushing during charging and discharging, which will seriously damage the integrity of the electrode. Here, we summarized the capacity failure mechanism of Si material during lithiation/delithiation process, as follows:

1. When the initial lithium insertion occurs, the basic structure of Si will change from the initial crystalline Si to amorphous Li_xSi. Then when the delithiation starts, the amorphous Li_xSi is further transformed into amorphous Si. And along with the whole charging and discharging process, the Si material will expand and shrink in volume, which will directly cause the pulverization of the Si material.
2. During the long-cycle charging–discharging test, the huge volume expansion will expose the surface of the fresh Si particles. This will consume limited lithium ions and the SEI will continue to form a lot of "dead" lithium. It is worth

noting that this process is irreversible, which causes the coulombic efficiency to be far below 100%.
3. During the charging and discharging process, the large volume expansion and contraction of Si particles will seriously affect the structural stability of the whole electrode, and a large amount of Si particles as active materials will lose effective electrical contact with the surrounding conductive agent and current collector.

7.3 Classification of Si-based composites
7.3.1 Si/carbon composites

Taking into account that carbon has natural softness and excellent electrical conductivity, it is considered to be the most suitable material for use in combination with Si in LIB. Researchers have done a lot of research work in this area. Carbon has a variety of allotropes, for example, through Si and amorphous carbon (AC) (He et al., 2018), carbon nanotube (CNT) (Kowase, Hori, Hasegawa, Momma, & Noda, 2017), and graphene (Mo, Lei, Rooney, & Sun, 2019) to construct a composite structure, excellent electrochemical performance is obtained. In the early stage of research on Si/carbon composites, Si particles were dispersed in the carbonaceous matrix (Dimov, Kugino, & Yoshio, 2003). Through this structural design, the electrical conductivity and volume expansion are effectively improved and buffered, thereby enhancing the electrochemical properties of the Si-based anodes to a certain extent Sohn et al. (2016). However, there are still two problems that are difficult to solve. The first is the agglomeration of Si particles, and the second is the instability of the SEI film, which will limit the further improvement of its electrochemical performance (Sohn et al., 2016).

Constructing a uniform carbon coating is considered another effective strategy and has been widely reported (Mery et al., 2019; Wang, Song, Zhao, Sun, & Du, 2020). The specific method of this strategy is to uniformly coat the surface of the Si particles with the thin carbon layer. This composite structure can provide a large number of electrical contacts and shorten the electron transmission path, which is beneficial to improve the conductivity and dynamics during charging and discharging (Liu et al., 2009, 2018). The carbon coating may avoid direct contact between the electrolyte and the surface of Si particles, which helps to reduce the consumption of limited lithium ions and inhibit the formation of SEI film, thereby effectively enhancing the coulombic efficiency of Si-based anodes (Zhang, Du, et al., 2014). Furthermore, this uniform carbon-coated technology can suppress and alleviate the volume change and pulverization and aggregation of Si particles during the lithiation/delithiation process, which is beneficial to maintain the stability of the composite structure and enhance the long-cycle performance of composite electrode (Kim et al., 2015). It is worth noting that if there is an uncoated part in the composite material, this will expose the Si surface to the electrolyte, which may lead to rapid capacity degradation after multiple charges and discharges (Guo, Sun, Chen, Wang, & Manivannan, 2011). Thus achieving uniform carbon coating is a key composite technology.

The different thickness of the carbon coating will also directly affect the electrochemical performance of the Si-based anode. Recently, Luo and other colleagues conducted a systematic study on the influence of the thickness of carbon coating on the electrochemical properties of Si-based anode in LIB (Luo et al., 2016). In this research work, the thickness of different carbon coatings is achieved through changing the concentration of the precursor solution. The sizes of Si particles prepared by this method are 0, 5 nm, 10 nm, and 15 nm, and are labeled as Si, Si@5 C, Si@10 C, and Si@15 C, respectively. Then their electrochemical performances are studied under the voltage range from 1.5 V to 0.005 V. Compared with other electrodes, the Si@5 C electrode exhibited the highest coulombic efficiency (85.8%, discharge capacity of 3774 mA h g^{-1}) under 0.5 A g^{-1}. This result can show that optimizing the thickness of carbon coating can effectively enhance the coulombic efficiency of Si-based composite electrode. When the thickness of the carbon coating is 5 nm, it shows the highest coulombic efficiency of Si-based composite electrode. The reasons are mainly two aspects. On the one hand, the thin carbon coating thickness can significantly increase the proportion of active material in the electrode; on the other hand, the thin carbon coating can effectively reduce the irreversible reaction between the surface of the silicon material and the electrolyte. Compared with other electrode materials, Si@10 C electrode shows the best cycle stability, especially the discharge capacity after 500 cycles of charge and discharge test at a current density of 0.5 A g^{-1} is still as high as 1006 mA h g^{-1}, which is several times higher than of other electrodes based on the same electrochemical testing. This result shows that when using extremely thin thicknesses, the carbon coating cannot maintain the integrity and stability of the electrode structure after hundreds of charging and discharging tests. On the contrary, when the carbon coating is too thick, it will limit the penetration of Li ions into the bulk material of the Si particles. It is worth noting that even under the high current density of 16.8 A g^{-1}, the discharge capacity of Si@10 C anode is still as high as 1209 mA h g^{-1}.

Recently, graphene is often used as a coating layer to composite with Si materials because of its excellent properties, such as light weight, large specific area and good mechanical flexibility (Feng et al., 2016; Qin et al., 2017). It is also important that graphene has excellent lithium storage performance and can be used as an active material in LIB (Mo, Li, et al., 2019). Therefore, a large amount of research works on graphene-coated Si material has been reported, and the composite electrode exhibits good electrochemical properties. Wen and other colleagues conducted a systematic study on the influence of graphene coating on the electrochemical performance of Si-based anode in LIB (Wen et al., 2013). The capacity calculation is based on the total mass of graphene and silicon. The reversible capacity of the composite electrode is as high as 1900 mA h g^{-1} under a charging–discharging rate of 1 C. In addition, the reversible specific capacity of composite electrode still maintains 1400 mA h g^{-1} over 120 cycles, and the capacity retention exceeds 70%. When the rate is further increased to 10 C, its reversible capacity still reaches 1000 mA h g^{-1}. Compared with bare Si electrode without graphene coating, the composite electrode shows better electrochemical performance in terms of cycle stability and rate performance. The reason is that the graphene coating in composite electrode may not only alleviate the volume expansion and contraction of Si particles during lithiation/delithiation, but also avoid the secondary agglomeration of Si particles under the long-cycle testing, thereby greatly improving the rate performance and cycle stability of composite electrode.

The above composite materials mainly use raw carbon materials to further enhance the electrochemical properties of Si-based composite electrode. This composite strategy can also be extended to the application of other materials in many research fields. With the continuous deepening of research, in addition to the research of original carbon materials, element doping in raw carbon materials is an effective strategy to enhance the electrochemical properties of composite electrode. In this regard, researchers have done a lot of research works in element-doped carbon materials mixed with Si materials in LIB. Among them, since the atomic radius of N is similar to that of C, most of the research is carried out on nitrogen-doped carbon materials (Tao, Huang, Fan, & Qu, 2013). It is worth noting that the nitrogen doping of carbon materials not only can enhance the electron and ion diffusivity, but also the composite electrode doped with nitrogen can further improve the mechanical strength and electrochemical activity (Liu et al., 2018; Park, Shim, Kim, & Kim, 2017; Tao et al., 2013). Thus the composite electrode doped with nitrogen element exhibits more excellent electrochemical properties. For example, Liu's research group prepared Si-based composites coated with nitrogen-doped porous carbon, which can produce a reversible capacity of 1565 mA h g^{-1} under 0.5 A g^{-1} over 100 cycles (Liu et al., 2018). However, under the same testing conditions, the capacity of pure Si electrode dropped rapidly from 3526.4 to 416 mA h g^{-1} over 30 charging–discharging cycles. Even after 400 cycles, the composite material still has a reversible capacity of 1086 mA h g^{-1} under 2 A g^{-1}. This also further shows that nitrogen-doped carbon materials can greatly increase the electrical conductivity of Si materials, thereby improving its rate performance. On the other hand, it also may effectively alleviate the huge volume change of Si materials, thereby enhancing its cycle stability.

7.3.2 Si /metal composites

Compared with carbon materials, metal materials, such as copper, silver, nickel, cobalt, iron, etc. which have better mechanical strength and electrical conductivity. Therefore it has also been composited with Si materials (Lee, Kim, & Yoon, 2019; Yoo, Lee, Ko, & Park, 2013) From the current research, copper and silver are the most studied metals, while other metals are less investigated. It is worth noting that the influencing mechanisms of these metals on the electrochemical properties of composite electrodes are generally similar. Similar to the composite method of carbon materials, Si particles are coated with a metal coating or dispersed in a metal matrix. In order to analyze the influence of metals on the electrochemical properties of composite electrodes, this section takes copper and silver as examples.

Among them, the method of copper (Cu) deposition on Si particles is mainly prepared by chemical deposition and magnesium thermal reduction. Recently, researchers reported that Cu nanoparticles with a particle size of 3–4 nm were uniformly deposited inside Si material with a mesoporous structure (Kwon, Kim, Kim, & Hong, 2019). In this composite material, the pores of Si materials are filled with Cu nanoparticles. It is well-known that Cu is electrochemically inert to lithium ions. The introduction of Cu nanoparticles in the composites can provide a good mechanical stability for Si materials under charging–discharging cycles. And the uniformly dispersed Cu nanoparticles in composites can also alleviate the volume expansion and contraction of Si materials, thereby effectively alleviating the problem of secondary agglomeration of particles. Compared with the original mesoporous Si particles, the presence of Cu nanoparticles can effectively inhibit the repeated generation of SEI film, thereby significantly improving the stability of SEI film. It is worth noting here that the uniformly dispersed Cu nanoparticles in composites have high electrical conductivity, which may accelerate the transmission of electrons, thereby exhibiting outstanding rate performance of Si-based composite electrode. Based on the above advantages, compared with mesoporous Si/carbon electrode and pure mesoporous Si electrode, mesoporous Si/Cu electrode shows the best electrochemical properties.

Specifically, the as-prepared mesoporous Si/Cu electrode has the first charge and discharge capacity of 1887 and 2100 mA h g^{-1} at 200 mA g^{-1}, respectively. After the current density increased to 1000 mA g^{-1}, the reversible specific capacity of the electrode was 1569 mA h g^{-1} over 200 charging–discharging cycles. On the contrary, under the same test conditions, the reversible specific capacities of pure Si electrode and Si/carbon electrode both showed a severe decrease under 200 charging–discharging cycles. More importantly, when the current density is further increased to 2000 mA g^{-1}, the reversible specific capacity of the as-prepared Si/Cu electrode is still as high as 1598 mA h g^{-1}, which further shows its excellent rate performance. On the contrary, when the current density is increased to 2000 mA g^{-1}, the reversible specific capacities of Si/C electrode and bare Si electrode are much lower, showing poor rate capability. This also further proves that the addition of Cu nanoparticles can comprehensively improve the rate performance and cycle stability of Si materials in LIB. Furthermore, the researchers also conducted a systematic study on the amount of Cu in the composite electrode, and found that the difference in the amount has a significant impact on the electrochemical properties. Through testing the cycle stability and rate performance of composite electrodes with different Cu concentrations from 18 wt.% to 56 wt.%, it is found that different Cu concentrations of 25% in the composite have the best electrochemical performance. This opens up a new path for the design of high-performance Si-based composite electrodes in LIB.

Compared to Cu, silver (Ag) exhibits the better electrical conductivity (Yoo et al., 2013). For this reason, researchers have done a lot of research work on Si/Ag electrodes in LIB, showing excellent electrochemical performance (Chen et al., 2012; Yang, Wen, Huang, Zhu, & Zhang, 2006; Yu et al., 2010). Among them, Yang and other researchers conducted a systematic study on the coating of Ag nanoparticles on the surface of Si materials and tested their electrochemical performance in LIB (Yang et al., 2006). This indicates that the coating layer of Ag nanoparticles on the surface of Si materials can provide a continuous conductive network, which is beneficial to promote the electron transport. In this strategy, achieving the uniformity of the Ag coating layer is the key to maintaining the stability of electrode architecture and inhibiting the secondary aggregation of Si particles during the long-term cycling process. The results show that the Si-based electrode with Ag coating layer has excellent electrochemical performance. In this research work, the weight of Si in the Ag-coated Si composite material is 90%. The cycle testing of composite electrode shows that the reversible specific capacity is about 800 mAh g^{-1} over 30 charging and discharging cycles. Moreover, the uniform Ag coating layer will enhance the long-cycle performance of Si-based composite electrode.

In the previous research work, no pore structure was formed in Si material. It is well-known that constructing a porous structure with void spaces in Si material is a common method to enhance the electrochemical properties of Si-based electrode in LIB. In this regard, in order to make full use of the porous Si architecture and Ag coating layer, Yu et al. successfully prepared the composite material by coating the Ag nanoparticles in the macroporous Si architecture (Yu et al., 2010). The mass ratios of Si and Ag in this composite are 92 wt.% and 8 wt.%, respectively. The results exhibit that uniform Ag coating layer on the surface of Si materials could be connected to each other to build a continuous conductive network, which helps to enhance electrode dynamics under the charging and discharging process. In the composite electrode, the macroporous structure may alleviate the volume expansion and contraction of Si materials during the lithiation and delithiation process, and play a role in reducing internal stress. Through the design of porous structure, on the one hand, the structural stability can be greatly improved. On the other hand, the internal pore structure of the Si material is also conducive to promoting the penetration of the electrolyte, thereby reducing the diffusion distance of lithium ions. Based on this porous Si structure, the Ag coating technology is further combined. Compared with commercial solid Si electrode and bare porous Si electrode, the Ag-coated porous Si composite electrode exhibits the best electrochemical properties in terms of lithium storage. Specifically, under a charging and discharging rate of 0.2 C, the reversible specific capacity and capacity retention rate of composite electrode were as high as 1163 mA h g^{-1} and 82% over 100 cycles, respectively. For comparison, under the same test conditions, the capacity retention rate of macroporous Si electrode is only 69%. In addition, the researchers also studied the rate performance of composite electrode. When the rate is increased to 4 C, the reversible specific capacity of the Ag-coated macroporous Si electrode can still reach 800 mA h g^{-1}. For comparison, the reversible specific capacity of porous Si electrode without Ag coating is only 590 mA h g^{-1} under the same test conditions.

7.3.3 Si/metal oxide composites

In addition to the above carbon and metal materials, metal oxide is also a material that can be used to composite with Si materials, which has great potential to enhance the electrochemical properties of Si-based composite electrode in LIB. So far, a large number of metal oxides and Si materials have been compounded, and the application of LIB has been studied. It is worth noting that almost all metal oxides can belong to transition metal oxides (TMO). Based on the conversion reaction mechanism of TMO, when it is used as anode for lithium storage, it is found that they can basically show a much better reversible specific capacity than commercial graphite (Zhang, Wu, & Lou, 2014). During the discharge process, the electrochemical reaction mechanism is that TMO undergoes a reduction reaction to transform into a transition metal (TM), and Li undergoes an oxidation reaction to form Li$_2$O. During the charging process, TM undergoes an oxidation reaction to transform into TMO, while Li$_2$O undergoes a reduction reaction to form Li. However, it is well-known that the second reaction is often partially irreversible. After multiple charge and discharge cycles, Li$_2$O and TM will remain in the electrode due to this irreversible situation. Additionally, although TMO also suffers a certain volume change under charging and discharging process, the volume change of TMO is much lower than that of Si. When combined with Si material, TMO can still alleviate the volume change of Si electrode during charging and discharging, thereby improving the structural stability (Hwa, Kim, Yu, Hong, & Sohn, 2013; Yue et al., 2017). Moreover, the uniform coating of TMO on the surface of Si material or using TMO as a matrix is beneficial to avoid direct contact between the Si material and electrolyte, thereby effectively inhibiting the occurrence of side reactions. During the discharge process, the TM formed in composite electrode also can significantly increase its electrical conductivity, thereby improving the electron diffusion kinetics of composite electrode. It is also worth noting that Li_2O is one of the main components of SEI film. Thus the production of Li_2O based on conversion reaction of TM will change the morphology of the SEI film on Si particle in the composite electrode.

Among all TMO materials, TiO_2 has aroused widespread research interest due to its outstanding structural stability under charging and discharging. TiO_2 not only has all the above advantages of TMO, but also has a very small volume expansion during the discharging process, which makes it an ideal material for compounding with Si (Lotfabad et al., 2013). Furthermore, TiO_2 has a variety of crystal structures with different electrochemical properties. Recently, Yang's research group has studied the influence of different TiO_2 crystal structures on the electrochemical performance of Si-based composite electrodes (Yang et al., 2017). The amorphous TiO_2 (a-TiO_2) and crystalline TiO_2 (c-TiO_2) were coated on the surface of Si material, respectively, and its electrochemical performance was systematically tested. The experimental results show that the elasticity of a-TiO_2 coating is much higher than that of c-TiO_2 in terms of structural integrity. After long-cycle charging and discharging testing, the a-TiO_2 coated Si-based electrode maintain good structural stability, while the c-TiO_2 coating was damaged under the same testing conditions. This will have adverse effects on the long-cycle performance of composite electrode in LIB. Compared with c-TiO_2, a-TiO_2 has the defects and disordered structure, which is beneficial to eliminate the plateau of the reaction between Li ion and TiO_2 and expand the potential range. In the composite electrode, the a-TiO_2 coating has a better ion and electron diffusivity than other coatings, which should be attributed to its open active diffusion channels and inherent isotropic characteristics. And due to the existence of defects, a-TiO_2 can provide excellent lithium ion conductivity.

It is worth noting here that different TiO_2 crystal structures will be different for the products after lithium insertion. In this research work, compared with c-TiO_2, a-TiO_2 will irreversibly transform into cubic $Li_2Ti_2O_4$ structure under the charging and discharging process. Compared with anatase (0.5 eV), the diffusion activation barrier of cubic $Li_2Ti_2O_4$ structure is only 0.257 eV, which indicates that the conductivity of a-TiO_2 is better than that of c-TiO_2 under the electrochemical testing. Based on the above reasons, the experimental results also show that the a-TiO_2-coated Si electrode exhibits the better rate performance and cycle stability than the c-TiO_2-coated Si electrode. Specifically, the results show that the initial discharge capacity of a-TiO_2-coated Si electrode is 3061 mA h g^{-1}, which is higher than that of c-TiO_2-coated Si electrode, which is 2200 mA h g^{-1} under the same test condition. The cycle performance has also been tested. After 200 charging and discharging cycles, a-TiO$_2$-coated Si electrode can provide a reversible capacity of 1720 mA h g^{-1} under 420 mA g^{-1}, which is significantly higher than the of c-TiO$_2$-coated Si electrode (150 mA h g^{-1}). When the current density is increased to 8.4 A g^{-1}, the reversible specific capacity of a-TiO$_2$-coated Si electrode is still as high as 812 mA h g^{-1}, which is also significantly higher than the corresponding value of c-TiO$_2$-coated Si electrode (33 mA h g^{-1}). In addition to the significant improvement in cycle stability and rate performance, more importantly, a-TiO$_2$ coating can effectively enhance the safety performance of the battery. Specifically, compared to carbon-coated Si electrode, the results show that the a-TiO$_2$ coating layer is more effective in suppressing any significant exothermic reaction under 285°C, thereby improving the safety performance of the battery.

The above metal oxides have good electrochemical reactivity in lithium storage. In addition to the above metal oxides, Shi's research group has studied the influence of ZrO$_2$ material on the electrochemical performance of Si-based composite electrodes, which shows that the ZrO$_2$ is inactive in lithium storage (Shi et al., 2006). In this research work, the Si nanoparticles are grown uniformly in a mesoporous ZrO$_2$ structure. In the composite, mesoporous ZrO$_2$ as a substrate can effectively alleviate the volume expansion and contraction of Si material during the lithiation/delithiation process, thereby ensuring the structural stability of composite electrode. In addition to buffering the volume change of Si material, the mesoporous structure can also effectively improve electrolyte penetration, thereby promoting the lithium ion diffusion. After 50 charging and discharging cycles, the reversible capacity of the composite electrode is up to 1500 mA h g^{-1} under 40 mA g^{-1}. This result shows that the existence of mesoporous ZrO$_2$ can enhance the long-cycle performance of Si-based composite electrode.

7.3.4 Si/polymer composites

In addition to the above inorganic materials, organic polymers have been used to compound with Si particles to improve their electrochemical performance due to their unique advantages. Among all organic polymers, conductive polymers have been extensively studied due to their excellent conductivity and mechanical flexibility. For instance, the polyaniline (PANI) as a conductive polymer is widely used in Si-based composite materials. Recently, researchers could greatly improve the electrochemical properties of Si-based composite electrodes by embedding Si particles in PANI substrate with a three-dimensional (3D) network architecture (Wu et al., 2013). Compared with the traditional electrode production, the composite material can be coated on the commercial copper foil when the electrode is made, and the whole process does not require any electrical conductors and adhesives. In this strategy, the composite of Si and PANI builds a 3D network structure using electrostatic interaction with positively charged polymers, and a thin layer of PANI coated on the surface of the Si particles. This porous polymer matrix contributes enough void space to buffer the volume expansion and contraction of Si particles during lithiation/delithiation. On the one hand, this composite structure not only provides 3D continuous network channels for electrons and ions, but also enhances the electrical contact of Si particles. On the other hand, although it is inevitable that some Si particles will crack in the long-cycle test, they still exist in the 3D continuous pore structure of PANI matrix, ensuring the good electrical integrity of composite electrode. The experimental results also showed that Si particles remained in the 3D continuous PANI matrix over 2000 charging and discharging cycles, and still maintained good electrochemical activity. In addition, it is inevitable that the PANI coating will crack over multiple cycles, but it will still maintain the good electrical contact between 3D continuous conductive network and Si particles.

Remarkably, the PANI coating on the surface of Si particles is conducive to the formation of a stable SEI film. According to all the above advantages, when the charge–discharge rate is 0.3 to 3.0 A g^{-1}, the reversible specific capacity of composite electrode is 2500 to 1100 mA h g^{-1}. This result shows that the 3D conductive network structure effectively improves the high-rate performance of composite electrode. More importantly, when the current density increases to 6 A g^{-1}, the reversible specific capacity of composite electrode is still as high as 550 mA h g^{-1} after 5000 ultralong cycles, and its capacity retention rate is calculated to exceed 90%. This result also further proves that the PANI substrate has a significant effect on alleviating the volume expansion and contraction of Si particles during the lithiation and delithiation.

In recent years, researchers have also reported a new type of poly(9,9-dioctylfluorene-co-fluorenone-co-methylbenzoic ester) material (CP), and coated it on the surface of Si particles to test its electrochemical performance (Gu et al., 2014). Similar to the effect of PANI, the CP coating layer can not only build more conductive networks for Si particles, thereby improving the conductivity of composite electrode; but also can effectively alleviate the volume expansion and contraction of Si particles under the lithiation and delithiation process, thereby enhancing the structural stability of composite electrode. In terms of electrode production, the composite can also be used directly as an electrode without any electrical conductors and adhesives. Under the low charge–discharge rate of 0.1 C, the reversible specific capacity of composite electrode is still as high as 1750 mA h g^{-1} over 938 cycles.

In short, the introduction of carbon materials, metals, metal oxides, or polymers into Si-based composites can not only provide good mechanical stability under charging and discharging process, thereby greatly improving its long-cycle performance; but also can construct a continuous conductive network, thereby effectively enhancing its high-rate capability.

7.4 Conclusions

From the perspective of structural design, this chapter describes in detail a variety of Si-based electrodes for high-performance LIB with enhanced cycle stability and rate performance, such as carbon materials, metals, metal oxides, and polymers. These materials have different effects on the electrochemical properties of Si-based composite electrodes, which are mainly attributed to their different electrochemical and physical properties. This provides a variety of design solutions for the construction of high-performance Si-based electrode architectures. According to the above research results, compared with pure Si electrodes, Si-based composite electrodes show better electrochemical properties. The main reason for the analysis is that on the one hand, the design of composite electrode may alleviate the volume expansion and contraction of Si particles under lithiation and delithiation process, thereby improving the structural stability; on the other hand, the coating layer on the surface of Si particles could significantly enhance the conductivity of composite electrode, thereby enhancing the electron diffusion kinetics.

Remarkably, the configuration of different composite structures has a great influence on the electrochemical performance of composite electrode. Among the various composite structures that have been reported, the core/shell structure with coupled material coated on the surface of Si material is considered to be the common and effective strategy to enhance the electrochemical properties of Si-based anodes in LIB. In this composite structure, on the one hand, the coating can effectively avoid the direct contact between the electrolyte and the surface of Si particles, thereby inhibiting the occurrence of harmful side reactions; on the other hand, it can provide more electrical contact to enhance the conductivity of composite electrode, thereby maintaining the electrochemical activity of cracking Si particles during long-term charging–discharging cycles.

Among the four coupled materials introduced above, carbon materials, especially graphene, are some of the best potential substitutes for realizing high-performance Si-based composite electrodes, due to their unique two-dimensional structure and excellent electrochemical and physical properties (Mo, Lei, Sun, & Rooney, 2014). However, the development of a low-cost and scalable graphene preparation technology is the key to commercialization in the future. Compared with carbon materials, metals have better structural stability and electrode dynamics of Si-based composites during charge and discharge cycles, which is attributed to their good mechanical strength and electrical conductivity. However, the current manufacturing process of Si/metal composite materials is relatively complicated, which is not conducive to scale-up. Compared with other coupled materials, metal oxides have most excellent electrochemical activity in lithium storage, and usually have a higher theoretical specific capacity. Nevertheless, most metal oxides as anodes have excessively high voltages of charging and discharging in LIB, which is the biggest obstacle to their practical application. It is worth noting here that the introduction of organic polymers can greatly simplify the manufacturing process of electrodes without any electrical conductors and adhesives, which can greatly promote the development of high-energy LIB devices. However, compared with other coupled materials, the lower conductivity of polymers still limits the high-rate capability of Si-based anodes. Based on the above analysis, researchers are currently actively exploring other types of coupled materials to enhance the electrochemical properties of Si-based anodes in LIB. Therefore this is a very promising and important research direction.

7.5 Outlook

In the design of Si-based composite structure, the main goal of combining Si particles with other coupled materials is to use the synergistic effect of various materials to overcome the shortcomings of Si, thereby improving the electrochemical properties of Si-based composite anodes in LIB. Therefore the development of new coupled materials based on their respective advantages and disadvantages is the key to achieving high-performance

Si-based anodes in LIB, and it is necessary to effectively alleviate the problems of low electrical conductivity and huge volume changes of Si materials. There are two main research directions in the development of new coupled materials. First, some coupled materials have excellent electrical conductivity. This requires more efforts to simplify the preparation process and enhance the integrity of composite structures. Second, some coupled materials have outstanding structural stability. There should be a focus on improving electrical conductivity through the design of polymer structures.

Additionally, the design of the multielement composite structure is also quite effective in greatly enhancing the rate performance and cycle stability of Si-based electrodes in LIB. It should be noted that compared to coupled materials, the theoretical specific capacity of Si material is much higher, so the use of coupled materials in composite will reduce the theoretical specific capacity of Si-based composite electrode to a certain extent. For this reason, designing a composite electrode with higher Si content means that the composite electrode has a higher reversible specific capacity. However, such composite electrodes often suffer from poor rate performance and short cycle life. On the contrary, designing a composite electrode with a lower Si content is conducive to achieve better cycle stability and rate performance. Nevertheless, such composite electrodes often suffer from the low reversible specific capacity. Therefore optimizing the mass ratio of the coupled materials in Si-based composite electrodes plays a very important role in balancing reversible specific capacity, rate performance, and cycle life.

So far, a series of significant progress has been made in improving the electrochemical properties of Si-based electrodes, but there are still some important issues to be resolved, such as large initial irreversible capacity, poor SEI stability, low tap density, complicated manufacturing process, and high production cost. Among these problems, the large initial irreversible capacity (i.e., low initial coulomb efficiency) is a key factor to limit the commercial application of Si-based electrodes. To realize the commercial application of anode materials in LIB, initial coulomb efficiency of anode materials needs to be higher than 95% (Buqa, Goers, Holzapfel, Spahr, & Novák, 2005; Yoshio, Wang, Fukud, Hara, & Adachi, 2000). Nevertheless, none of the high-capacity Si-based electrodes reported at this stage can meet this requirement. Therefore it is necessary to increase research efforts to solve the above problems. In order to solve the above problems more effectively, structural engineering and compositional design are the important research directions in the next research stage. It is strongly recommended to study multicomponent composite materials with structural engineering. The main reason for the analysis is that the synergistic effects of different structures and coupled materials could greatly enhance electrochemical performance of Si-based composite electrode. So far, many Si-based composites have been reported, but there are still many areas worthy of in-depth study. On the one hand, the electrochemical performance of composite electrodes may be further improved by combining Si particles with multiple coupled materials to combine the design of composite structure. On the other hand, researchers need to develop new coupled materials with excellent electrical conductivity and structural stability, which may effectively improve the electrochemical properties of high-capacity Si-based composite electrode. In addition to the above strategies, more efforts can be made in the optimization of conductive additives, binders, and electrolytes, which may further enhance the cycle stability and rate performance of Si-based composite electrode in LIB.

In addition, in order to realize the rapid development of Si-based composite materials as anodes in LIB, it is necessary to link and distinguish practical applications with academic research. In the process of basic academic research, it is important to increase efforts to study the practical application of Si-based composite materials. Under the guidance of basic academic research, it is necessary to pay more attention to ensuring safety, reducing costs, simplifying fabrication processes and increasing manufacturing efficiency for practical applications. In terms of basic academic research, the development of new composite structure and coupled material is the main research direction of high-performance Si-based electrode in the future. In terms of practical applications, Si/carbon composites with low silicon content have made positive progress in the commercial application of lithium-ion battery. The next step is to increase the Si content in composites to achieve high energy densities of Si-based electrode in LIB. The types and structures of Si/carbon composite materials can be studied in depth through coupling Si particles with different carbon structures and materials. Likewise, there are also many directions worthy of in-depth research in other noncarbon coupled materials. Based on the abovementioned advantages, the introduction of noncarbon coupled materials is also beneficial to improve the electrochemical properties of Si-based electrode, thereby satisfying the requirements of high-energy LIB with long cycle life and high power density. More importantly, the composition optimization of carbon coupled materials or noncarbon coupled materials in Si-based electrodes must be well resolved to achieve excellent electrochemical properties. In the aspect of electrode manufacturing, more effort is needed to develop a new type of manufacturing that is simple and efficient. Although there are still some difficulties that limit the commercial application of high-capacity Si-based materials in LIB, people should still have the confidence and ability to solve these difficulties in order to promote the commercial application of Si-based electrodes in high-energy LIB.

References

Buqa, H., Goers, D., Holzapfel, M., Spahr, M. E., & Novák, P. (2005). High rate capability of graphite negative electrodes for lithium-ion batteries. *Journal of the Electrochemical Society, 152*, A474–A481.

Chen, D., Mei, X., Ji, G., Lu, M., Xie, J., Lu, J., & Lee, J. Y. (2012). Reversible lithium-ion storage in silver-treated nanoscale hollow porous silicon particles. *Angewandte Chemie International Edition, 51*, 2409–2413.

Choi, J. W., & Aurbach, D. (2016). Promise and reality of post-lithium-ion batteries with high energy densities. *Nature Reviews Materials, 1*, 16013.

Dimov, N., Kugino, S., & Yoshio, M. (2003). Carbon-coated silicon as anode material for lithium ion batteries: Advantages and limitations. *Electrochimica Acta, 48*, 1579–1587.

Dunn, B., Kamath, H., & Tarascon, J. M. (2011). Electrical energy storage for the grid a battery of choice. *Science (New York, N.Y.), 334*, 928–935.

Feng, K., Ahn, W., Lui, G., Park, H. W., Kashkooli, A. G., Jiang, G., ... Chen, Z. (2016). Implementing an in-situ carbon network in Si/reduced graphene oxide for high performance lithium-ion battery anodes. *Nano Energy, 19*, 187–197.

Forney, M. W., Ganter, M. J., Staub, J. W., Ridgley, R. D., & Landi, B. J. (2013). Prelithiation of silicon-carbon nanotube anodes for lithium ion batteries by stabilized lithium metal powder. *Nano Letters, 13*, 4158–4163.

Gu, M., Xiao, X. C., Liu, G., Thevuthasan, S., Baer, D. R., Zhang, J. G., ... Wang, C. M. (2014). Mesoscale origin of the enhanced cycling-stability of the Si-conductive polymer anode for Li-ion batteries. *Scientific Reports, 4*, 3684.

Guo, J., Sun, A., Chen, X., Wang, C., & Manivannan, A. (2011). Cyclability study of silicon-carbon composite anodes for lithium-ion batteries using electrochemical impedance spectroscopy. *Electrochimica Acta, 56*, 3981–3987.

He, Y., Xiang, K., Zhou, W., Zhu, Y., Chen, X., & Chen, H. (2018). Folded-hand silicon/carbon three-dimensional networks as a binder-free advanced anode for high-performance lithium-ion batteries. *Chemical Engineering Journal, 353*, 666–678.

Hwa, Y., Kim, W. S., Yu, B. C., Hong, S. H., & Sohn, H. J. (2013). Enhancement of the cyclability of a Si anode through Co_3O_4 coating by the Sol-Gel method. *Journal of Physical Chemistry C, 117*, 7013–7017.

Iaboni, D. S. M., & Obrovac, M. N. (2016). $Li_{15}Si_4$ formation in silicon thin film negative electrodes. *Journal of the Electrochemical Society, 163*, 255–261.

Kim, H., Seo, M., Park, M. H., & Cho, J. (2010). A Critical size of silicon nano-anodes for lithium rechargeable batteries. *Angewandte Chemie International Edition, 49*, 2146–2149.

Kim, J. S., Pfleging, W., Kohler, R., Seifert, H. J., Kim, T. Y., Byun, D., ... Lee, J. K. (2015). Three-dimensional silicon/carbon core-shell electrode as an anode material for lithium-ion batteries. *Journal of Power Sources, 279*, 13−20.

Kowase, T., Hori, K., Hasegawa, K., Momma, T., & Noda, S. (2017). A-few-second synthesis of silicon nanoparticles by gas-evaporation and their self-supporting electrodes based on carbon nanotube matrix for lithium secondary battery anodes. *Journal of Power Sources, 363*, 450−459.

Kwon, S., Kim, K. H., Kim, W. S., & Hong, S. H. (2019). Mesoporous Si-Cu nanocomposite anode for a lithium ion battery produced by magnesiothermic reduction and electroless deposition. *Nanotechnology, 30*, 405401.

Lee, S. Y., Kim, S. I., & Yoon, S. (2019). Si nanoparticles coated with Co-containing N-doped carbon: preparation and characterization as Li-ion battery anode materials. *Journal of Nanoscience and Nanotechnology, 19*, 7753−7757.

Lin, Z., Liu, T. F., Ai, X. P., & Liang, C. D. (2019). Aligning academia and industry for unified battery performance metrics. *Nature Communications, 10*, 5262.

Liu, N., Mamat, X., Jiang, R., Tong, W., Huang, Y., Jia, D., ... Hu, G. (2018). Facile high-voltage sputtering synthesis of three-dimensional hierarchical porous nitrogen-doped carbon coated Si composite for high performance lithium-ion batteries. *Chemical Engineering Journal, 343*, 78−85.

Liu, Y., Wen, Z. Y., Wang, X. Y., Hirano, A., Imanishi, N., & Takeda, Y. (2009). Electrochemical behaviors of Si/C composite synthesized from F-containing precursors. *Journal of Power Sources, 189*, 733−737.

Lotfabad, E. M., Kalisvaart, P., Cui, K., Kohandehghan, A., Kupsta, M., Olsen, B., & Mitlin, D. (2013). ALD TiO_2 coated silicon nanowires for lithium ion battery anodes with enhanced cycling stability and coulombic efficiency. *Physical Chemistry Chemical Physics: PCCP, 15*, 13646−13657.

Luo, W., Wang, Y., Chou, S., Xu, Y., Li, W., Kong, B., ... Yang, J. (2016). Critical thickness of phenolic resin-based carbon interfacial layer for improving long cycling stability of silicon nanoparticle anodes. *Nano Energy, 27*, 255−264.

Manj, R. Z. A., Zhang, F., Rehman, W. U., Luo, W., & Yang, J. (2020). Toward understanding the interaction within silicon-based anodes for stable lithium storage. *Chemical Engineering Journal, 385*, 123821.

Mery, A., Bernard, P., Valero, A., Alper, J. P., Boime, N. H., Haon, C., ... Sadki, S. (2019). A polyisoindigo derivative as novel n-type conductive binder inside Si@C nanoparticle electrodes for Li-ion battery applications. *Journal of Power Sources, 420*, 9−14.

Mo, R. W., Lei, Z. Y., Sun, K. N., & Rooney, D. (2014). Facile synthesis of anatase TiO_2 quantum-dot/graphene-nanosheet composites with enhanced electrochemical performance for lithium-ion batteries. *Advanced Materials, 26*, 2084−2088.

Mo, R. W., Lei, Z. Y., Rooney, D., & Sun, K. N. (2019). Anchored monodispersed silicon and sulfur nanoparticles on graphene for high-performance lithiated silicon-sulfur battery. *Energy Storage Materials, 23*, 284−291.

Mo, R. W., Li, F., Tan, X. Y., Xu, P. C., Tao, R., Shen, G. R., ... Lu, Y. F. (2019). High-quality mesoporous graphene particles as high-energy and fast-charging anodes for lithium-ion batteries. *Nature Communications, 10*, 1474.

Mo, R. W., Tan, X. Y., Li, F., Tao, R., Kong, D. J., Xu, J. H., ... Lu, Y. F. (2020). Tin-graphene tubes as anodes for lithium-ion batteries with high volumetric and gravimetric energy density. *Nature Communications, 11*, 1374.

Park, M. H., Kim, M. G., Joo, J., Kim, K., Kim, J., Ahn, S., ... Cho, J. (2009). Silicon nanotube battery anodes. *Nano Letters, 9*, 3844–3847.

Park, S. W., Shim, H. W., Kim, J. C., & Kim, D. W. (2017). Uniform Si nanoparticle-embedded nitrogen-doped carbon nanofiber electrodes for lithium ion batteries. *Journal of Alloys and Compounds, 728*, 490–496.

Qin, J., Wu, M., Feng, T., Chen, C., Tu, C. Y., Li, X. H., ... Wang, D. X. (2017). High rate capability and long cycling life of graphene-coated silicon composite anodes for lithium ion batteries. *Electrochimica Acta, 256*, 259–266.

Shi, D. Q., Tu, J. P., Yuan, Y. F., Wu, H. M., Li, Y., & Zhao, X. B. (2006). Preparation and electrochemical properties of mesoporous Si/ZrO_2 nanocomposite film as anode material for lithium ion battery. *Electrochemistry Communications, 8*, 1610–1614.

Shi, F. F., Song, Z. C., Ross, P. N., Somorjai, G. A., Ritchie, R. O., & Komvopoulos, K. (2016). Failure mechanisms of single-crystal silicon electrodes in lithium-ion batteries. *Nature Communications, 7*, 11886–11893.

Sohn, H., Kim, D. H., Yi, R., Tang, D., Lee, S. E., Jung, Y. S., & Wang, D. H. (2016). Semimicro-size agglomerate structured silicon-carbon composite as an anode material for high performance lithium-ion batteries. *Journal of Power Sources, 334*, 128–136.

Tao, H. C., Huang, M., Fan, L. Z., & Qu, X. (2013). Effect of nitrogen on the electrochemical performance of core-shell structured Si/C nanocomposites as anode materials for Li-ion batteries. *Electrochimica Acta, 89*, 394–399.

Wang, F., Song, C. S., Zhao, B. X., Sun, L., & Du, H. B. (2020). One-pot solution synthesis of carbon-coated silicon nanoparticles as an anode material for lithium-ion batteries. *Chemical Communications, 56*, 1109–1112.

Wen, Y., Zhu, Y., Langrock, A., Manivannan, A., Ehrman, S. H., & Wang, C. (2013). Graphene-bonded and -encapsulated Si nanoparticles for lithium ion battery anodes. *Small (Weinheim an der Bergstrasse, Germany), 9*, 2810–2816.

Wu, H., & Cui, Y. (2012). Designing nanostructured Si anodes for high energy lithium ion batteries. *Nano Today, 7*, 414–429.

Wu, H., Yu, G., Pan, L., Liu, N., McDowell, M. T., Bao, Z., & Cui, Y. (2013). Stable Li-ion battery anodes by in-situ polymerization of conducting hydrogel to conformally coat silicon nanoparticles. *Nature Communications, 4*, 1943.

Yang, J., Wang, Y., Li, W., Wang, L., Fan, Y., Jiang, W., ... Zhao, D. (2017). Amorphous TiO_2 shells: A vital elastic buffering layer on silicon nanoparticles for high-performance and safe lithium storage. *Advanced Materials, 29*, 1700523.

Yang, X., Wen, Z., Huang, S., Zhu, X., & Zhang, X. (2006). Electrochemical performances of silicon electrode with silver additives. *Solid State Ionics, 177*, 2807–2810.

Yoo, S., Lee, J. I., Ko, S., & Park, S. (2013). Highly dispersive and electrically conductive silver-coated Si anodes synthesized via a simple chemical reduction process. *Nano Energy, 2*, 1271–1278.

Yoon, T., Nguyen, C. C., Seo, D. M., & Lucht, B. L. (2015). Capacity fading mechanisms of silicon nanoparticle negative electrodes for lithium ion batteries. *Journal of the Electrochemical Society, 162*, A2325–A2330.

Yoshio, M., Wang, H. Y., Fukud, K. J., Hara, Y., & Adachi, Y. (2000). Effect of carbon coating on electrochemical performance of treated natural graphite as lithium-ion battery anode material. *Journal of the Electrochemical Society, 147*, 1245–1250.

Yu, Y., Gu, L., Zhu, C., Tsukimoto, S., van Aken, P. A., & Maier, J. (2010). Reversible storage of lithium in silver-coated three-dimensional macroporous silicon. *Advanced Materials, 22*, 2247–2250.

Yue, L., Tang, J., Li, F., Xu, N., Zhang, F., Zhang, Q., ... Zhang, W. (2017). Enhanced reversible lithium storage in ultrathin $W_{18}O_{49}$ nanowires entwined Si composite anode. *Materials Letters, 187*, 118–122.

Zhang, L., Wu, H. B., & Lou, X. W. (2014). Iron-oxide-based advanced anode materials for lithium-ion batteries. *Advanced Energy Materials, 4*, 1300958.

Zhang, R., Du, Y., Li, D., Shen, D., Yang, J., Guo, Z., ... Zhao, D. (2014). Highly reversible and large lithium storage in mesoporous Si/C nanocomposite anodes with silicon nanoparticles embedded in a carbon framework. *Advanced Materials, 26*, 6749–6755.

Nano silicon carbon hybrid particles and composites for batteries: Fundamentals, properties and applications

Yohan Oudart[1], Rudy Guicheteau[1], Jean-Francois Perrin[1], Raphael Janot[2,3], Mathieu Morcrette[2,3], Mariana Gutierrez[1,2], Laure Monconduit[3,4] and Nicolas Louvain[3,4]

[1]*Nanomakers, Rambouillet, France* [2]*Laboratoire de Réactivité et Chimie des Solides, Université de Picardie Jules Verne, Hub de l'énergie, Amiens, France* [3]*Réseau sur le Stockage Electrochimique de l'Energie (RS2E), Hub de l'Energie, Amiens, France* [4]*ICGM, The University of Montpellier, CNRS, ENSCM, Montpellier, France*

8.1 Introduction

Silicon nanoparticles have interesting properties for Li-ion batteries. Today, batteries use graphite (natural or synthetic) as negative electrodes. Graphite has been used as an anode active material since the very beginning of Li-ion batteries, but, compared to the positive active materials, its evolution has remained low and is mainly centered on the interface with electrolyte for improved durability at high temperature. Graphite is particularly interesting for its good stability during electrochemical cycling, its low cost, its large availability, and its relatively high capacity (350–375 mAh/g based on the LiC_6 formation vs. 180–250 mAh/g for cathode active materials). Its main drawback is the quasi-absence of improvement perspectives for capacity. Other materials offer potentially much larger capacities, among them silicon and lithium appear to be the most promising with storage capacities of 3579 mAh/g (for $Li_{15}Si_4$ phase) and 3860 mAh/g (Whittingham, 1976), respectively. They both have very good capacities, large availabilities (in contrast to germanium for instance) and good techno-economical potential. However, lithium faces large technological challenges which have not been solved yet: anode volume expansion with loss of contact with the current collector, with dendrites formation, and unreacted lithium are the main challenges (Fang, Wang, & Meng, 2019).

8.2 Nanosilicon for batteries

8.2.1 Silicon generalities

Silicon is of huge interest for energy storage, due to its ability to form alloys with lithium. This property has been largely studied (Okamoto, 1998; Wen & Huggins, 1981) and the main alloys at room temperature are $Li_{13}Si_4$; Li_7Si_3; $Li_{12}Si_7$, LiSi, and $Li_{22}Si_4$, as found in the Li-Si binary phase diagram. The $Li_{22}Si_5$ alloy is the richest in lithium and has a theoretical capacity of 4200 mAh/g, but its formation is not observed under normal electrochemical conditions. Their electrochemical domains are illustrated below along with the corresponding volume changes and particles size increases.

In situ XRD studies indicate that the lithiation of crystalline silicon is made progressively, leading to a succession of amorphous alloys (Chon et al., 2011; Key et al., 2009; Li, 2000; Morris et al., 2013; Schott et al., 2017).

By in situ NMR spectroscopy, some lithium silicon alloys have been observed along discharge and charge cycling and have been attributed to amorphous Li_2Si, $Li_{3.5}Si$, and crystalline $Li_{3.75}Si$ ($Li_{15}Si_4$) (Key et al., 2009; Morris et al., 2013; Ogata et al., 2014). At the end of discharge below 50 mV, the crystallized $Li_{15}Si_4$ phase is observed (Hatchard & Dahn, 2004; Obrovac & Christensen, 2004; Okamoto, 1998). For this alloy, the volumic expansion is 270% (Touidjine, 2016; Wen & Huggins, 1981; Okamoto, 1990). For spherical particles, it means a particle diameter increase of 55%. (Fig. 8.1)

Fig. 8.2 shows that reduction is progressive from the separator side (Hatchard & Dahn, 2004). After delithiation, silicon is amorphous.

In the case of particles, lithiation occurs at the surface of the silicon particles. Uncompleted lithiation can leave a crystalline core surrounded by a lithiated surface.

8.2.2 Nanosizing

Due to the high volumetric expansion during cycling (up to 270% for the $Li_{15}Si_4$ phase), decrepitation is observed for particles larger than 150 nm (Chen et al., 2019; Kasavajjula et al., 2007; Obrovac & Chevrier, 2014). This leads to electrode

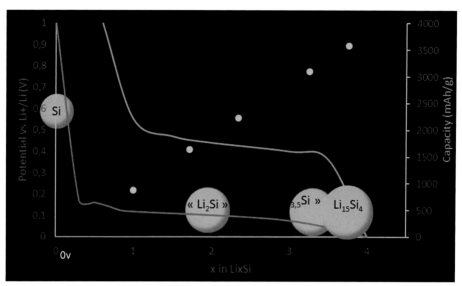

Figure 8.1 Silicon particle size evolution and LixSi alloy stoichiometry during lithiation (corresponding electrochemical capacities in yellow dots) (Wen & Huggins, 1981; McDowell et al., 2013; Ogata et al., 2014; Chon et al., 2011; Schott et al., 2017).

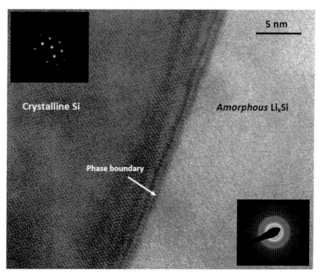

Figure 8.2 TEM image of silicon electrode after partial lithiation (Chon et al., 2011).

Figure 8.3 SEM picture of a silicon wafer, showing fractures after delithiation (Chon et al., 2011).

cracking at large scale as shown below with a silicon wafer (Hatchard & Dahn, 2004; McDowell et al., 2013). (Fig. 8.3)

During cycling, the particles bigger than 150 nm fracture progressively and this has two detrimental effects on the electrochemical cycling stability: electrical contact within the electrode materials is partially lost and the particles' surface is drastically increased (Chon et al., 2011; McDowell et al., 2013).

A way to avoid swelling and cracking is the use of only a small fraction of the silicon by reducing the voltage range (especially by increasing the low cut-off voltage) to limit lithiation and swelling. This has been done with micrometric particles (Andersen et al., 2019). However, in those types of formulation, the silicon capacity is reduced to 1000 mAh/g, meaning that 72% of the silicon is unused.

8.2.3 Various forms of nanosilicon

To get particles below 150 nm, the main approach is bottom-up particles preparation. Indeed, the top-down approach is usually based on crushing particles, leading to large size distribution, especially with difficulties to avoid some large particles (McDowell et al., 2013). Moreover, such low particle size requires usually ball-milling in liquid media with solvent under inert atmosphere and high energy.

The bottom-up approach enables a much better control on the particle morphology and size distribution. The different forms are:

— Nanofilms (Graetz et al., 2003)
  • Despite good specific capacities, the overall anode surface capacity is limited as the electrode thickness is very low. Thickness increase leads to a sharp decrease of the electrochemical performances (Maranchi et al., 2003).
— Nanowires (Cui et al., 2009)
  • Those 1D material includes nanotubes (Wu et al., 2012) and nanowires (Chen et al., 2019). Their interest is that they can grow directly on the current collector with a limited use of conductive additives and binders. Energy densities can be high but as the length of the tubes and wires is limited, the anode overall surface capacity is usually low (Obrovac & Chevrier, 2014; Sourice, 2015).
— Spherical particles and string of spherical particles (McDowell et al., 2013)

Laser pyrolysis is a well-known technique for nanosilicon synthesis. It was reported for the first time in 1980 by Marra (Cannon et al., 1982). Similar to most bottom-up approaches, the precursor is a gas, usually silane or chlorosilane. The silicon–hydrogen bonds are excited and broken by a CO_2 laser leading to the direct formation of silicon and dihydrogen (with silane). The use of a laser enables a very good control of the reaction zone and therefore of the particle size distribution (Herlin-Boime et al., 2004). This method has been scaled up to industrial size (Sourice, 2015). Powders obtained by this process have spherical primary particles aggregated as strings of a few particles. As a negative electrode the powder state enables various formulations and high loadings (Escamilla-Pérez et al., 2019) (Fig. 8.4).

8.2.4 General behavior and SEI

Silicon nanoparticles solve the cracking problem, but they develop high surface areas: 150 nm spherical particles have a surface area of 17 m^2/g for instance, and 65 m^2/g for 40 nm spherical particles. During complete silicon lithiation, a low potential is reached and the electrolyte salt and solvent react with the particles and form the well-documented SEI (Müller, 2011). This usually occurs at potential below 0.8 V versus Li/Li^+ (Philippe et al., 2012; Verma et al., 2010). An irreversible consumption of lithium and

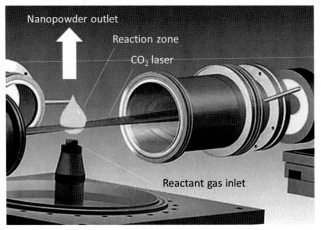

Figure 8.4 Principle of laser pyrolysis (Herlin-Boime et al., 2004).

silicon occurs. This layer has a good ionic conductivity but a poor electronic conductivity (Obrovac & Christensen, 2004; Pereira-Nabais et al., 2013; Verma, Maire, & Novák, 2010). To form a stable layer, a "formation cycle" is usually run at a slower C-rate than the following cycles or higher temperature. During cycling, the volume expansion can leave new bare silicon surfaces. The SEI is formed at the interface and causes an increase in irreversible capacity upon subsequent cycling. Many works are under way to improve the electrolyte in order to form a more stable SEI (Aurbach, 1996; Obrovac & Christensen, 2004, Arreaga-Salas et al., 2012; Xu, 2004):

$$LiPF_6 \rightarrow LiF + PF_5$$

$$PF_6^- + 2\,e^- + 3\,Li^+ \rightarrow 3\,LiF + PF_3$$

$$PF_5 + H_2O \rightarrow 2\,HF + POF_3$$

Around silicon particles, the usual surface layer is made of silicon oxide (Si-O-Si) and silanol groups (Si-OH). This oxide can react with the decomposed PF_6^- anion to SiO_xF_y species and HF can also form SiF_6^{2-} anions. Those compounds form cracks at the Si particles surface. Interestingly, those cracks are less important for Si particles smaller than 150 nm. (Obrovac and Christensen, 2004)

The classical solvents of the electrolyte are alkyl carbonates that can also react with the silicon and the fluorinated SEI (containing species like LiF, LiP_xF_y, and $LiPO_yF_z$), thus increasing irreversible capacities. This phenomenon is more important with small particles and larger specific surface areas. (Obrovac and Christensen, 2004)

The SEI containing organic and inorganic species is formed upon the first lithiation but also evolves upon subsequent cycling in the case of Si due to volumetric expansion, crackings, and exposure of fresh surfaces. This is the main drawback of silicon as a negative electrode for Li-ion batteries. Two main complementary methods are investigated as workarounds:
- modifying the electrolyte formulation to make it more stable upon cycling by the use of additives such as FEC and VC (Li et al., 2018a, 2018b). This is not the topic of this chapter.
- protecting the silicon to reduce and stabilize the SEI formation. This chapter describes the most documented approaches which is the use of carbon.

8.3 Carbon

Carbon is the most used material for Li-ion batteries. The chemical stability and specific capacity of graphite make it the most efficient material in batteries. This paved the way for the first Li-ion batteries commercialization by Sony in 1991 based on Yoshino et al.'s patent (Yoshino & Nakajima, 1985). {{{ Table 8.1 }}}.

To combine the stability of carbon with the capacity of silicon, the mix of silicon and carbon (mainly graphite) has been tested. This method is simple but requires the development and use of specific binders, which are not always easily or economically available. This is why nanoparticles of silicon hybridized with carbon have been developed.

One important feature is the thermodynamically favorable reaction of silicon with carbon to form silicon carbide which is not active for lithium storage (Kim, Blomgren, & Kumta, 2004) and thus its formation should be avoided:

Table 8.1 Comparison of the performances of graphites and silicon (Pilot, 2018).

	Artificial graphite	Natural graphite (Pilot, 2018)	Si
Capacity (mAh/g)	365	350	Up to 3576
First irreversible capacity	90%–94%	90%	75%–80% without C protection and usual formulation, up to 90% with coating
Metric tonnage 2017	70 000	50 000	<10 t

$$Si + C \rightarrow \beta\text{-SiC} \quad \Delta H_f = -73.2 \text{ kJ/mol for } \beta\text{-SiC at 298 K}$$

(Varadachari et al., 2009)

However, the activation energy for this reaction is high and it does not occur below 900°C–1000°C. On the other hand, graphite is the best carbon phase for lithium storage but its formation requires hours of crystallization at temperatures above 2400°C (Coulon et al., 1994). Therefore, graphite/silicon composites can only be made by mixing and not by synthesis, without direct bonds between them. Some teams succeeded in forming graphene coating (Son et al., 2015) on silicon thanks to a very special CVD parameters set: CH_4 and CO_2 at 1000°C leads to graphene growth on silicon, whereas at 900°C graphene is not well formed and at 1100°C the silicon oxide layer was too thick. Without CO_2, SiC is readily formed. In full cells configuration, capacity drops by 28% after 200 cycles, with an initial capacity of 970 mAh/g. These results were achieved with 100 nm silicon powder.

However, other forms of carbon can store lithium and can be useful for silicon hybridization.

8.3.1 Carbon forms

Two properties are important for carbon as anodes materials: electrical conductivity and lithium insertion/intercalation. Both of those properties are favored with the presence of conjugated aromatic cycles (Graczyk-Zajac et al., 2010). In the specific case of graphite, plans of graphene are perfectly aligned and this favors electric transfer. Aromatic content and graphitization are favored by temperature, graphite being formed above 2400°V (see Fig. 8.5). Lithium intercalation is made between the graphene sheets yielding its good storage capacity (372 mAh/g). This reversible intercalation is usually summarized as:

$$Li + 6 \text{ C} - \rightarrow LiC_6.$$

Lithium diffusion is important for lithiation and delithiation processes. A well-structured phase will enable a good permeability and a good intercalation of lithium. As a general principle, the more structured carbon, the more stable the lithiation capacity: up to 372 mAh/g for graphite, 220 mAh/g for carbon black (Super P) (Attia et al., 2019), 290 mAh/g for petroleum pitch (Rutgers 250 M) (Gutierrez) with very disordered species but high aromatic contents. For cokes, compositions are various, initial capacities range from 500 mAh/g to 800 mAh/g after treatment at 800°C but decrease below 200 mAh/g after a few tens of cycles. Oxidation can improve stability around 400 mAh/g but with 200 mAh/g capacity loss during the first

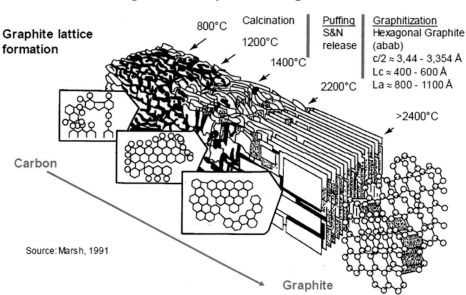

Figure 8.5 Carbon structure evolution from mesophase depending on temperature (Marsh, 1991).

20 cycles (Concheso et al., 2009). Moreover, the higher the temperature at which thermal treatment is performed, the better is the graphitization.

For carbon treated at temperature up to 1000°C, the structure and the lithiation property will also depend on the precursor type, mainly soft and hard carbon with the main difference being that soft carbon can be transformed into graphite at high temperature and this is not the case for hard carbons due to a high content of structural defects. For example, soft carbons can be made from pitch at 900–1000°C. Those structures can have capacities higher than graphite as they have many structural defaults and microporosity along with small graphitic domains (Buiel & Dahn, 1999; Yong & Fan, 2013). They however have large irreversible capacity at first cycle and high polarization between charge and discharge (Buiel and Dahn, 1999) (Fig. 8.6).

Hard carbons are obtained from oxygen or sulfur-rich precursors at temperature superior to 1000°C. Oxygen and sulfur enable partial graphitization, but along with reticulation, disable full graphitization. Typical precursors are cellulose and sucrose. Oxygen control and precursor oxygen content is rather important.

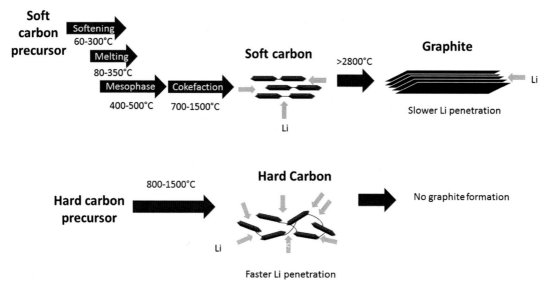

Figure 8.6 Comparison between different carbons (Müller, 2011).

Nanosilicon (<80 nm) is oxidized slowly from 500°C and rapidly from 800°C (Sourice, 2015). Silicon oxidation into silicon dioxide is thermodynamically favored and exothermic ($\Delta H_{f\ liq} = -902.7$ kJ/mol) and is one of the most stable forms of silicon (Nist data, 2020).

As silicon oxide capacity and also initial irreversible loss are less good than for pure silicon, its oxidation should be avoided by the absence of oxygen or oxygen precursors (oxygen-rich carbons) during thermal treatments. Carbon oxidation can occur earlier and independently from silicon oxidation (see thermogravimetric analysis for carbon-coated Si particles in Fig. 8.7).

To avoid oxygen, nitrogen can be used; however, the nitridation of silicon from N_2 gas is also thermodynamically very favorable ($\Delta H_f = -733$ kJ/mol for the reaction $3Si + 2N_2 = Si_3N_4$) but occurs only at high temperature, above 1350°C, close to silicon's melting point (1410°C) (Xiaohan, 2019).

The temperature value has to be a compromise as higher temperature increases graphitization and improves carbon properties but inactive SiC can be formed (Datta & Kumta, 2007; Graczyk-Zajac et al., 2010; Obrovac and Chevrier, 2014; Obrovac and Chevrier, 2014) For instance, SiC formation has been observed at a temperature as low as 800°C for petroleum pitch and polysilane. However, this low temperature is probably due to the silicon precursor reactivity. With nanosilicon and pitch

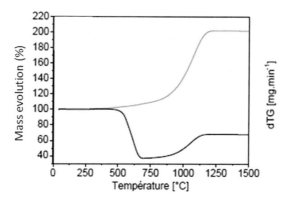

Figure 8.7 Thermogravimetric analysis of pure nanosilicon particles (red) and carbon-coated silicon nanoparticles containing 60% carbon (black) (Sourice et al., 2015).

blends, SiC formation with 12% Si (atomic) has been observed at 1000°C (Saint, 2005).

Along with pyrolysis temperature, physical properties change; Young's modulus increases up to 1000°C and then decreases: elasticity of the carbon host matrix is the worst at about 1000°C. From 1500°C, thermal and electrical conductivities both increase due to the middle- and long-distance graphitization process (also illustrated in Fig. 8.8) (Coulon et al., 1994). Up to 1300°C–1400°C, carbon dehydrogenation is observed along with crystalline thickness decrease (Endo et al., 2000). The first cycles storage capacity also decreases. Then a temperature increase leads to graphitization and crystalline thickness increases, to end up with graphite, the most stable material (Dahn et al., 1995).

8.3.2 Pitch as precursor

Different carbon precursors have been tested (Li et al., 2018a, 2018b): graphite, PVC, pitches (coal or petroleum), PVA, PAN, PVP, phenolic resins, saccharides, and lignin (Du et al., 2018). The most often used in academic literature are pitches and PVC. At larger scale, PVC is not used as its mass loss during pyrolysis is very high (83%) and chloride emissions are generated and have to be treated (Saint et al., 2007; Saint, 2005). The interest of PVC is the good elasticity of the resulting carbons and the final Si/C composites have low specific surface areas.

Pitch is a carbonaceous mixture of organic by-products resulting from distillation of coal or petroleum, not readily useful for energy production (combustion). Its main characteristics are its high boiling points (>200°C) and its high content in aromatic molecules. Its main application is road covering, and some special grades are used as amorphous carbon coating for graphite electrodes of Li-ion devices.

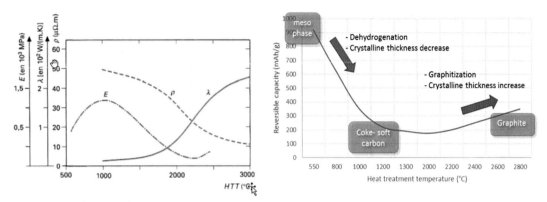

Figure 8.8 Left: Influence of pyrolysis temperature on physical properties (E = Young's modulus, λ = thermal conductivity; ρ = resistivity) (Coulon et al., 1994); Right: Specific reversible capacity in Li-ion batteries of carbon materials ("soft carbon family") versus temperature (Dahn et al., 1995; Endo et al., 2000).

During pyrolysis of pitch, hydrogen is produced due to aromatization of the molecules. It only explains a minor part of the weight loss which is due to the gasification of carbon material (Dang et al., 2016). This usually ends between 500°C and 600°C depending on the precursor. Between 350°C and 450°C, polycyclic hydrocarbons of about 10 cycles face condensation reaction and their size increases (Dupupet, 2008; European, 2016). At temperature of 420°C–500°C, a liquid crystal phase called mesophase appears (cf. Figures 8.44 et 45) (Savage, 2012) (Fig. 8.9).

From temperature over 700°C and in all cases at 900°C, aromatics are the main molecules. A further temperature increase will mainly improve crystallinity, instead of aromaticity (Dang et al., 2016). The pitch structure contains aliphatic and aromatic compounds in various concentrations. The C-C and C-H bond breaking during aromatization requires energy (420 kJ/mol to break an aromatic C-H bond and 325 kJ/mol for a C-H bond in methyl group). Thus aliphatic groups will require more energy than aromatic groups to fully decompose in aromatic cycles (Dang et al., 2016). On the other hand, the more aromatic the compound, the higher activation energy to initiate C-H bonds breaking (which is normal as those products are thermodynamically more stable) (Savage, 2012).

Therefore carbon chemistry is rather complex and difficult to characterize. In most cases, silicon will be associated with noncrystalline carbons even though some graphitic contribution can be observed. For instance, Raman spectroscopy analysis has been done on 30 nm silicon coated with carbon (19% C in weight): the spectra show two bands, the one at 1600 cm^{-1} called G being

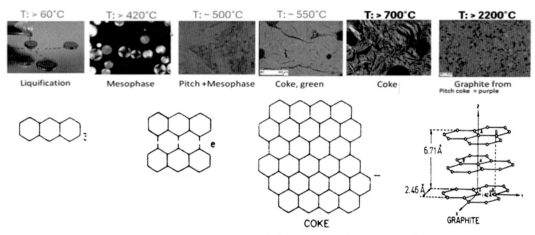

Figure 8.9 Phase formed during thermal treatment of pitch Savage, 2012; European, 2016).

attributed to organized graphitic carbon (Desrues, 2019), whereas the "D" band corresponds to disorganized carbon. This carbon is made from ethylene decomposed by a CO_2 laser.

So apparently, it is not possible to combine all the positive properties of carbon and silicon due to temperature limitations. However various approaches have been developed to combine the most interesting properties of each, that is to say silicon capacity and carbon stability.

8.4 Carbon coating of silicon

Many protocols have been used to make a carbon coating on silicon, from solid precursor (mechanosynthesis), carbon dissolved in solution (sucralose, PVC), or from gas, mainly acetylene.

8.4.1 Mechanical milling

Powders are crushed together in a high energy system with hard milling tools (usually hardened steel or tungsten carbide) (Bridel, 2010; Tarascon et al., 2005; Wen & Huggins, 1981). Depending on the apparatus and the targeted particle size, operating milling times range from 1 h to 100 h. A temperature increase is usually observed but it remains below 200°C (Costa, 2001). During crushing, two main mechanisms are encountered: pulverization which tends to form small particles (Costa, 2001), and aggregation which has the opposite effect to pulverization in terms of mean particle size (Tarascon et al., 2005).

This method has been successfully used with various carbon precursors, including graphite, to obtain Si/C composites. It could be followed by thermal treatments. Coating is usually observed and irreversible capacity is decreased even though it is difficult to identify the effect of silicon particle size decrease and coating only itself (Costa, 2001; Kasavajjula, Wang, & Appleby, 2007). Despite the low temperature, SiC formation has been reported for highly energetic milling processes (Obrovac and Chevrier, 2014). Silicon particle size distribution is usually large and possible contamination by the reactor materials such as iron or tungsten, and their related carbides, can be found (up to few wt.%) (Costa, 2001; Endo, Kim, Nishimura, Fujimo, & Miyashita, 2000).

8.4.2 Gas-phase synthesis

The gaseous way is advantageously a single-step process as carbon is directly decomposed onto silicon particles. Acetylene is the main raw material as it is cheap, easily available, and highly exothermic (-243 kJ mol^{-1} at 900K for the reaction $C_2H_2 = 2C + H_2$) (Nist data, 2020) as well as PECVD (plasma enhanced chemical vapor deposition; Chaukulkar et al., 2014) (Fig. 8.10).

Acetylene decomposition starts at the temperature of 700°C (Hurd, 1934). Some silicon surfaces can also catalyze this reaction and lower this temperature down to 600°C (Silvestrelli et al., 2001). For homogeneity reasons, it is preferable that the hot silicon particles meet the colder acetylene gas in order not to decompose acetylene alone and form carbon nanoparticles (Desrues, 2019; Sourice, 2015).

To avoid particle handling and a step of aerosol formation, coating in line with silicon nanoparticle production is of strong interest. This has been exemplified by laser pyrolysis with two different systems: firstly, acetylene gas injection just after the silicon formation zone (Tenegal, 2012). This method has been

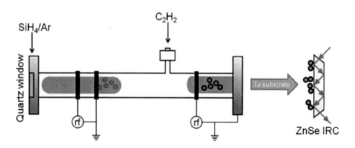

Figure 8.10 Scheme of plasma reactor for carbon-coated silicon particles (Chaukulkar et al., 2014).

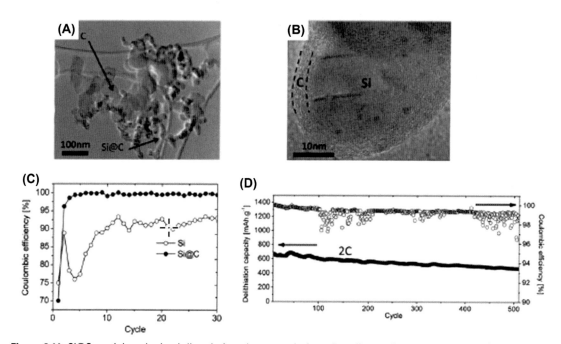

Figure 8.11 Si@C particles obtained directly from laser pyrolysis and cycling performances obtained (Sourice et al., 2015).

demonstrated at several kg/hours and electrochemical performances are shown below.

Another embodiment is the use of a second reaction zone where the flame is formed via the interaction of ethylene and CO_2 laser. A radiation flame is observed (Sourice, 2015). Those approaches validate the interest of carbon coating for silicon surface protection (Fig. 8.11).

8.4.3 Liquid process

This method is made in four steps:
1. Carbon precursor dissolution in a solvent.
2. Silicon nanoparticles dispersion in the slurry.
3. Solvent removal (usually evaporation).
4. Thermal treatment.

A key step is the homogeneity of dispersion of nanosilicon as agglomerates can be formed (Alavarez Barragan et al., 2018). High shear system (Dispermat, Ultra-Turrax) can be used and, for small samples preparation, ultrasonication is the most used technique even though it should not be used with flammable solvent (cavitation can cause local temperature increase and ignite inflammation). After the solvent removal, the carbon precursor should be thermally treated to have a good protection of silicon. The effect of thermal treatment is described in the composite section.

Sucrose can be used to obtain a coating on the particles with limited success in terms of stability (Alavarez Barragan et al., 2018). The most complex carbon structures have been obtained through the liquid phase route using templates to create voids between silicon and carbon for instance. The idea is to have enough void to accommodate silicon particles swelling during cycling without breaking the carbon shell. This has improved significantly the silicon stability at rather high capacity values (1000 mAh/g). The preparation includes several steps, the key one being the covering of particles with a surfactant which is then covered by a silica crust. After surfactant removal, the hollow silica crust is used as a template for carbon deposition. After thermal treatment, this silica layer is removed by HF leaching (Yang et al., 2015) (Fig. 8.12).

A pomegranate structure has then been obtained with "grains" of silicon inside a hollow carbon shell. Performances are also very stable with high capacities (around 700 mAh/g) for thousands of cycles, despite high irreversible losses during the very first cycle (50%) (Peng Guan et al., 2018) (Fig. 8.13).

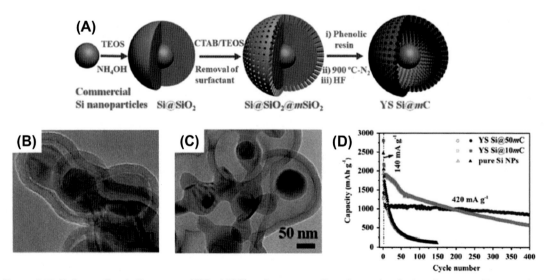

Figure 8.12 Hollow yolk–shell structure (Si@void@C) and corresponding electrochemical performance (Yang et al., 2015).

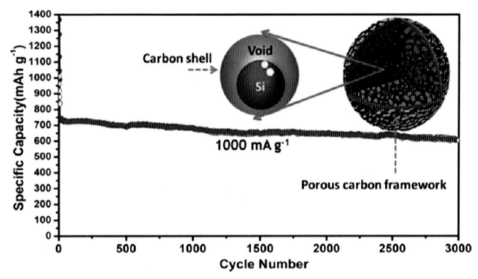

Figure 8.13 Pomegranate assembly of Si@void@C particles and corresponding chemical performances (Peng Guan et al., 2018).

This validates the general assumption that swelling side-effects could be limited by having enough space for it. Those preparations would be difficult to scale up economically but validate that silicon can be cycled under stable conditions if the system can accommodate volume expansion. This can be done also using elastic binders in the electrode formulation.

8.4.4 Si/C composites

The previous section mainly detailed the preparation of thin carbon coating around silicon particles. Another approach is to embed silicon into a carbon matrix to form a Si/C composite. The main advantage of this approach is to obtain micrometric particles similar in size to graphite (10–40 μm) ready and easy to use. Moreover, carbon protection around silicon is much thicker and tunable (Kim et al., 2016; Li et al., 2018b; Mochida et al., 1991).

The main routes are the liquid process (see thin coating part) with the same steps, the main difference being the higher carbon content. For some precursors, especially pitches, it is possible to take advantage of the low temperature melting point to disperse silicon into hot fused carbon, before a thermal step. During solvent evaporation and thermal treatment, some micropores can be created, thus increasing specific surface area and, therefore, irreversible capacity (Mochida et al., 1991).

Moreover graphite and some other additives can be added in the composite to enhance properties (electronic conductivity, stability). This method can for instance yield Si/C/graphite composite with specific capacity of 80% after 100 cycles with 17 wt% Si and a current density of 130 mA/g (Dang et al., 2016; Escamilla-Pérez et al., 2019; Li et al., 2018a). The carbon matrix act as an intermediate between electrolyte and Si nanoparticles. It avoids direct contact between them and creates a chemically and mechanically stable SEI (Li et al., 2018a, 2018b; Obrovac & Christensen, 2004; Paireau et al., 2015; Shen et al., 2017).

A key step for hybridization is the thermal treatment and it is developed here using pitches as carbon precursors (Fig. 8.14).

Various precursors can be used. One important feature is the mass loss during thermal treatment. At 900°C, some precursors have very large mass losses (from more than 90% for cellulose and sucrose, down to 44% for some pitch) (Li et al., 2016), which makes them difficult to use due to the large gas treatment/energy recovery unit which has to be implemented:

Among other criteria, the aromatic content of the pitch is interesting as it will favor crystalline carbon. Oxygen content is important as high content will favor the reticulation and the formation of hard carbons.

Silicon in carbon matrix is well documented (Desrues, 2019), (Dahn et al., 1995; Obrovac & Christensen, 2004; Escamilla-Pérez et al., 2019; Xing, 1997; Yong & Fan, 2013). Carbon matrix enables ionic and electric conductivities and helps to form a more stable SEI (Dahn et al., 1995; Obrovac & Christensen, 2004).

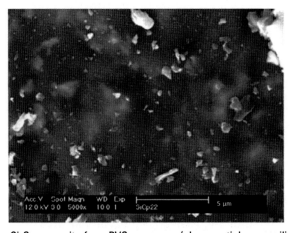

Figure 8.14 SEM image of a Si-C composite from PVC precursor (clear particles are silicon) (Saint, 2005).

Initial coulombic efficiencies (ICE) of Si/C composites are very variable. PVC and pitches as precursors coupled with nanosilicon have the highest values (78%–80%) (Escamilla-Pérez et al., 2019; Liu et al., 2005; Saint et al., 2007). Lignin yields lower coulombic efficiency due to its higher oxygen content (71.6% ICE) (Li et al., 2018a, 2018b).

A good summary of all those approaches has been made by combining several of them. Hu et al. mixed nanosilicon particles with micrometric graphite and CMC as binder. The blend was then dried and pyrolyzed at 900°C: silicon nanoparticles cover the graphite surface (Hu et al., 2018). Then, pitch dissolved in THF is added and pyrolyzed at 900°C: a pitch-based carbon layer is formed around the silicon on graphite particles. Then, the particles are placed in a CVD equipment and a carbon layer coming from acetylene decomposition at 700°C is deposited on the particles to form graphite@Si-$C_{ex\text{-}CMC}$@$C_{ex\text{-}pitch}$@$C_{ex\text{-}CVD}$. Those complex particles have carbon coming from four different sources and forming four successive layers, one of them containing also silicon. Liquid and gaseous processes are used (we can note that the first Si-graphite mixing could have been done by powder mixing). This material has a first discharge capacity of 762 mAh/g and then the capacity stabilized around 600 mAh/g from cycle 80 to 140. This demonstrates that very good electrochemical performances can be obtained with silicon and complex carbon structure/texture. The challenge now is to obtain very good electrochemical performances with a more simple structure and this implies a good knowledge of carbon and silicon chemistries.

8.5 Conclusion

Silicon has a very high specific capacity for Li-ion batteries anodes (3579 mA h g^{-1} based on the $Li_{15}Si_4$ formation) and do not form dendrites, which is a strong advantage over lithium, the other potential high capacity anode material. Silicon has two disadvantages: cracking during electrochemical cycling, which can be solved by using particles smaller than 150 nm; and limited cycling stability due to the continuous formation of SEI (solid electrolyte interphase). SEI is due to silicon surfaces reaction with electrolyte. Those reactions are limited with carbon, especially graphite, but carbon-based material have capacities limited to 372 mA h g^{-1}. Silicon nanoparticles hybridization with carbon can benefit from both interests: high silicon capacity and high carbon stability. This can be made possible by making a carbon shell around the silicon. Formation of silicon carbide should be avoided and, therefore, limits the type of carbon to "cokes" that can be obtained below 1000°C. Various ways have been tested: solid, liquid, and gaseous carbon precursors, usually followed by a thermal treatment to increase carbon crystallinity. Simple coating and complex structure (carbon hollow structure around Si, pomegranate) have been demonstrated at the laboratory scale. Very good performances can be obtained with silicon and complex structure. It validates some principles to improve silicon stability: use Si particles with diameter below 150 nm, protect the silicon surface with a carbon coating, and enable swelling without electrode cracking. One challenge for academic studies is to increase the overall active material stability. For larger use, the challenge now is to obtain very good performances with simple and cheap structure and this implies a good knowledge of both carbon and silicon chemistries.

Direct carbon coating techniques in line with silicon particles synthesis seems to be the most efficient way, among them laser pyrolysis has demonstrated large-scale capacities. Integration of nanosilicon in carbon matrix is also simple and exhibits good results, especially when using pitch as the carbon precursor.

References

Alavarez Barragan, A., Nava, G., Wagner, N. J., & Mangolini, L. (2018). Silicon-carbon composites for lithium-ion batteries: A comparative study of different carbon deposition approaches. *Journal of Vacuum Science & Technology B, 36*, 011402.

Andersen, H. F., Foss, C. E. L., Voje, J., Tronstad, R., Mokkelbost, T., Vullum, P. E., ... Mæhlen, J. P. (2019). Silicon-carbon composite anodes from industrial battery grade silicon. *Nature Scientific Reports, 9*, 14814.

Arreaga-Salas, D. E., Sra, A. K., Roodenko, K., Chabal, Y. J., & Hinkle, C. L. (2012). Progression of solid electrolyte interphase formation on hydrogenated amorphous silicon anodes for lithium-ion batteries. *Journal of Physical Chemistry C, 116*(16), 9072–9077.

Attia, P. M., Supratim Das, S. J., Harris., Martin, Z. B., & William, C. (2019). Chueh, Electrochemical kinetics of SEI growth on carbon black: Part I. experiments. *Journal of the Electrochemical Society, 166*(4), E97–E106.

Aurbach, D. (1996). A comparative study of synthetic graphite and Li electrodes in electrolyte solutions based on ethylene carbonate-dimethyl carbonate mixtures. *Journal of the Electrochemical Society, 143*(12), 3809.

Bridel, J.-S. PhD thesis, Optimisation de composites silicium-polymère-carbone pour électrodes négatives d'accumulateurs lithium-ion, Université de Picardie Jules Vernes, 2010.

Buiel, E., & Dahn, J. R. (1999). Li-insertion in hard carbon anode materials for Li-ion batteries. *Electrochimica Acta, 45*(1–2), 121–130.

Cannon, W. R., Danforth, S. C., Flint, J. H., Haggerty, J. S., & Marra, R. A. (1982). Sinterable ceramic powders from laser-driven reactions: I, process description and modeling. *Journal of the American Ceramic Society*, 65(7), 324−330.

Chaukulkar, R. P., de Peuter, K., Stradins, P., Pylypenko, S., Bell, J. P., Yang, Y., & Agarwal, S. (2014). Single-step plasma synthesis of carbon-coated silicon nanoparticles. *ACS Applied Materials & Interfaces*, 6(21), 19026−19034.

Chen, H., He, S., Hou, X., Wang, S., Chen, F., Qin, H., ... Zhou, G. (2019). Nano-Si/C microsphere with hollow double spherical interlayer and submicron porous structure to enhance performance for lithium-ion battery anode. *Electrochimica Acta*, 312, 242−250.

Chon, M. J., Sethuraman, V. A., McCormick, A., Srinivasan, V., & Guduru, P. R. (2011). Real-time measurement of stress and damage evolution during initial lithiation of crystalline silicon. *Physical Review Letters*, 12.

Concheso, A., Santamaria, R., Menendez, R., Jimenez-Mateos, J. M., Alcantara, R., Ortiz, G. F., ... Tirado, J. L. (2009). Effect of oxidation on the perfroamnces of low temperature petroleum cokes as anodes in lithium ion batteries. *Journal of Applied Electrochemistry*, 38, 899−906.

Costa, P. (2001). Nanomatériaux - Structure et Élaboration. *Technique de l'ingénieur*. Available from: https://www.techniques-ingenieur.fr/base-documentaire/archives-th12/archives-etudes-et-proprietes-des-metaux-tiamb/archive-1/nanomateriaux-m4026/.

Coulon, M., Reynvaan, C., & Maire, J. (1994). *Le carbone en électrotechnique*, 33.

Cui, L.-F., yang, Y., Hsu, C.-M., & Cui, Y. (2009). Carbon − silicon core − shell nanowires as high capacity electrode for lithium ion batteries. *American Chemical Society*, 9(9), 3370−3374.

Dahn, J. R., Zheng, T., Liu, Y., & Xue, J. S. (1995). Mechanisms for lithium insertion in carbonaceous materials. *Science (New York, N.Y.)*, 270, 590−593.

Dang, A., Li, H., Li, T., Zhao, T., Xiong, C., Zhuang, Q., ... Ji, X. (2016). Preparation and pyrolysis behavior of modified coal tar pitch as C/C composites matrix precursor. *Journal of Analytical and Applied Pyrolysis*, 119, 18−23.

Datta, M. K., & Kumta, P. N. (2007). Silicon, graphite and resin based hard carbon nanocomposite anodes for lithium ion batteries. *Journal of Power Sources*, 165(1), 368−378.

Desrues, A. PhD thesis, Matériaux Composites Si@C Nanostructurés Pour Anodes de Batterie Li-Ion à Haute Densité d'énergie. Relations Entre Structure/Morphologie et Mécanismes de Dégradation, Université Paris Sud, 2019.

Dupupet, G. (2008). Fibres de carbone. *Technique de l'ingénieur*, 22. Available from: https://www.techniques-ingenieur.fr/base-documentaire/materiaux-th11/materiaux-composites-presentation-et-renforts-42142210/fibres-de-carbone-am5134/.

Du, L., Wu, W., Luo, C., Zhao, H., Xu, D., Wang, R., & Deng, Y. (2018). Lignin derived Si@C composite as a high performance anode material for lithium ion batteries. *Solid State Ionics*, 319, 77−82.

Endo, M., Kim, C., Nishimura, K., Fujimo, T., & Miyashita, K. (2000). Recent development of carbon materials for Li-ion batteries. *Carbon*, 38, 183−197.

Escamilla-Pérez, A. M., Roland, A., Giraud, S., Guiraud, C., Virieux, H., Demoulin, K., ... Monconduit, L. (2019). Pitch-based carbon/nano-silicon composite, an efficient anode for Li-ion batteries. *RSC Advances*, 9(19), 10546−10553.

European Carbon and Graphite association. Status of Coal Tar Pitch, High Temperature (CTPht) as an Intermediate in the Manufacture of Carbon and Graphite Products. 9; 2016.

Fang, C., Wang, X., & Meng, Y. S. (2019). Key issues hindering a practical lithium metal anode. *Trends in Chemistry, 1*(2), 152–158. Available from: https://doi.org/10.1016/j.trechm.2019.02.015.

Graczyk-Zajac, M., Mera, G., Kaspar, J., & Riedel, R. (2010). Electrochemical studies of carbon-rich polymer-derived SiCN ceramics as anode materials for lithium-ion batteries. *Journal of the European Ceramic Society, 30*(15), 3235–3243.

Graetz, J., Ahn, C. C., Yazami, R., & Fultz, B. (2003). Highly reversible lithium storage in nanostructured silicon. *Electrochemical and Solid-State Letters, 6*(9), A194.

Gutierrez, M., unpublished result.

Hatchard, T. D., & Dahn, J. R. (2004). In Situ XRD and electrochemical study of the reaction of lithium with amorphous silicon. *Journal of the Electrochemical Society, 151*(6), A838.

Herlin-Boime, N., Mayne-L'Hermite, M., & Reynaud, C. (2004). Synthesis of covalent Nanoparticules by CO2 laser. *Encyclopedia of Nanoscience and Nanotechnology, 10*, 1–26.

Hurd, C. D. (1934). Pyrolysis of unsaturated hydrocarbons. *Industrial & Engineering Chemistry, 26*, 50–55.

Hu, X., Huang, S., Hou, X., et al. (2018). A double core-shell structure silicon carbon composite anode material for a lithium ion battery. *Silicon, 10*, 1443–1450.

Kasavajjula, U., Wang, C., & Appleby, A. J. (2007). Nano- and bulk-silicon-based insertion anodes for lithium-ion secondary cells. *Journal of Power Sources, 163*(2), 1003–1039.

Key, B., Bhattacharyya, R., Morcrette, M., Seznec, V., Tarascon, J. M., & Grey, C. P. (2009). Real-time NMR investigations of structural changes in silicon electrodes for lithium-ion batteries. *Journal of the American Chemical Society, 131*, 9239–9249.

Kim, I.-S., Blomgren, G. E., & Kumta, P. N. (2004). Tiltel XX. *Journal of Power Sources, 130*, 275.

Kim, S. Y., Lee, J., Kim, B.-H., Kim, Y.-J., Yang, K. S., & Park, M.-S. (2016). Facile synthesis of carbon-coated silicon/graphite spherical composites for high-performance lithium-ion batteries. *ACS Applied Materials & Interfaces, 8*(19), 12109–12117.

Li, H. (2000). The crystal structural evolution of nano-Si anode caused by lithium insertion and extraction at room temperature. *Solid State Ionics, 135*(1–4), 181–191.

Liu, Y., Matsumura, T., Imanishi, N., Hirano, A., Ichikawa, T., & Takeda, Y. (2005). Preparation and characterization of Si/C composite coated with polyaniline as novel anodes for Li-ion batteries. *Electrochemical and Solid-State Letters, 8*(11), A599.

Li, Y., Hu, Y.-S., Li, H., Chen, L., & Huang, X. (2016). A superior low-cost amorphous carbon anode made from pitch and lignin for sodium-ion batteries. *Journal of Materials Chemistry A, 4*(1), 96–104.

Li, Y., Liu, W., Long, Z., Xu, P., Sun, Y., Zhang, X., ... Jiang, N. (2018a). Si@C microsphere composite with multiple buffer structures for high-performance lithium-ion battery anodes. *Chemistry—A European Journal, 24*(49), 12912–12919.

Li, P., Zhao, G., Zheng, X., Xu, X., Yao, C., Sun, W., & Dou, S. X. (2018b). Recent progress on silicon-based anode materials for practical lithium-ion battery applications. *Energy Storage Materials, 15*, 422–446.

Maranchi, J. P., Hepp, A. F., & Kumta, P. N. (2003). High capacity, reversible silicon thin-film anodes for lithium-ion batteries. *Electrochemical and Solid-State Letters, 6*(9), A198.

Marsh, H. (1991). A tribute to Philip L Walker. *Carbon, 29*(6), 703–704.

McDowell, M. T., Lee, S. W., Nix, W. D., & Cui, Y. (2013). 25th anniversary article: Understanding the lithiation of silicon and other alloying anodes for lithium-ion batteries. *Advanced Materials, 25*(36), 4966–4985.

Mochida, I., Zeng, S.-M., Korai, Y., Hino, T., & Toshima, H. (1991). The introduction of a skin-core structure in mesophase pitch fibers through a successive stabilization by oxidation and solvent extraction. *Carbon, 29*(1), 23–29.

Morris, A. J., Needs, R. J., Salager, E., Grey, C. P., & Pickard, C. J. (2013). Lithiation of silicon via lithium zintl- defectcomplexes from first principles. *Physical Review B: Condensed Matter and Materials Physics, 87*(17), 6–9.

Müller, D.R. Artificial graphite for lithium ion batteries. P35, London presentation 2011.

Nist data base https://webbook.nist.gov/cgi/cbook.cgi?ID = C74862&Mask = 1# Thermo-Gas, consulted on 06/11/2020.

Obrovac, M. N., & Chevrier, V. L. (2014). Alloy negative electrodes for Li-ion batteries. *Chemical Reviews, 114*(23), 11444–11502.

Obrovac, M. N., & Christensen, L. (2004). Structural changes in silicon anodes during lithium insertion/extraction. *Electrochemical and Solid-State Letters, 7*(5), A93.

Ogata, K., Salager, E., Kerr, C. J., Fraser, A. E., Ducati, C., Morris, A. J., ... Grey, C. P. (2014). Revealing lithium-silicide phase transformations in nanostructured silicon-based lithium ion batteries via in situ NMR spectroscopy. *Nature Communications, 5*, 3217.

Okamoto, H. (1990). *The Li-Si (lithium-silicon) system, 11*(3), 7.

Okamoto, H. (1998). Lithium-silicon. Section III: Supplemental literature review. *Journal of Phase Equilibria, 19*(5), 486–486.

Paireau, C., Jouanneau, S., Ammar, M.-R., Simon, P., Béguin, F., & Raymundo-Piñero, E. (2015). Si/C composites prepared by spray drying from cross-linked polyvinyl alcohol as Li-ion batteries anodes. *Electrochimica Acta, 174*, 361–368.

Peng Guan, J., Li, T., Lu, T., Guan, Z., Ma, Z., Peng, X., & Zhu, L. (2018). Zhang, facile and scalable approach to fabricate granadilla-like porous-structured silicon-based anode for lithium ion batteries. *ACS Applied Materials & Interfaces, 10*(40), 34283–34290.

Pereira-Nabais, C., Światowska, J., Chagnes, A., Ozanam, F., Gohier, A., Tran-Van, P., ... Marcus, P. (2013). Interphase chemistry of Si electrodes used as anodes in Li-ion batteries. *Applied Surface Science, 266*, 5–16.

Philippe, B., Dedryvère, R., Allouche, J., Lindgren, F., Gorgoi, M., Rensmo, H., ... Edström, K. (2012). Nanosilicon electrodes for lithium-ion batteries: interfacial mechanisms studied by hard and soft x-ray photoelectron spectroscopy. *Chemistry of Materials: a Publication of the American Chemical Society, 24*(6), 1107–1115.

Pilot, C., Avicenne presentation, 7th March Cleveland, Ohio, 2018.

Saint, J., PhD thesis, Matériaux d'électrode négative pour accumulateurs à ions lithium: Etude des systèmes binaires Li-Ga et Li-B et des composites silicium-carbone, Université de Picardie Jules Vernes, 2005.

Saint, J., Morcrette, M., Larcher, D., Laffont, L., Beattie, S., Pérès, J.-P., ... Tarascon, J.-M. (2007). Towards a fundamental understanding of the improved electrochemical performance of silicon–carbon composites. *Advanced Functional Materials, 17*(11), 1765–1774.

Savage, G. (2012). *Carbon-carbon composites*.

Schott, T., Robert, R., Pacheco Benito, S., Ulmann, P. A., Lanz, P., Zürcher, S., ... Trabesinger, S. (2017). Cycling behavior of silicon-containing graphite electrodes, part b: effect of the silicon source. *Journal of Physical Chemistry C, 121*(46), 25718–25728.

Shen, T., Xia, X., Xie, D., Yao, Z., Zhong, Y., Zhan, J., ... Tu, J. (2017). Encapsulating silicon nanoparticles into mesoporous carbon forming pomegranate-structured microspheres as a high-performance anode for lithium ion batteries. *Journal of Materials Chemistry A, 5*(22), 11197–11203.

Silvestrelli, P. L., Toigo, F., & Ancilotto, F. (2001). Acetylene on Si(100) from first principles: Adsorption geometries, equilibrium coverages and thermal decomposition. *Journal of Chemical Physics, 114*, 19.

Son, I., Hwan Park, J., Kwon, S., et al. (2015). Silicon carbide-free graphene growth on silicon for lithium-ion battery with high volumetric energy density. *Nature Communications, 6*, 7393.

Sourice, J., Quinsac, A., Leconte, Y., Sublemontier, O., Porcher, W., Haon, C., ... Reynaud, C. (2015). One-step synthesis of Si@C nanoparticles by laser pyrolysis: High-capacity anode material for lithium-ion batteries. *ACS Applied Materials & Interfaces, 7*(12), 6637–6644.

Sourice, J. PhD thesis, Synthèse de nanocomposites cœur-coquille silicium carbone par pyrolyse laser double étage: application à l'anode de batterie lithium-ion, Université Paris Sud, Paris XI, 2015.

Tarascon, J.-M., Morcrette, M., Saint, J., Aymard, L., & Janot, R. (2005). on the benefits of ball milling within the field of rechargeable Li-based batteries. *Comptes Rendus Chimie, 8*(1), 17–26.

Tenegal, F., Method for producing multilayer submicron particles by laser pyrolysis patent EP2872444 (B1), 2012.

Touidjine, A. PhD thesis, Optimisation de l'électrode négative à base de silicium pour les batteries lithium-ion, Université de Picardie Jules Vernes, 2016.

Varadachari., et al. (2009). *Modelling and Simulation in Materials Science and Engineering, 17*, 075006.

Verma, P., Maire, P., & Novák, P. (2010). A Review of the features and analyses of the solid electrolyte interphase in Li-ion batteries. *Electrochimica Acta, 55*(22), 6332–6341.

Wen, C. J., & Huggins, R. A. (1981). Chemical diffusion in intermediate phases in the lithium-silicon system. *Journal of Solid State Chemistry, 37*(3), 271–278.

Whittingham, M. S. (1976). Electrical energy storage and intercalation chemistry. *Science (New York, N.Y.), 192*(4244), 1126 6.

Wu, H., Chan, G., Choi, J. W., Ryu, I., Yao, Y., McDowell, M. T., ... Cui, Y. (2012). Stable cycling of double-walled silicon nanotube battery anodes through solid–electrolyte interphase control. *Nature Nanotechnology, 7*(5), 310–315.

Xiaohan W, PhD thesis, University of New south wales, 2019.

Xing, W. (1997). Pyrolysed Pitch-polysilane blends for use as anode materials in lithium ion. *batteries. Solid State Ionics, 93*(3–4), 239–244.

Xu, K. (2004). Nonaqueous liquid electrolytes for lithium-based rechargeable batteries. *Chemical Reviews, 104*(10), 4303–4418.

Yang, J., Wang, Y. X., Chou, S. L., Zhang, R., Xu, Y., Fan, J., ... Dou, S. X. (2015). Yolk-shell silicon mesoporous carbon anode with compact solid electrolute interphase film for superior lithium-ion batteries. *Nano Energy, 18*, 133–142.

Yong, Y., & Fan, L.-Z. (2013). Silicon/carbon nanocomposites used as anode materials for lithium-ion batteries. *Ionics, 19*(11), 1545–1549.

Yoshino, A.; Nakajima,T.; Patent JP9769585 Secondary battery, 1985.

9

Nanostructured silicon for energy applications

Tenzin Ingsel and Ram K. Gupta
Department of Chemistry, Kansas Polymer Research Center, Pittsburg State University, Pittsburg, KS, United States

9.1 Introduction

The global annual energy consumption is estimated to rise by 50% between 2004 to 2030 (Hochbaum & Yang, 2010). Therefore, the field of energy conversion research is full of opportunities for socially essential and practical applications. Hydrogen as a green energy can be produced via electrocatalytic or photocatalytic decomposition of water. Out of all the available materials for electrocatalysis and photocatalysis applications, silicon (Si) is an attractive choice because of its outstanding photoelectric behavior, optimizable electrical property, high crystal abundance, increased stability, low-cost, and benign nature (Zhang, Jie, Zhang, Ou, & Zhang, 2017). Silicon has also become an essential part of smart devices, smartphones, computers, transistors, and among many other electronics. In recent decades, interest in nanosized materials has gained momentum where thorough studies have been carried out to understand the properties of nanostructured silicon, such as their high surface area, quantum size effect, and robust light-trapping capabilities, etc. The merits of nanostructured silicon's desirable properties have revealed its roles in energy fields, such as lithium-ion batteries, solar cells, supercapacitors, and catalysis (Kabashin, Singh, Swihart, Zavestovskaya, & Prasad, 2019; Zhang et al., 2017).

When it comes to nanostructured silicon synthesis, the different synthetic routes can be grouped into the well-known synthesis methods of the two paradigms: top-down and bottom-up (Hu, Li, & Yu, 2010). Another way to classify fabrication types of silicon is based on physical, chemical, electrochemical, and physicochemical methods. Wet chemical synthesis routes used in obtaining silicon have been known to produce luminescent silicon nanocrystals; however, one of the most significant drawbacks seen in wet chemical synthesis of silicon is harmful toxic components etching and reducing agents and organosilicon precursors (Kabashin et al., 2019). Some of the wet chemical synthesis methods may not produce the desired reduced crystalline silicon material. On the other hand, there are dry fabrication methods like chemical vapor deposition, microwave plasma, and thermal annealing of silicon oxide-based polymers, among many others (Kabashin et al., 2019).

Silicon is an important semiconductor material in modern electronic devices and contemporary energy-related applications. Numerous breakthroughs in silicon technology have been linked to significant shifts in various application sectors requiring silicon. Silicon has drawn attention for its use in advanced energy conversion systems that are either solar-driven or voltage-driven. Numerous researches in low-cost renewable resources other than silicon are also explored to mitigate nonrenewable fossil fuel sources' depletion and climate change. Fig. 9.1 sheds light on the

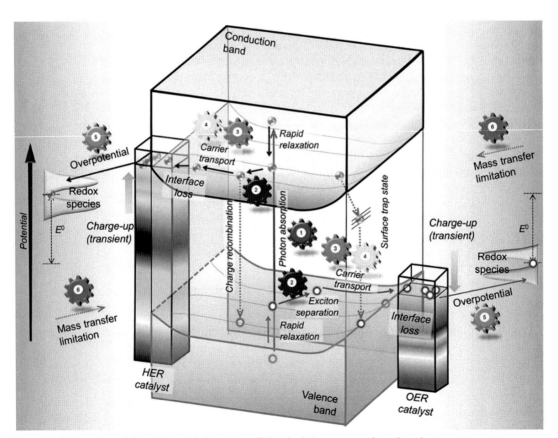

Figure 9.1 Schematics of the photocatalytic water splitting in the presence of semiconductors. Adapted from Takanabe, K. (2017). Photocatalytic water splitting: Quantitative approaches toward photocatalyst by design. ACS Catalysis, 7, 8006–8022. Copyright (2017) American Chemical Society.

fundamental mechanisms and semiconductors' role (e.g., silicon) in electrocatalysis and photocatalysis applications (Takanabe, 2017). The following subsections will discuss silicon in photoelectrochemical and photovoltaic devices, lithium-ion batteries, supercapacitors, and the hydrogen evolution reaction.

9.2 Silicon for energy applications

Due to its unique morphology, suitable bandgap, and electrochemical behavior, silicon-based materials are widely used for energy applications. In the following sections, energy applications of silicon are briefly covered.

9.2.1 Applications of silicon in photoelectrochemical and photovoltaic devices

The use of solar energy is one of the most promising options for tackling global energy shortages and environmental issues. Devices that can convert solar energy to chemical or electrical energy through photochemical or photovoltaic processes are a boon to humankind. Numerous interdisciplinary, cutting-edge researches in photoelectrochemical and photovoltaic devices are being carried out to improve their energy conversion efficiencies and cost-efficiency for commercial use. Silicon is an attractive choice for solar energy applications and has been included in all three generations of solar cell technologies. Silicon-based solar cells of different structures and designs have been exploited for decades (Zhang et al., 2017). For instance, planar silicon material can be structurally patterned for improved light trapping efficiency and antireflection characteristics, enabling the fabrication of ultrathin layered silicon solar cells (Seo et al., 2013). Such nanolevel texturing or patterning concept has helped investigators study nanocone, nanodome, and nanowire structured silicon for their light absorption capabilities (Jeong et al., 2012). Let us look at various silicon-based solar cells.

Silicon–organic heterojunction solar cells, a subset of silicon-based solar devices, can merge the benefits of high-performance crystalline silicon with cheap organic materials to deliver high conversion efficient silicon-based devices. In a study, as shown in Fig. 9.2, two naphthalene di-imide (NDI)-based conjugated polymers, N2200 and its fluorinated equivalent F-N2200, were utilized to coat/passivate the silicon surface (Han et al., 2017). The silicon used in this study was texturized to obtain an inverted pyramid-shaped surface known for being beneficial for

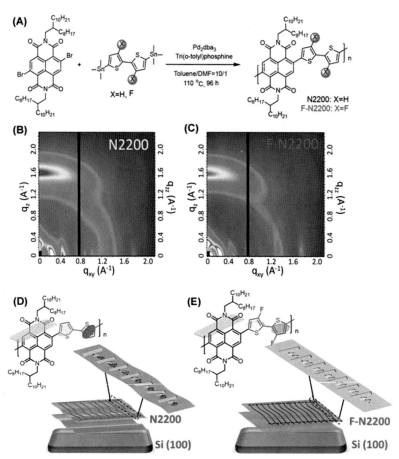

Figure 9.2 (A) Synthesis of N2200 and F-N2200 polymers. In situ 2D grazing incidence wide-angle X-ray scattering (GIWAXS) of (B) N2200 and (C) F-N2200 films on a silicon substrate. Schematic representation of how (D) N2200 and (E) F-N2200 were stacked on a silicon substrate. Adapted from Han, Y., et al. (2017). Naphthalene diimide-based n-type polymers: Efficient rear interlayers for high-performance silicon-organic heterojunction solar cells. ACS Nano, 11, 7215–7222. Copyright (2017) American Chemical Society.

surface recombination and light-trapping. Fig. 9.3A shows the schematic image of silicon/organic solar cell configuration. Fig. 9.3B shows the current density and voltage (J-V) relationship of the silicon/organic solar cells with different conjugated polymers under air mass (AM) 1.5 G at 100 mW/cm^2. Solar cells designed with F-N2200 exhibited the highest power conversion efficiency (PCE) of 14.5%, open-circuit voltage (V_{oc}) of 0.635 V, fill factor (FF) of 0.733, and short-circuit current density (J_{sc}) of 31.1 mA/cm^2. The enhanced F-N2200 silicon-based solar cell

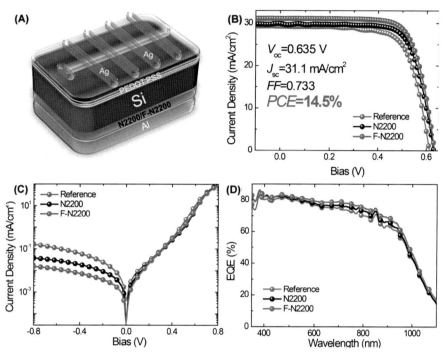

Figure 9.3 (A) Silicon/organic material solar cell design. (B) Current density versus voltage relations measured under standard simulated AM 1.5 G illumination at 100 mW/cm^2. Adapted from Han, Y., et al. (2017). Naphthalene diimide-based n-type polymers: Efficient rear interlayers for high-performance silicon-organic heterojunction solar cells. ACS Nano, 11, 7215–7222. Copyright (2017) American Chemical Society.

performance is attributed to F-N2200 improved morphology and crystallinity (Han et al., 2017).

Silicon with microwire or nanowire geometry can provide an edge over conventional wafer-based silicon by lowering the cost of silicon and the volume requirement. SiNWs can be synthesized using vapor–liquid–solid methods like chemical vapor deposition and physical vapor deposition techniques such as electron beam evaporation and molecular beam epitaxy (Sivakov et al., 2009). Silicon nanowires (SiNWs) possess outstanding optical properties; the array arrangement aids in further entrapping incident light, improving antireflective properties, and providing combined light scattering effects among the silicon nanowires in the array. Theoretical and experimental studies have been carried out to understand how changing different structural parameters in SiNWs affect their light-harvesting properties to potentially strategize to enhance SiNWs' light-trapping abilities (Zhang et al., 2017). In a simulation study, the effects of structural parameters such as

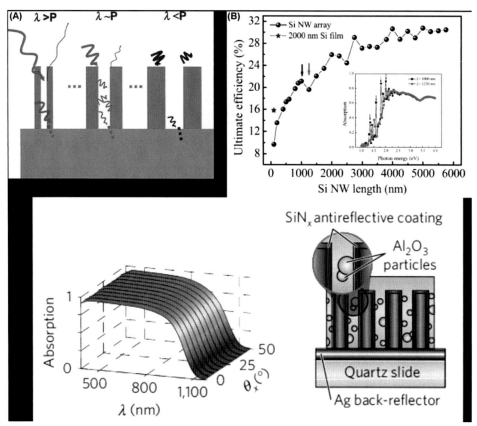

Figure 9.4 (A) The schematic describes three distinct processes that could happen between the SiNW arrays with distinctive structural parameters and the incident light. (B) SiNW wire length versus calculated UE (Li et al., 2012). (C) Schematic diagram and absorption measurement of SiNW array composed of SiN_x antireflective coating and Al_2O_3 light scatterers anchored on Ag back-reflector. Adapted with permission from Li J., Yu H., Li Y. (2012) Solar energy harnessing in hexagonally arranged Si nanowire arrays and effects of array symmetry on optical characteristics. Nanotechnology, 23:194010. Copyright (2012) IOP Publishing Ltd. Adapted from Kelzenberg, M.D., et al. (2010). Enhanced absorption and carrier collection in Si wire arrays for photovoltaic applications. Nature Materials, 9, 239–244. Copyright (2010) Springer Nature.

periodicities, diameters, and lengths on the material's optoelectronic characteristics were studied. Fig. 9.4A shows the schematics representing how the incident light from the point of view of wave optics interact with the SiNW arrays. The representation shows that when the wavelength of the light is more significant than the periodicity of the SiNW arrays, these incident lights won't interact as much with the arrays; this phenomenon is not desirable in light-harvesting applications. When the wavelength of the incident light is comparable to that of the array's periodicity, improved light scattering with enhanced light absorption was observed. However, in the case where the wavelength of the incident light was so much higher than the periodicity, most light was reflected. Fig. 9.4B showcases SiNW lengths as an independent variable and the calculated ultimate efficiency (UE) as the dependent value. The general trend of the UE versus SiNWs length suggests that the UE is increasing with the wire's length until saturation is reached. Note that the SiNW with a length 1000 nm has a UE yield higher than that of a 2000-nm thick silicon nanofilm (Li, Yu, & Li, 2012; Zhang et al., 2017). Other than optimizing structural parameters in SiNWs, using antireflective coatings like silicon nitride-based (SiN_x) on SiNW strategy is widely known to enhance SiNW's light-absorbing abilities. As shown in Fig. 9.4C, on top of the SiN_x coating, Al_2O_3 nanoparticles were filled between SiNWs with an Ag back-reflector incorporated to scatter light (Kelzenberg et al., 2010; Zhang et al., 2017).

Here are a few more unique SiNW arrays in their applications in solar cells. A SiNW array coated with carbon quantum dots (CQD) showed potential in optoelectronic devices' applications. The CQD/SiNW array photovoltaic device fabricated was of a low-cost and unique heterostructure. When these SiNW devices were compared based on photovoltaic device characteristics like the fill factor (FF), open voltage (V_{oc}), and short-circuit current density (J_{sc}), improved photovoltaic performances were observed with few layers of CQD. In particular, five-layered CQD coated on SiNW array device showcased ∼7% increase in power conversion efficiency (PCE) from SiNW array without CQD (Xie et al., 2014). Here are brief mentions of metal oxide-based silicon photocathodes for water-splitting applications. Three-dimensional structured zinc oxide/silicon branched nanowire photoelectrodes were studied for their applications in light-driven water splitting. Zinc oxide/silicon branched NW material's photoelectrochemical stability was further improved when a thin layer of TiO_2 was coated on the photocathode by the well-known atomic-layer deposition method (Kargar et al., 2013).

WO_3 thin films/SiNW core−shell morphology photoanodes were fabricated. The cores were made up of Si microwires and the shells of a WO_3 layer. What was observed was that by improving the porosity in the WO_3 thin film, the photocurrent density of the material nearly doubled from the initial photocurrent density value (Coridan, Arpin, Brunschwig, Braun, & Lewis, 2014) Fig. 9.5 shows an emerging device, transparent solar cells (TSCs), that takes advantage of the transparent property and light for electricity conversion. The fabrication process of TSC and its morphologies are described in Fig. 9.5. (Kang et al., 2019). Now that silicon-based solar cells have been briefly introduced, silicon-based materials for battery, supercapacitor,

176 Chapter 9 Nanostructured silicon for energy applications

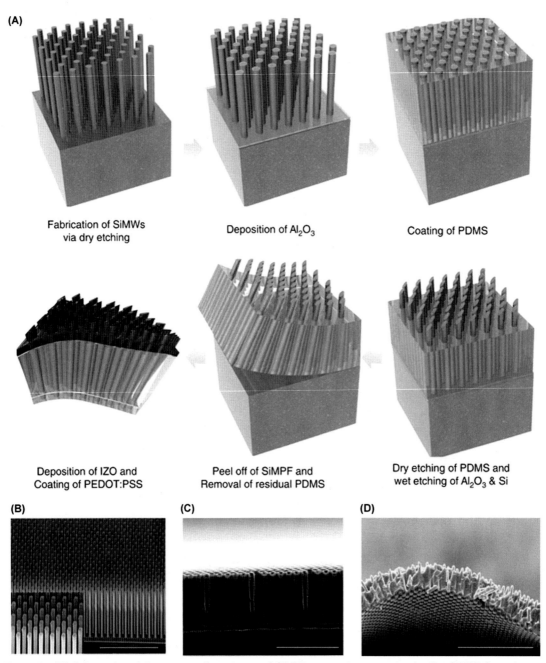

Figure 9.5 (A) Schematics of the construction process of SiMW array polymer composite film SiMPF. Scanning electron microscopy images of (B) SiMW array with scale bar as 50 μm. (C) SiMW array with Si microwire array filled with poly(dimethyl siloxane), which is surface treated at a scale bar as 50 μm. (D) SiMPF peeled off from silicon parent at scale bar 100 μm. Adapted from Kang, S.B., et al. (2019). Stretchable and colorless freestanding microwire arrays for transparent solar cells with flexibility. Light: Science & Applications, 8. Copyright (2019) Springer Nature. The article was printed under a CC-BY license.

and hydrogen evolution reaction applications will be discussed in the next subsections.

9.2.2 Silicon for lithium-ion batteries

An increase in global energy consumption, high dependence on portable devices, and environmental concerns have driven cutting-edge research in energy storage systems. Rechargeable lithium-ion batteries (LIBs), an important subset of energy storage systems, have been extensively used in portable electronic devices. The LIBs are highly anticipated promising materials that could be used to replace petroleum fuel in automobiles and in large-scale energy storage systems that can store renewable energy. Presently, the commercial LIBs show limited storage capacity where the standard graphite anode is known to have a maximum theoretical gravimetric capacity of 372 mAh/g (Franco Gonzalez, Yang, & Liu, 2017). If lithium metal was to be used as the anode, their theoretical gravimetric capacity is 3800 mAh/g (Franco Gonzalez et al., 2017). Therefore, new safe anode materials are explored to replace conventional graphite anode. Here are a few basics about lithium-ion batteries. To keep in mind, oxidation happens on the anode and reduction on the cathode. For simplicity, the electrodes are named cathode or anode based on their function during the discharge cycle. When LIBs are charged, the graphite anode goes through reduction, where lithium ions intercalate between the layers of graphite, as shown in Eq. (9.1):

$$Li + 6C\ LiC_6 \qquad (9.1)$$

This reaction happens in 1:6 ratios with one lithium to six carbon. On the other hand, when LIBs are discharged, the oxidation of graphite and deintercalation of lithium liberates stored energy.

As discussed earlier, the replacement of potentially better-performing anode material requires exploring and actualizing materials with higher theoretical capacity. Silicon is an attractive anode material that can alloy with lithium electrochemically to form $Li_{22}Si_5$ with a theoretical gravimetric capacity higher than 4000 mAh/g (Franco Gonzalez et al., 2017). However, at room temperature, it is observed that the lithium–silicon alloying results in a low level of lithiation, forming $Li_{15}Si_4$ with a theoretical gravimetric capacity less than 4000 mAh/g. Despite all the favorable silicon properties, a determining factor that impedes the commercialization of silicon for anode material is its inferior cycling performance. Fig. 9.6 shows

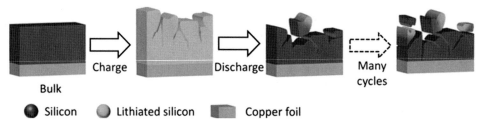

Figure 9.6 Schematic image of volumetric changes observed in bulk silicon on electrochemical cycling, leading to disintegration of silicon material from the current collector. Adapted from Franco Gonzalez, A., et al., (2017). Silicon anode design for lithium-ion batteries: Progress and perspectives. Journal of Physical Chemistry C, 121, 27775–27787. Copyright (2017) American Chemical Society.

a mechanical strain that the bulk silicon experiences from volumetric expansion during electrochemical cycling, which leads to the disintegration of silicon and eventually it gets disconnected from the current collector (Franco Gonzalez et al., 2017). Volume changes of electrode material also affect anode/electrolyte interphase. During the lithiation process, solid electrolyte interphase (SEI) is formed on the surface of silicon due to reduced solvents and salts of the electrolyte. The SEI film may help make the anode chemically stable in the electrolyte through passivation and hence enhance the battery performance. Conversely, the SEI layer can break or disintegrate when volume change happens in the silicon; eventually, the SEI layer formation and its disintegration will lead to the consummation of lithium with excessive growth of ionically insulating thick SEI, and in turn degrading battery capacity. Therefore there are various strategies utilized in improving silicon anode's performance in Li-ion batteries by tackling three main objectives: improving silicon's conductivity, enhancing the mechanical properties of silicon electrodes, and improving the chemical stability of the electrolyte-electrode interphase (Franco Gonzalez et al., 2017).

The following are a few literature examples of different methods undertaken for silicon anode's lithium-ion battery performance. Nanostructuring or nanosized morphological fabrication of silicon has allowed LIB technology to face common problems faced by Si-based electrodes, such as their poor ionic and electronic conductivities, combined with their volume expansion issues. Therefore investigators have delved into synthesizing silicon of morphologies like nanopowder, nanotubes, nanorods, nanofibers, and nanocomposites with graphite (Profatilova, Stock, Schmitz, Passerini, & Winter, 2013). The first in situ transmission electron microscopy (TEM) analysis on silicon cells helped find 150 nm as the critical diameter size for silicon nanostructure before it fractures, as shown

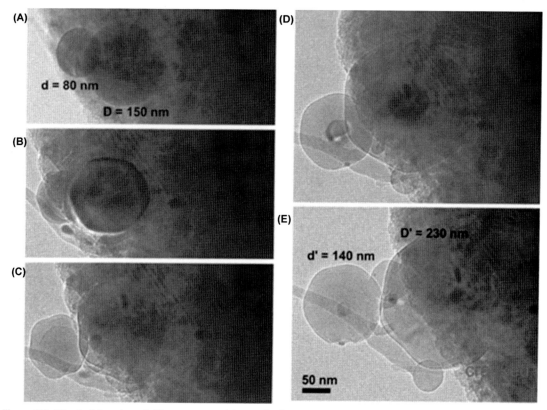

Figure 9.7 Effect of the size of silicon nanoparticles on its fracture during the lithiation process. (A) Pure silicon nanoparticle before lithiation with smaller particle (d = 80 nm) and the bigger (D = 150 nm). (b—d) Silicon nanoparticles during lithiation. (E) The smaller silicon nanoparticle still intact, while the bigger one has cracked upon full lithiation. Adapted from Franco Gonzalez, A., et al., (2017). Silicon anode design for lithium-ion batteries: Progress and perspectives. Journal of Physical Chemistry C, 121, 27775—27787. Copyright (2017) American Chemical Society.

in Fig. 9.7. The smaller silicon nanoparticle (d = 80 nm to d' = 140 nm) did not fracture (Fig. 9.7E), while silicon nanoparticle (D = 150 nm to D' = 230 nm) did, upon the completion of lithiation process (Liu & Huang, 2011).

Even though silicon's nanosize morphology yields beneficial merits like a large surface for improved conductivities, this could also cause excessive SEI growth and impeded battery capacities after multiple cycles (Franco Gonzalez et al., 2017). A break-through study challenged the silicon nanowire diameter upper limit, which was known to be 300 nm. Silicon wire anodes with a large diameter of 1 μm showcased a stable capacity of 3150 mAh/g for 100 cycles with a current density of 2.1 A/g (Quiroga-González, Carstensen, & Föll, 2013). Fig. 9.8 delineates the differences in the abilities

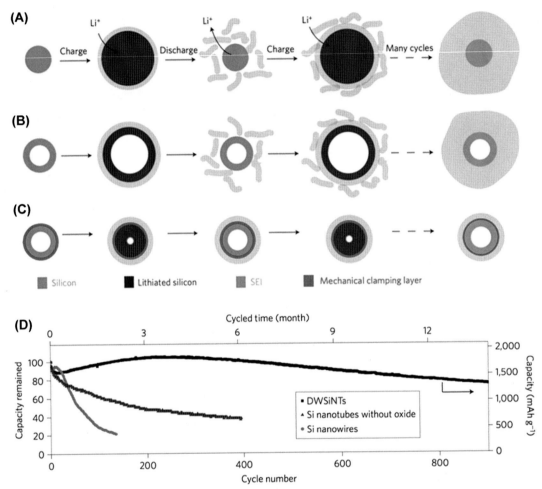

Figure 9.8 Schematic images of SEI formation on different silicon surfaces. (A) SEI formation on the silicon nanowires upon lithiation. (B) SEI formation on silicon nanotube during lithiation. (C) Double-walled silicon nanotubes surface handling healthy growth of SEI during lithiation process. (D) Galvanostatic cycling of silicon nanotubes, silicon nanowires, and double-walled silicon nanotubes. Adapted from Wu, H., et al., (2012). Stable cycling of double-walled silicon nanotube battery anodes through solid-electrolyte interphase control. Nature Nanotechnology, 7, 310–315. Copyright (2012) Springer Nature.

maintained by silicon nanotubes, nanowires, and double-walled silicon (Si-SiO$_x$) nanotubes (DWSiNTs). Fig. 9.8 illustrates SEI formation during the lithiation process on silicon nanowire, silicon nanotube, DWSiNTs. Unlike the silicon nanotube and nanowire anodes, the DWSiNT anode is not prone to excessive, unnecessary SEI formation, and such a phenomenon is attributed to the SiO$_2$ protective outer layer (Wu et al., 2012).

Porous silicon structures are created to lessen mechanical strain in silicon's form by feeding space to silicon for volume expansion. Microporous silicon synthesis was innovated wherein the silica is reduced via magnesiothermic reaction, as shown in Eq. (9.2), and magnesia is removed with the help of 1 M HCl (Bao et al., 2007).

$$2Mg(g) + SiO_2(s) 2MgO(s) + Si(s) \qquad (9.2)$$

Porous silicon was obtained via the abovementioned magnesiothermic reduction of silica. Here, rice husk is utilized as silica, and Fig. 9.9 sheds light on a potential cost-effective green production route for nanosilicon. The nanosilicon recovered from rice husk exhibited a capacity of 2650 mAh/g after 200 cycles at a current rate of 2100 mA/g (Liu, Huo, McDowell, Zhao, & Cui, 2013).

Generally, intermetallic silicon composites including magnesium, aluminum, copper, iron, silver, calcium, and many others

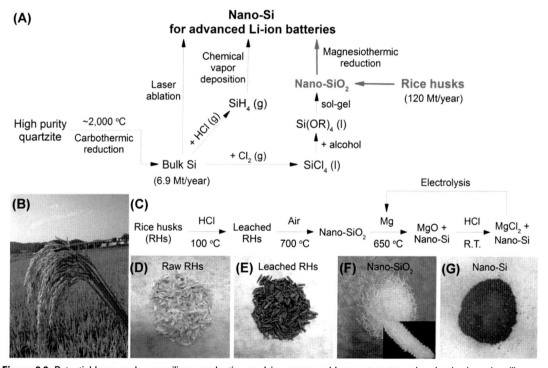

Figure 9.9 Potential large-scale nanosilicon production applying green and low-cost routes using rice husk as the silica source. (A) Different nanostructured silicon synthesis routes, the red highlighted route is through the utilization of rice husks. (B) Ripe rice panicle. (C) Flow chart showcasing silicon nanoparticles recovered from rice husks. (D–G) Digital pictures of intermediate materials for silicon nanoparticles. Adapted from Liu, N., et al., 2013. Rice husks as a sustainable source of nanostructured silicon for high performance Li-ion battery anodes. Scientific Reports, 3, 1–7. Copyright (2013) Springer Nature.

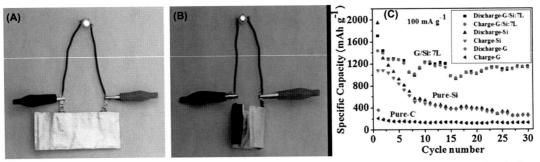

Figure 9.10 (A,B) Flexible battery made up of silicon multilayer and graphene with liquid electrolyte and cathode material LiCoO$_2$, powering a LED. (C) Cycling performance of pure silicon, pure carbon, and silicon/graphene at a current density of 100 mA/g and a voltage window of 1.2–0.02 V. Adapted from Mori, T., et al., (2015). High specific capacity retention of graphene/silicon nanosized sandwich structure fabricated by continuous electron beam evaporation as anode for lithium-ion batteries. Electrochimica Acta, 165, 166–172. Copyright (2015) Elsevier.

can enhance the electrochemical performance of LIBs by potentially improving the low innate conductivity of silicon (Wang, Xu, Huang, Li, & Ma, 2016). A tin-doped silicon nanowire retained a capacity above 1000 mAh/g for 100 cycles at a current density of 2.8 A/g (Bogart, Lu, Gu, Wang, & Korgel, 2014). Composites of silicon and carbon are attractive choices because of their enhanced conductivity and minor volumetric change on lithiation. Silicon/graphene was fabricated via an electron beam evaporation method (Mori et al., 2015). As shown in Fig. 9.10A,B, silicon/graphene electrode (anode), LiCoO$_2$ (cathode), liquid electrolyte, and separator are combined to make a battery able to power a light-emitting diode. Fig. 9.10C exhibits the cycling performances of 750-nm-thick pristine silicon and carbon and the silicon/graphene composite. The silicon/graphene composite proves its superiority when cycling performances (Mori et al., 2015).

Unusual novel designs have helped achieve excellent electrochemical performance in silicon-based anodes. For instance, a pomegranate-inspired design of silicon nanoparticle coated with carbon maintained a decent capacity of 1160 mA/g at a current density of 0.635 A/g for 1000 cycles. This particular design provides space between the carbon coating and the silicon nanoparticle for the electrode's volume change during lithiation (Liu et al., 2014). Binders are generally used to hold the conductive species and active material together on top of a current collector. Some binders can subdue volume expansion as they can enhance mechanical properties (Wu et al., 2019). In summary, the methods used in improving silicon anode material, namely, porous structures, nanostructures, composites, and coatings, have both positive and negative effects on their electrochemical battery performances.

9.2.3 Supercapacitors based on nanostructured silicon

Silicon, a most basic material used in electronics, is chemically versatile and is manufactured cheaply. Silicon is widely researched for applications in energy storage devices like supercapacitors because of its high surface area, decent conductivity, and 1D electron transport nature. Supercapacitors being an essential group of energy storage devices, are divided into three different types based on the charge storage mechanism: (a) electrochemical double-layer capacitors (EDLCs) enable the charge separation at the electrode/electrolyte interface for charge storage; (b) pseudocapacitors store charge by using reversible and fast surface reduction and oxidation reactions; and (c) hybrid capacitors as the name suggests, consist of both pseudocapacitor and EDLC in the same device (Hu, Pei, & Ye, 2015).

With the rapid expansion and growth in nanotechnology, numerous researches have been stimulated in fabricating miniaturized electronic devices. These miniaturized devices have advantageous properties such as low volume and low weight, playing an essential role in electronics applications in space, medical devices, microrobots, and sensors. Micropower sources for these portable electronic devices are heavily researched as they require high power density and long lifetime. Microsupercapacitors possess unique properties such as long life span, fast charge–discharge, and high power density. Microsupercapacitors can play a role in either complementing or replacing commercially available energy storage devices such as microbatteries or thin-film batteries as these microbatteries face limited lifetime and low power densities (Hu et al., 2015). Now that the importance of supercapacitors and microsupercapacitors is discussed, the following are hybrid silicon-based supercapacitor/microsupercapacitor examples of silicon and polymer, silicon and carbon, and silicon and metal oxides.

It is widely known that the surface of silicon exhibits rapid oxidation in aqueous electrolytes and dissolution in saline solutions, limiting the application of silicon nanowires in microsupercapacitors (μ-SCs). Therefore, coating SiNWs with thin films of highly capacitive materials is strategically crucial to improve the stability of the SiNWs in different chemicals. In the literature, the deposition of silicon carbide (SiC), ultrathin porous carbon, graphene nanosheets at high temperature, metal oxides, and conductive polymers on silicon have been discussed (Devarapalli, Szunerits, Coffinier, Shelke, & Boukherroub, 2016). In a study, SiNWs were coated with thin glucose film, and with a facile hydrothermal

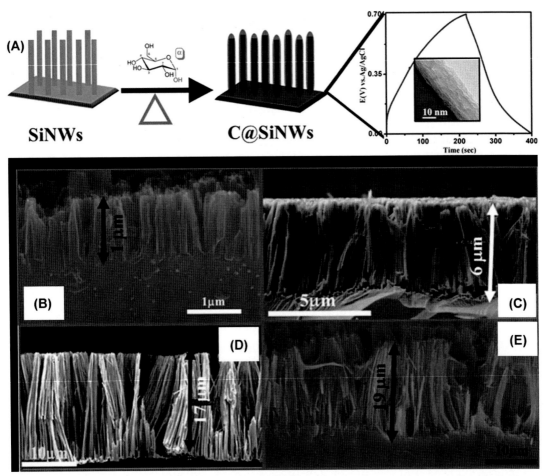

Figure 9.11 (A) Schematic of a hydrothermal synthesis process using glucose as a precursor of carbon for the carbon coating of vertically aligned silicon nanowire arrays. (B–E) Cross-sectional SEM images of C@SiNWs formed by varying the concentration of the etching solution. Adapted from Devarapalli, R.R., et al. (2016). Glucose-derived porous carbon-coated silicon nanowires as efficient electrodes for aqueous micro-supercapacitors. ACS Applied Materials & Interfaces, 8, 4298–4302. Copyright (2016) American Chemical Society.

process, the glucose underwent pyrolysis that generated activated carbon uniformly deposited on SiNWs (Fig. 9.11A) (Devarapalli et al., 2016). The cross section SEM images of the nanowires synthesized using a varying concentration of etching solution are shown in Fig. 9.11(B–E). The carbon-based SiNWs exhibited a high specific capacity of 25.6 mF/cm^2 at a current density of 0.1 mA/cm^2. Another carbonaceous/polymer-based SiNWs device showcased an areal density of 8.5 mF/cm^2, power density of 1.3 mW/cm^2, and an energy density of 26 mJ/cm^2 with excellent cycling stability (Aradilla et al., 2016). This microsupercapacitor is a diamond-coated SiNW with poly(3,4-(ethylenedioxy)thiophene) (PEDOT) electrochemically deposited on top.

Additionally, a commercial PEDOT: PSS solution was used to drop-cast on aluminum-coated SiNWs (Valero, Mery, Gaboriau, Gentile, & Sadki, 2019). The resulting composite exhibited a specific capacitance of 3.4 mF/cm^2 at a current density of 2 A/g and an energy density of 8.2 mJ/cm^2 and a power density of 4.1 mW/cm^2. Direct gas-phase graphene passivation was carried out in a nanostructured silicon study to improve its electrochemical stability in corrosive liquid media (Chatterjee et al., 2014). As mentioned earlier, the examples demonstrate the role of carbon-based materials in enhancing the electrochemical surface area and improving the chemical stability in silicon-based materials.

Microsupercapacitors are presently known to possess low power capabilities and specific capacitance as their solid electrolytes show low ionic conductivities. An optimized interelectrode distance and electrode thickness is required in microsupercapacitors for enhanced energy density properties and prevention of potential unwanted short circuit incidents. In a study, hydrothermal growth of carbon and MnO_2 on silicon interdigits was carried out to obtain a three-dimensional silicon/carbon/MnO_2 (Si/C/MnO_2) hybrid electrode for enhanced electrochemical properties (Wang et al., 2020). Fig. 9.12 depicts the fabrication process of solid-state three-dimensional silicon/carbon/MnO_2 hybrid material for microsupercapacitor applications. Fig. 9.13 shows the electrochemical behavior of silicon/carbon (Si/C) and Si/C/MnO_2 solid-state supercapacitors. As seen in Fig. 9.13(B,C), the cyclic voltammograms of Si/C showed close to rectangular behavior while Si/C/MnO_2 showed little deviation from this behavior which could be due to the Faradic reaction at the highly porous electrode surfaces. The energy storage capacity of Si/C/MnO_2 solid-state supercapacitor was observed to be much higher than that of Si/C (Fig. 9.13D) and charge-discharge characteristics also showed improvement in discharge time for Si/C/MnO_2 solid-state supercapacitor (Fig. 9.13(E,F)). Such a hybrid solid-state device showcased a power density of 117.82 $\mu W/cm^2$, a specific capacitance of 29.45 mF/cm^2, and an energy density of 2.62 $\mu Wh/cm^2$ at 10 mV/s. The hybrid structure of Si/C/MnO_2 was able to take advantage of EDLC and pseudocapacitive charge storage mechanisms. Such studies pave the road for favorable applications of three-dimensional energy storage devices.

In Fig. 9.14, the coating of In_2O_3 film is carried out on mesoporous silicon (Zhu et al., 2019). The In_2O_3 film acts as an active electrode material that enhances the electrochemical

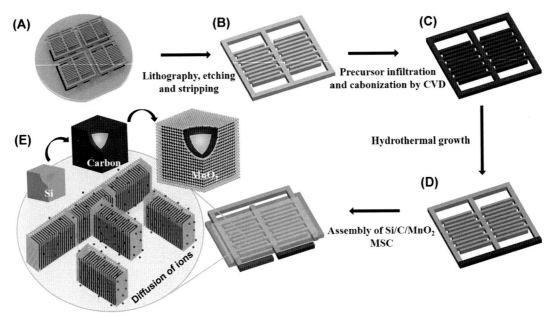

Figure 9.12 (A) Schematic of silicon framework synthesis. (B) A stand-alone three-dimensional silicon framework. (C) Functionalization of silicon interdigits by carbon film coating. (D) MnO_2 nanosheets deposited on the carbon-coated silicon electrodes. (E) Workings of ion diffusion principles in (Si/C/MnO_2) hybrid electrode. Adapted from Wang, Y., et al. (2020). Silicon-based 3D all-solid-state micro-supercapacitor with superior performance. ACS Applied Materials & Interfaces, 12, 43864–43875. Copyright (2020) American Chemical Society.

performance of the resulting microsupercapacitor. In general, metal oxides such as NiO, RuO_2, FeO_3, MoO_3, MnO_2, and SnO_2 have been studied for their pseudocapacitive contributions as electrode material. In_2O_3 is a cost-effective suitable choice for its atomic layer deposition on mesoporous silicon due to its favorable properties such as high electrical conductivity and wide bandgap (2.9–3.75 eV), and high optical transmittance. The on-chip microsupercapacitors displayed a specific capacitance of 1.36 mF/cm^2 at a scan rate of 10 mV/s (Zhu et al., 2019). In summary, electrochemically active carbon, pseudocapacitive inorganic components, and hybrids are described and explained above in their role in silicon-based electrode materials for applications in supercapacitors and microsupercapacitors.

9.2.4 Silicon as electrocatalyst for hydrogen evolution reaction

In general, the hydrogen evolution reaction (HER) process can be part of a solar-driven and electrically driven water-splitting

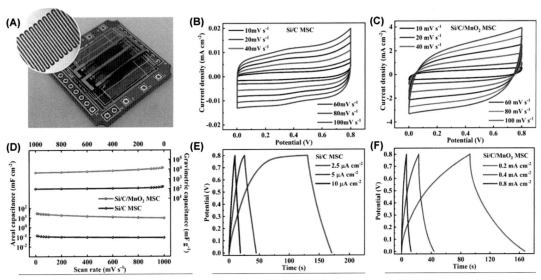

Figure 9.13 (A) Optical images of the all-solid-state Si/C/MnO$_2$ MSC integrated on a PCB. CV curves for the all-solid-state MSC based on (B) Si/C and (C) Si/C/MnO$_2$ electrodes. (D) Comparison of the specific capacitances as a function of the scan rates. GCD curves for (E) Si/C-based MSC and (f) Si/C/MnO$_2$-based MSC. Adapted from Wang, Y., et al. (2020). Silicon-based 3D all-solid-state micro-supercapacitor with superior performance. ACS Applied Materials & Interfaces, 12, 43864–43875. Copyright (2020) American Chemical Society.

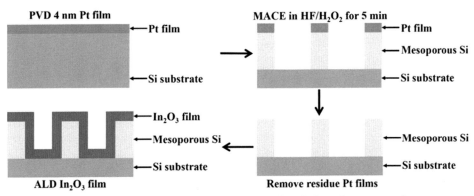

Figure 9.14 Fabrication schematic of the single electrode of the PS/In$_2$O$_3$-based supercapacitors. Adapted from Zhu, B., et al. (2019). High-performance on-chip supercapacitors based on mesoporous silicon coated with ultrathin atomic layer-deposited in 2O 3 films. ACS Applied Materials & Interfaces, 11, 747–752. Copyright (2019) American Chemical Society.

system. One approach uses solar energy in semiconductors' presence to convert water to hydrogen, while the other utilizes electrical power. Silicon is an attractive material for water electrocatalysis applications, specifically as a HER electrocatalyst, and for photocatalysis applications, specifically as photoanodes for hydrogen generation. When discussing Si's photocatalysis applications, multiple pieces of literature mention a few different geometries of Si photoanode materials, such as planar, nanowires/microwires, micro pyramids, nanoporous, nanobelts, and nanoholes (Thalluri et al., 2019; Warren, Atwater, & Lewis, 2014).

Silicon microwire array, a subset of Si-based materials for energy, is known for improved surface area, low-cost growth techniques, flexibility, useful radial carrier collection, and antireflection effect. However, to take full advantage of Si microwires' desirable qualities for their use as photocathodes, other catalysts are required to improve Si microwires' surface catalytic activity toward HER. Without such cocatalysts, the surface of Si and the electrolyte results in ineffective charge transfer kinetics. The cocatalyst can also act as a protector that can prevent the formation of an oxide layer on the Si and hence prevent the reduction of photocatalytic activities (Warren et al., 2014). In general, noble metals like Pt, Ag, and Au are used as cocatalysts; however, the push toward fabricating cost-effective photocathodes calls for the employment of cheap nonnoble cocatalysts. In a study, the impact of surface area improvement on the catalytic behavior was analyzed. Ni, Ni-Mo, and Pt catalytic materials were electrodeposited on planar and microwire geometries of Si-based substrate. It was observed that catalysts deposited onto the Si-based microwires substrate exhibited improved HER performance. Additionally, Ni-Mo loaded Si microwires showed potential as an alternative to noble metal-based semiconductor-assisted photocathodes (McKone et al., 2011).

Other interesting Si geometries like Si-inverted pyramid and Si pyramids with cocatalysts like doped transition metal chalcogenides and metal phosphides are discussed in the literature to strategically improve the capabilities of the Si photoabsorber in receiving electrons that are light-induced and enhance the adsorption/desorption of hydrogen on the catalyst surface for enhanced photoelectrochemical reactions. Figs. 9.15 and 9.16 shows the schematic images of how Si-based photocathodes of different geometries participate in photoelectrochemical response and their corresponding microstructural images. The Co_2P coated p-Si photocathode, under AM 1.5 G illumination, was able to deliver a high density of photocurrent of 35.2 mA/cm^2 at 0 V versus. the reversible hydrogen electrode (RHE). The photocathode was able to catalyze the HER above 30 mA/cm^2 for more than 150 hours with minimal degradation. The Co_2P cocatalyst not only helped prevent the photocathode from corrosion but also enhanced the photoelectrochemical HER process. A cobalt-based transition metal chalcogenide, CoS_x doped with phosphorus, was used as a cocatalyst in a TiO_2 thin film-modified Si pyramids textured-

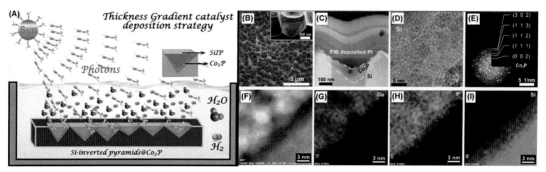

Figure 9.15 (A) Schematic image of Co$_2$P coated inverted pyramid textured Si in photon-induced hydrogen generation. Microstructural, morphological, and compositional properties of inverted pyramid textured p-Si with Co$_2$P photocathode. (B) SEM image. (C) Cross-sectional TEM image. (D) Crystalline structure of Co$_2$P view by high-resolution TEM. (E) FFT-ED pattern of Co$_2$P. (F) Annular dark-field imaging of Si-Co$_2$P interface. (G–I) Elemental mappings of Co, P, and Si. Adapted from Thalluri, S.M., et al. (2019). Inverted pyramid textured p-silicon covered with Co2P as an efficient and stable solar hydrogen evolution photocathode. ACS Energy Letters, 4, 1755–1762. Copyright (2019) American Chemical Society.

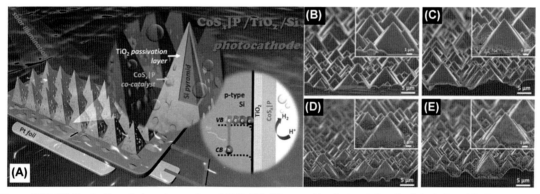

Figure 9.16 (A) Illustration of phosphorus-doped CoS$_x$/TiO$_2$/Si pyramids photocatalyst for light-induced hydrogen evolution. SEM images of (B) pristine Si. (C) CoS$_x$ coated on Si. (D) Si pyramid decorated by P-doped CoS$_x$ (S/P = 1). (E) Si pyramid decorated by P-doped CoS$_x$ (S/P = 7). Adapted from Chen, C.J., et al. (2018). Amorphous phosphorus-doped cobalt sulfide modified on silicon pyramids for efficient solar water reduction. ACS Applied Materials & Interfaces, 10, 37142–37149. Copyright (2018) American Chemical Society.

photocathode. An appropriate amount of phosphorous dopant in CoS$_x$ brought the photocathode nearer to the thermoneutrality of hydrogen adsorption. The photocathode with the most appropriate P-doping quantity showcased -20.6 mA/cm^2 at 0 V (Chen et al., 2018; Thalluri et al., 2019). Another metal chalcogenide-based cocatalyst, molybdenum sulfide (MoS$_x$) quantum dots, was introduced onto a silicon black (bSi) surface. Such manipulation resulted in these two constituents working in synergism towards enhancing the activity for hydrogen evolution reaction. The

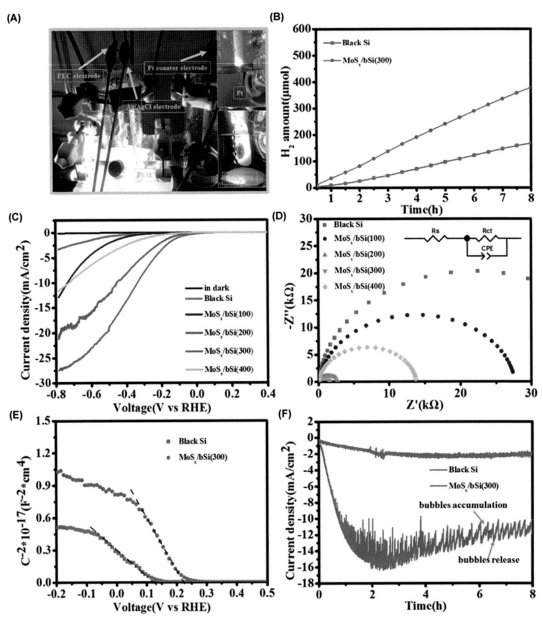

Figure 9.17 (A) Photoelectrochemical (PEC) device. (B) Hydrogen generation versus time catalyzed by MoS$_x$/bSi (300) and pristine bSi at 0 V versus RHE under 100 mW/cm^2 illumination. (C) Linear sweep voltammetry (LSV) curves of photocathodes. (D) Nyquist plots of the photocathodes. (E) Bare black Si and MoS$_x$/bSi (300) photocathodes at 1 kHz measured in the dark. (F) The photocathodes' stability was measured under 100 mW/cm^2 illumination at 0 V versus RHE. Adapted from Wang, B., et al. (2019). MoS x quantum dot-modified black silicon for highly efficient photoelectrochemical hydrogen evolution. ACS Sustainable Chemistry & Engineering, 7, 17598–17605. Copyright (2019) American Chemical Society.

resulting onset potential of HER was a low 0.255 V versus RHE. with an obtained photocurrent density of 12.2 mA/cm². As shown in Fig. 9.17, photoelectrochemical characterization of bare black Si and MoS$_x$/bSi photocathodes showcases that the best-synthesized photocathode material was MoS$_x$/bSi (300)

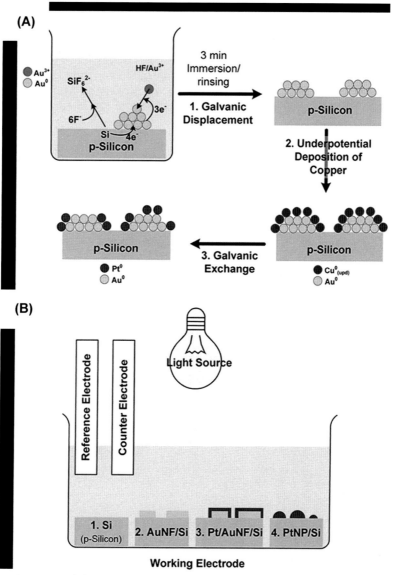

Figure 9.18 Schematic diagram: (A) P-silicon Pt-Au nanofilm. (B) Photochemical H$_2$ measurement on p-Si with different catalysts. Adapted from Kye, J., et al. (2013). Platinum monolayer electrocatalyst on gold nanostructures on silicon for photoelectrochemical hydrogen evolution. ACS Nano, 7, 6017–6023. Copyright (2013) American Chemical Society.

photocathodes with 300 μL MoS$_x$ quantum dot solution concentration (Wang et al., 2019).

In literature, various thin film-based metal catalysts anchored on planar p-type silicon are also used for light-driven hydrogen generation, as shown in Fig. 9.18 (Kye et al., 2013). On top of that, as discussed earlier, the hydrogen can be generated electrochemically or through photochemical reactions. Here are some examples of Si-based catalysts that participate in electrocatalytic reactions. Oxides of different elements such as MoO$_2$, CeO$_2$, SiO$_2$, and TiO$_2$ have been shown to enhance various catalytic materials' catalytic properties. When a silicon monoxide (SiO) was utilized as a reducing agent in a Pt-modified graphene nanocomposite synthesis, the resulting electrode delivered current density higher

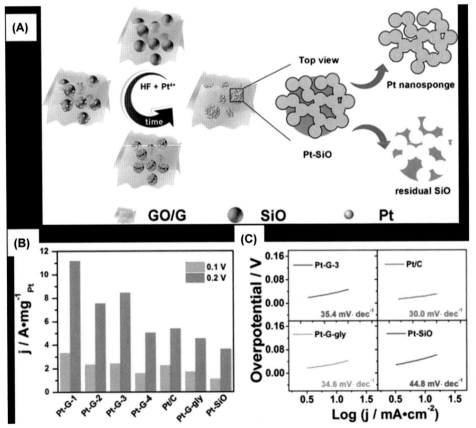

Figure 9.19 (A) Schematic illustration of the synthesis of nanosponge Pt-SiO. (B) Mass activities of the catalysts. (C) Tafel plots of catalysts. Adapted from Liao, F., et al. (2018). Nanosponge Pt modified graphene nanocomposites using silicon monoxides as a reducing agent: high efficient electrocatalysts for hydrogen evolution. ACS Sustainable Chemistry & Engineering, 6, 15238−15244. Copyright (2018) American Chemical Society.

than that of the commercial Pt-based electrocatalyst at high overpotentials. The reason being that SiO escalates the hydrogen desorption rate at high overpotentials. SiO not only enhances the catalytic activity of Pt-graphene (Pt-G) nanocomposite, it could help augment the stability of the catalyst by potentially preventing the restacking of graphene and agglomeration of Pt. Fig. 9.19A shows the schematics of the usage of SiO for the fabrication of Pt-G. Fig. 9.19B,c illustrates the superior HER electrocatalytic activity of Pt-G-3, facilitated by the synergistic effect of Pt, residual SiO, and graphene (Liao et al., 2018).

In another study, Ir/Si binary nanowire catalyst was designed to test density functional theory calculations, which theorized that two different strategies are in play synergistically. The strategies were that iridium nanoparticle sizes were reduced to ~2.2 nm. Additionally, as shown in Fig. 9.20, the three-stepped hydrogen-generation process is happening on the two catalysts where adsorption of H^+ is happening on iridium (Volmer reaction), diffusion of hydrogen to silicon is occurring, and finally, the desorption of H_2 from silicon is taking place either through Heyrovsky or Tafel reaction. With such use of the binary metal/SiNW catalyst system, hydrogen adsorption and desorption

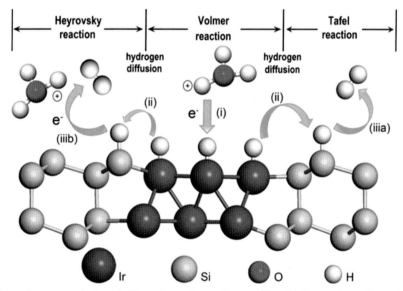

Figure 9.20 Schematic representation of different reaction pathways: (i) on iridium surface, the protons are reduced to hydrogen atoms; (ii) from iridium atoms to the Si atoms, the hydrogen atoms get diffused; hydrogen molecules are formed either through Tafel (iii a) or Heyrovsky (iii b). Adapted from Sheng, M., et al. (2019). Approaching the volcano top: iridium/silicon nanocomposites as efficient electrocatalysts for the hydrogen evolution reaction. ACS Nano, 13:2786–2794. Copyright (2019) American Chemical Society.

were enabled onto two different surfaces. This in turn aided in combating limitations posed by Sabatier's principle of catalysis (Sheng et al., 2019).

9.3 Conclusion

Silicon nanostructured materials are an eminent part of the modern microelectronics industry. Silicon materials are being explored for their applications in optics, analytical detection, photodiodes, biomarkers, solar cells, and lithium-ion batteries. Silicon is an attractive choice owing to its excellent semiconductor properties, optimizable conductivity, biocompatibility, and abundance. In recent decades, investigators have witnessed nanosilicon showcasing enhanced properties as anodic material in lithium-ion batteries, potential in therapeutic modalities, luminescent behavior in the near-infrared, and large surface area for hydrogen generation from water. Research in nanostructured silicon and its applications is booming, and this chapter has attempted to distill to readers the existing and emerging silicon technologies.

References

Aradilla, D., Gao, F., Lewes-Malandrakis, G., Müller-Sebert, W., Gentile, P., Boniface, M., . . . Bidan, G. (2016). Designing 3D multihierarchical heteronanostructures for high-performance on-chip hybrid supercapacitors: poly(3,4-(ethylenedioxy)thiophene)-coated diamond/silicon nanowire electrodes in an aprotic ionic liquid. *ACS Applied Materials & Interfaces, 8*, 18069−18077.

Bao, Z., Weatherspoon, M. R., Shian, S., Cai, Y., Graham, P. D., Allan, S. M., . . . Sandhage, K. H. (2007). Chemical reduction of three-dimensional silica micro-assemblies into microporous silicon replicas. *Nature, 446*, 172−175.

Bogart, T. D., Lu, X., Gu, M., Wang, C., & Korgel, B. A. (2014). Enhancing the lithiation rate of silicon nanowires by the inclusion of tin. *RSC Advances, 4*, 42022−42028.

Chatterjee, S., Carter, R., Oakes, L., Erwin, W. R., Bardhan, R., & Pint, C. L. (2014). Electrochemical and corrosion stability of nanostructured silicon by graphene coatings: Toward high power porous silicon supercapacitors. *Journal of Physical Chemistry C, 118*, 10893−10902.

Chen, C. J., Liu, C. W., Yang, K. C., Yin, L. C., Wei, D. H., Hu, S. F., & Liu, R. S. (2018). Amorphous phosphorus-doped cobalt sulfide modified on silicon pyramids for efficient solar water reduction. *ACS Applied Materials & Interfaces, 10*, 37142−37149.

Coridan, R. H., Arpin, K. A., Brunschwig, B. S., Braun, P. V., & Lewis, N. S. (2014). Photoelectrochemical behavior of hierarchically structured Si/WO$_3$ core-shell tandem photoanodes. *Nano Letters, 14*, 2310−2317.

Devarapalli, R. R., Szunerits, S., Coffinier, Y., Shelke, M. V., & Boukherroub, R. (2016). Glucose-derived porous carbon-coated silicon nanowires as efficient electrodes for aqueous micro-supercapacitors. *ACS Applied Materials & Interfaces, 8,* 4298–4302.

Franco Gonzalez, A., Yang, N. H., & Liu, R. S. (2017). Silicon anode design for lithium-ion batteries: Progress and perspectives. *Journal of Physical Chemistry C, 121,* 27775–27787.

Han, Y., Liu, Y., Yuan, J., Dong, H., Li, Y., Ma, W., … Sun, B. (2017). Naphthalene diimide-based n-type polymers: Efficient rear interlayers for high-performance silicon-organic heterojunction solar cells. *ACS Nano, 11,* 7215–7222.

Hochbaum, A. I., & Yang, P. (2010). Semiconductor nanowires for energy conversion. *Chemical Reviews, 110,* 527–546.

Hu, H., Pei, Z., & Ye, C. (2015). Recent advances in designing and fabrication of planar micro-supercapacitors for on-chip energy storage. *Energy Storage Materials, 1,* 82–102.

Hu, X., Li, G., & Yu, J. C. (2010). Design, fabrication, and modification of nanostructured semiconductor materials for environmental and energy applications. *Langmuir: the ACS Journal of Surfaces and Colloids, 26,* 3031–3039.

Jeong, S., Garnett, E. C., Wang, S., Yu, Z., Fan, S., Brongersma, M. L., … Cui, Y. (2012). Hybrid silicon nanocone-polymer solar cells. *Nano Letters, 12,* 2971–2976.

Kabashin, A. V., Singh, A., Swihart, M. T., Zavestovskaya, I. N., & Prasad, P. N. (2019). Laser-processed nanosilicon: a multifunctional nanomaterial for energy and healthcare. *ACS Nano, 13,* 9841–9867.

Kang, S. B., Kim, J. H., Jeong, M. H., Sanger, A., Kim, C. U., Kim, C. M., & Choi, K. J. (2019). Stretchable and colorless freestanding microwire arrays for transparent solar cells with flexibility. *Light: Science & Applications, 8.*

Kargar, A., Sun, K., Jing, Y., Choi, C., Jeong, H., Jung, G. Y., … Wang, D. (2013). 3D Branched nanowire photoelectrochemical electrodes for efficient solar water splitting. *ACS Nano, 7,* 9407–9415.

Kelzenberg, M. D., Boettcher, S. W., Petykiewicz, J. A., Turner-Evans, D. B., Putnam, M. C., Warren, E. L., … Atwater, H. A. (2010). Enhanced absorption and carrier collection in Si wire arrays for photovoltaic applications. *Nature Materials, 9,* 239–244.

Kye, J., Shin, M., Lim, B., Jang, J. W., Oh, I., & Hwang, S. (2013). Platinum monolayer electrocatalyst on gold nanostructures on silicon for photoelectrochemical hydrogen evolution. *ACS Nano, 7,* 6017–6023.

Li, J., Yu, H., & Li, Y. (2012). Solar energy harnessing in hexagonally arranged Si nanowire arrays and effects of array symmetry on optical characteristics. *Nanotechnology, 23,* 194010.

Liao, F., Shen, W., Sun, Y., Li, Y., Shi, H., & Shao, M. (2018). Nanosponge Pt modified graphene nanocomposites using silicon monoxides as a reducing agent: high efficient electrocatalysts for hydrogen evolution. *ACS Sustainable Chemistry & Engineering, 6,* 15238–15244.

Liu, N., Huo, K., McDowell, M. T., Zhao, J., & Cui, Y. (2013). Rice husks as a sustainable source of nanostructured silicon for high performance Li-ion battery anodes. *Scientific Reports, 3,* 1–7.

Liu, N., Lu, Z., Zhao, J., McDowell, M. T., Lee, H.-W., Zhao, W., & Cui, Y. (2014). A pomegranate-inspired nanoscale design for large-volume-change lithium battery anodes. *Nature Nanotechnology, 9,* 187–192.

Liu, X. H., & Huang, J. Y. (2011). In situ TEM electrochemistry of anode materials in lithium ion batteries. *Energy & Environmental Science, 4,* 3844–3860.

McKone, J. R., Warren, E. L., Bierman, M. J., Boettcher, S. W., Brunschwig, B. S., Lewis, N. S., & Gray, H. B. (2011). Evaluation of Pt, Ni, and Ni-Mo electrocatalysts for hydrogen evolution on crystalline Si electrodes. *Energy & Environmental Science, 4*, 3573–3583.

Mori, T., Chen, C. J., Hung, T. F., Mohamed, S. G., Lin, Y. Q., Lin, H. Z., ... Liu, R. S. (2015). High specific capacity retention of graphene/silicon nanosized sandwich structure fabricated by continuous electron beam evaporation as anode for lithium-ion batteries. *Electrochimica Acta, 165*, 166–172.

Profatilova, I. A., Stock, C., Schmitz, A., Passerini, S., & Winter, M. (2013). Enhanced thermal stability of a lithiated nano-silicon electrode by fluoroethylene carbonate and vinylene carbonate. *Journal of Power Sources, 222*, 140–149.

Quiroga-González, E., Carstensen, J., & Föll, H. (2013). Good cycling performance of high-density arrays of Si microwires as anodes for Li ion batteries. *Electrochimica Acta, 101*, 93–98.

Seo, K., Yu, Y. J., Duane, P., Zhu, W., Park, H., Wober, M., & Crozier, K. B. (2013). Si microwire solar cells: Improved efficiency with a conformal SiO_2 layer. *ACS Nano, 7*, 5539–5545.

Sheng, M., Jiang, B., Wu, B., Liao, F., Fan, X., Lin, H., ... Shao, M. (2019). Approaching the volcano top: iridium/silicon nanocomposites as efficient electrocatalysts for the hydrogen evolution reaction. *ACS Nano, 13*, 2786–2794.

Sivakov, V., Andrä, G., Gawlik, A., Berger, A., Plentz, J., Falk, F., & Christiansen, S. H. (2009). Silicon nanowire-based solar cells on glass: Synthesis, optical properties, and cell parameters. *Nano Letters, 9*, 1549–1554.

Takanabe, K. (2017). Photocatalytic water splitting: Quantitative approaches toward photocatalyst by design. *ACS Catalysis, 7*, 8006–8022.

Thalluri, S. M., Wei, B., Welter, K., Thomas, R., Smirnov, V., Qiao, L., ... Liu, L. (2019). Inverted pyramid textured p-silicon covered with Co_2P as an efficient and stable solar hydrogen evolution photocathode. *ACS Energy Letters, 4*, 1755–1762.

Valero, A., Mery, A., Gaboriau, D., Gentile, P., & Sadki, S. (2019). One step deposition of PEDOT-PSS on ALD protected silicon nanowires: Toward ultrarobust aqueous microsupercapacitors. *ACS Applied Energy Materials, 2*, 436–447.

Wang, B., Wu, H., Xu, G., Zhang, X., Shu, X., Lv, J., & Wu, Y. C. (2019). MoS x quantum dot-modified black silicon for highly efficient photoelectrochemical hydrogen evolution. *ACS Sustainable Chemistry & Engineering, 7*, 17598–17605.

Wang, J., Xu, T., Huang, X., Li, H., & Ma, T. (2016). Recent progress of silicon composites as anode materials for secondary batteries. *RSC Advances, 6*, 87778–87790.

Wang, Y., Sun, L., Xiao, D., Du, H., Yang, Z., Wang, X., ... Lu, B. (2020). Silicon-based 3D all-solid-state micro-supercapacitor with superior performance. *ACS Applied Materials & Interfaces, 12*, 43864–43875.

Warren, E. L., Atwater, H. A., & Lewis, N. S. (2014). Silicon microwire arrays for solar energy-conversion applications. *Journal of Physical Chemistry C, 118*, 747–759.

Wu, H., Chan, G., Choi, J. W., Ryu, I., Yao, Y., Mcdowell, M. T., ... Cui, Y. (2012). Stable cycling of double-walled silicon nanotube battery anodes through solid-electrolyte interphase control. *Nature Nanotechnology, 7*, 310–315.

Wu, Z.-H., Yang, J.-Y., Yu, B., Shi, B.-M., Zhao, C.-R., & Yu, Z.-L. (2019). Self-healing alginate–carboxymethyl chitosan porous scaffold as an effective binder for silicon anodes in lithium-ion batteries. *Rare Metals, 38*, 832–839.

Xie, C., Nie, B., Zeng, L., Liang, F. X., Wang, M. Z., Luo, L., ... Yu, S. H. (2014). Core-shell heterojunction of silicon nanowire arrays and carbon quantum dots for photovoltaic devices and self-driven photodetectors. *ACS Nano, 8*, 4015–4022.

Zhang, B., Jie, J., Zhang, X., Ou, X., & Zhang, X. (2017). Large-scale fabrication of silicon nanowires for solar energy applications. *ACS Applied Materials & Interfaces, 9*, 34527−34543.

Zhu, B., Wu, X., Liu, W. J., Lu, H. L., Zhang, D. W., Fan, Z., & Ding, S. J. (2019). High-performance on-chip supercapacitors based on mesoporous silicon coated with ultrathin atomic layer-deposited in 2O 3 films. *ACS Applied Materials & Interfaces, 11*, 747−752.

Application of silicon-based hybrid nanoparticles in catalysis

Pratibha and Jaspreet Kaur Rajput
Department of Chemistry, Dr. B.R Ambedkar National Institute of Technology, Jalandhar, Punjab, India

10.1 Introduction

Silicon is well-explored amongst the several abundant elements present on Earth. The properties of silicon generally resemble those of carbon as they belong to the same group of the periodic table. Silicon is usually considered as a chemically inert element but has strong affinity toward oxygen. A wide range of varieties of Si-O compounds have been known to date. The most common among them are silica, clay, polyhedral oligomeric silsesquioxanes (POSS), and silicates (Gogoi, Barua, Khan, & Karak, 2019). In recent times, silicon and its oxide has been fabricated in various size ranges varying from micro to nano. The key aspect of such materials is their high surface area to volume ratio. Due to this feature, the nanoparticles of silicon and silicon incorporated materials have been found to be active in diverse fields from electronics to catalysis to biomedical applications (McInnes & Voelcker, 2009). In addition, the hybrid functional materials can also be constructed by incorporating silicon with some inorganic and organic components. Such hybrid nanomaterials have been found to exhibit significant roles in catalysis because of the enhanced thermal and chemical stability and tunable properties. Thus the following section describes the significance of the nanocatalysis and the potential application of the numerous silicon-based hybrid nanoparticles in various catalytic reactions.

In order to reduce the energy consumption and thus to achieve the objectives of the sustainability, the need for catalysts emerges in the modern science and technology. In today's world, approximately 90% of chemical processes occur in the presence of catalysts. Catalysts are essential in numerous fields such as industrial production of organic chemicals, pharmaceutical industry, production of food, fuels, and goods (Cuong et al., 2016). Moreover, the goal of attaining the sustainability prompts the researchers to focus on the "Green Chemistry" concepts, which rely on developing a catalytic process without the use of any toxic reagent, volatile organic compounds, harsh reaction conditions, and minimizing the waste production, thus making the catalyst highly active, selective, stable, energy efficient, and environmentally benign in nature.

Broadly, the catalytic system can be categorized into two different types, namely homogeneous and heterogeneous. In the former case, the catalytic species shows high activity and selectivity owing to excellent interaction between the catalyst and reactant species present in the same phase. So in previous literature, the use of wide variety of homogeneous catalysts such as Bronsted acids, Lewis acids, metal ions, biomolecules, metal complexes, etc. in various organic synthetic transformations has been reported (Singh & Rajput, 2017). But the major aspect that limits the industrial commercialization of the homogeneous catalyst is the difficulty of its recovery as such from the mixture after accomplishment of the chemical process. On the other side, although the heterocatalytic system offers the effortless separation of the catalyst owing to the different phase of reactants and catalyst, the catalytic activity and selectivity is comparatively lesser because of the limited active catalytic surface area than the homogeneous catalytic system. Thus in order to combine the advantageous properties of both types of catalytic system, that is, facile separation like heterogeneous system and high activity like homogeneous system, the methodology of heterogenization of active homogeneous species on the solid support comes up with an alternative way (Sharma, Sharma, Dutta, Zboril, & Gawande, 2015). However, the outcomes were not as satisfactory because of the fact that the grafting was only carried out on the exterior exposed face of a solid support, as a consequence of which the available active catalytic sites gets decreased compared to pure homogeneous system, thus resulting in the overall decrease in reactivity. In addition to this, the catalyst instability further imposes restrictions on the industrial use of these catalysts as most of the active catalytic molecules get leached away from the solid surface with the passage of time (Gawande, Monga, Zboril, & Sharma, 2015). Thus the desire of generating an efficient, stable, highly active, and selective catalytic system indulges the researchers to make use of nanoparticulates in the catalytic system. The key benefit of using the nanoranged particles as catalyst is their small size which offers a highly active catalytic surface area. Moreover, the decrease in particle size resulted in the increase in proportion of surface active atoms. As a whole, the quantum confinement effect, dispersibility, durability, stability, and the comparative high active catalytic surface area are the key factors possessed by the nanoparticulates which are responsible for constructing a highly active, energy efficient, and environment-friendly path for catalytic systems; thus following the "Green Chemistry" concepts (Chaturvedi, Dave, & Shah, 2012). In addition to this, the surface properties of the nanoparticles (NPs) can also be tuned according to the desire by functionalization with various organic (polymers, surfactant) and inorganic (carbon, silica, metals, and metal oxides) moieties.

10.2 Types of silicon-based nanoparticles

Since, silicon possesses strong affinity toward oxygen, different varieties of Si-O compounds can be prepared (Weinhold & West, 2011). The familiar among them is silica or silicon dioxide (SiO_2). The first ever method for the production of silica NPs was introduced by Stöber and Fink in 1968 (Stöber, Fink, & Bohn, 1968). After their work, different methodologies like sol–gel method, hydrothermal synthesis, microemulsion, microwave assisted, sonochemical synthesis, etc. have been adopted for the preparation of various NPs varying in size and shape. In addition to silica, nanoclay, polyhedral oligomericsilsesquioxanes (POSS), and silicates are some other representatives of the SiO family. Silica NPs have been tremendously utilized in organic synthetic transformations because of their excellent thermal, chemical, and mechanical stability. Owing to the fact that both the exterior and interior pore surface of silica NPs can be functionalized with various moieties, much consideration has been centered around the use of modified silica NPs in diverse fields (Ghosh, 2019).

Nanosized silica consisting of mesopores (pores having diameter from 2 to 50 nm) are generally recognized as mesoporous silica NPs (MSNs). Besides having applications in different fields like biomedical, biosensing, catalytic, drug delivery, etc. the distinct pore structure of mesoporous silica can also work as molecular sieves (Davidson et al., 2018). Moreover, the properties of mesoporous silica can also be tuned by changing the pore size. Thus depending upon the pore size, various kinds of mesoporous silica nanomaterials have been identified. The most common among them are MCM-41 (Mobil Crystalline Materials) and SBA-15 (Santa Barbara Amorphous type materials). In addition to this, FDU-2, MCM-50, SBA-16, KIT-5, COK-12, HMM-33, etc. are some other types of mesoporous materials that has been reported in the previous literature (Narayan, Nayak, Raichur, & Garg, 2018).

Nanoclays are formed from the layered silicates consisting of tetrahedrally bound Si atoms with an octahedrally shared edge of Mg$(OH)_2$ or $Al(OH)_3$. The diverse morphology of nanoclay has been synthesized by varying the chemical composition of the constituting elements. Some of them are halloysite, kaolinite, illite, bentonite, nacrite, beidellite, talc, kenyaite, and brucite (Faheem, 2008). Besides having catalytic applications in organic transformations, the nanocomposites of clay can also serve as excellent adsorbent material for the removal of heavy metals, dyes, and other organic pollutants (Cavallaro et al., 2019; Martínez, Volzone, & Huck, 2015).

The other class of the Si-O family are the polyhedral oligomeric silsesquioxanes NPs. Silsesquioxanes comprise Si-O-Si linkages connected in a manner to construct a cage-shaped structure having tetrahedral Si vertices and organic groups. Generally, this can be viewed as a three-dimensional structure of a Si-O framework having a silica-like core surrounded by a shell of organic moieties. Furthermore, the properties of silsesquioxanes can be varied by varying the composition and the type of organic groups on the silicon atoms (Tolinski, 2009).

Beyond this, the incorporation or grafting of various types of organic and inorganic moieties onto the silica leads to the construction of various hybrid nanomaterials. Broadly, these can be categorized as silicon/metal-based and silicon/carbon-based hybrid nanosystems. Various metals and metal oxide NPs such as Pd, Cu, Ag, Au, Fe, zinc oxide, and iron oxide, and carbonaceous materials, such as CNTs, graphene, and graphene oxide, have been utilized in order to prepare the silicon-based hybrid nanomaterials possessing different properties which further can be used in various applications (Sugimoto, Fujii, & Imakita, 2016; Wang & Kumta, 2010). In other words, we can say that the methodology of modification of surface properties of silica NPs by functionalization with a wide variety of groups provides a way to synthesize suitable nanostructured materials required for various catalytic applications.

With this foundation, the catalytic activity of the various silicon-based hybrid nanomaterials in different types of organic transformations is explored in this chapter. Moreover, the major aspects highlighting the future prospects of these hybrid nanomaterials are also thoroughly discussed in the present study.

10.3 Silicon-based hybrid nanoparticles as catalyst for organic conversions

The exclusive properties possessed by the nanosized materials, like smaller size, high surface area to volume ratio, thermal and chemical stability, make them appropriate aspirants for catalytic purposes. Each kind of organic reaction demands a specific catalyst depending upon the requirement of the reacting partners involved in the reaction. Usually a catalyst works by increasing the selectivity of a reaction, by making it move in a particular direction.

As mentioned previously, the basic requirement for any catalyst is its chemical stability, easy recoverability, and recyclability. In order to make the task of recoverability more feasible, the production of magnetically active nanocatalyst was initiated and further various types of magnetic nanocatalyst have been utilized in organic transformations. Moreover, the coating of functionalized silica NPs onto the surface of the magnetic iron oxide NPs resulted in the construction of magnetic/silica hybrid NPs consisting of magnetic core and silica shell. Therefore considering all these facts, the importance of the various types of hybrid NPs incorporating silicon in diverse organic reactions such as coupling reactions, multicomponent reactions, addition reactions, oxidation and reduction reactions, addition reactions, and many more is illustrated in the following section.

10.3.1 Coupling reactions

The coupling reactions play a significant role in the synthetic organic chemistry. The industrial importance of these types of reactions lies in the synthesis of carbon–carbon and carbon–heteroatom bonds. The literature studies disclose that diverse types of silicon incorporated nanohybrids have been utilized in the coupling reactions; a few of them are discussed here:

10.3.1.1 Carbon–Carbon coupling reactions

Cai et al. introduced a very simple and convenient approach for the preparation of -41 supported sulfur palladium(0) complex [-41-S-Pd(0)] by the immobilization of 3-(2-cyanoethyl sulfanyl) propyltriethoxysilane on -41, proceeded by reaction with palladium chloride and then subsequent reduction in the presence of hydrazine hydrate (Fig. 10.1(A)). The overall sulfur and palladium content was observed to be 0.72 and 0.35 mmol/g,

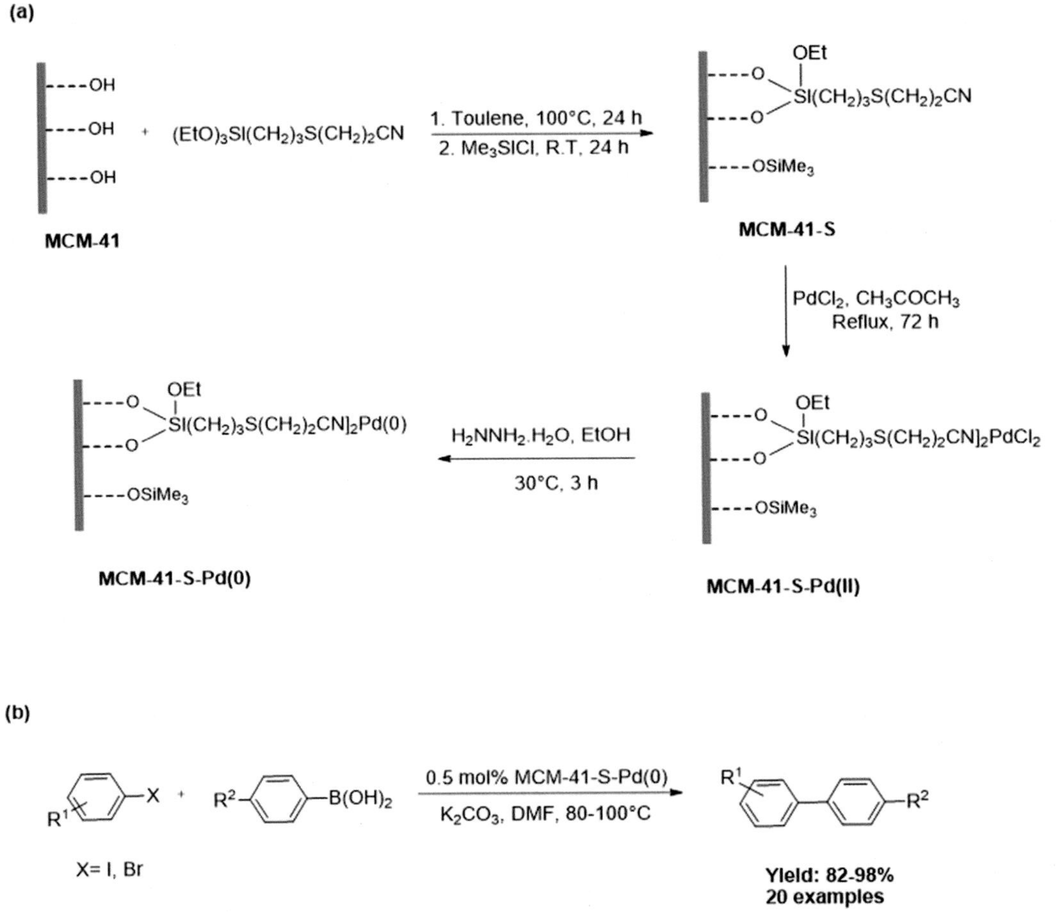

Figure 10.1 Schematic illustration of the (A) synthesis of MCM-41-S-Pd(0); (B) Suzuki reaction catalyzed by MCM-41-S-Pd(0).

respectively in the resulted complex. The catalytic activity of the synthesized complex was further investigated in the Suzuki coupling reaction of aryl boronic acids with aryl halides (Fig. 10.1(B)). The various biphenyl products were isolated in the range of 82%–98% of yield. It was also demonstrated that catalyst possesses the excellent stability and reusability without having any noticeable variation in yield or in turnover number (Cai, Xu, & Huang, 2007).

However, in the same year, another heterogeneous catalyst based on -41 and containing both S and Pd was introduced. As depicted in Fig. 10.2, the mercapto-functionalized -41 anchored palladium (0) complex [MCM-41-SH-Pd(0)] was prepared using

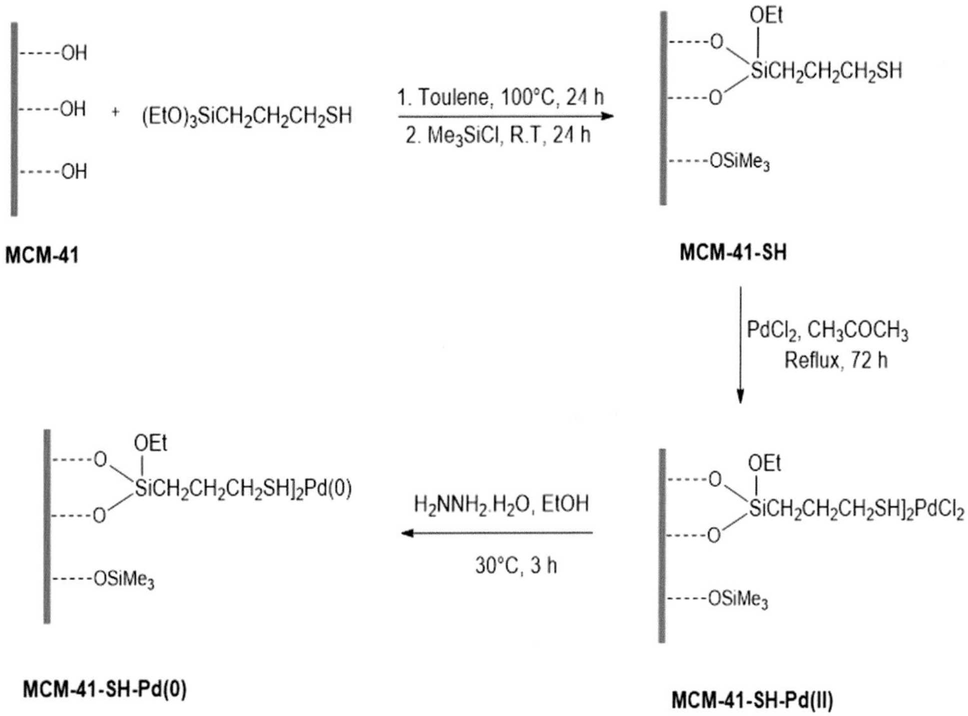

Figure 10.2 Schematic illustration of the synthesis of MCM-41-SH-Pd(0).

a similar route as above and was further successfully utilized as a heterogeneous catalyst for Suzuki coupling reaction (Xu, Hao, & Cai, 2007). Thiel et al. fabricated the mesoporous magnetic nanocomposites (Mag-MSN) so as to enhance recoverability of catalyst from the Suzuki–Miyaura cross-coupling reaction. The silica-coated magnetic nanoparticles (MNPs) were embedded in mesoporous silica matrix in the presence of template surfactant (CTAB). Further, in order to make the catalyst active for the C-C coupling reaction, the covalent grafting of the Pd complex of the type $(L)_2PdCl_2$ [L = Si(OMe)$_3$ functionalized PPh$_3$] was carried out on the accessible pore channels of the synthesized magnetic mesoporous material. The variously substituted aryl halides were utilized as substrate for the coupling in the presence of Mag-MSN, thus producing the biphenyls in sufficient yields (Shylesh, Wang, Demeshko, & Thiel, 2010).

Sarkar and research group introduced a nanostructured catalyst, SBA-16 supported Pd-complex, for Heck, Suzuki, and Sonogashira coupling reactions. The Pd was coordinatively attached to four N atoms of 1,2 diaminocyclohexane immobilized on mesoporous silica SBA-16 using 3-(chloropropyl) triethoxysilane as linker (Fig. 10.3). They had also investigated the heterogeneity of the fabricated complex. It was thus demonstrated that leaching of Pd from the manufactured complex was even less than 0.18 ppm, thus it was confirmed that active catalytic species present in the reaction medium were of a heterogeneous nature instead of the homogeneous Pd. Moreover, the recovery of the catalyst and the purification of products was easily carried out using these heterogeneous mesoporous silica-based Pd complexes. Thus these types of catalytic reactions serve as an alternative to the original Heck, Suzuki, and Sonogashira reaction which usually utilize the homogeneous palladium catalyst (Sarkar, Rahman, & Yusoff, 2015). Fig. 10.4 represents the general scheme for the SBA-16 supported Pd-complex catalyzed cross-coupling reactions.

Ngnie, Dedzo, and their research group designed the modified kaolinite by anchoring triethanolamine (TEA) and 1-(2-hydroxyethyl)-3-methylimidazolium (ImIL) separately on its inner surface. These modifications allowed the uniform deposition of Pd NPs onto the platelets of modified kaolinite in contrast to the unmodified one, which showed the comparative low abundance and random arrangement of Pd NPs on its surface. As expected, the modified catalysts (K-TEA/Pd and K-ImIL/Pd) showed the higher activity toward the Suzuki–Miyaura and Heck coupling compared to the unmodified kaolinite (K/Pd). This was due to the insufficiency of Pd NPs in the latter case. In addition to this, the comparison between the catalytic activities of the two modified kaolinite catalysts revealed that the yield of the products obtained in the presence of K-ImIL/Pd was less (40% in case of Suzuki Coupling and only 5% for Heck reaction) than the other (K-TEA/Pd). This was supposed to be because of the existence of residual BH_4^- ions (counter anions of the grafted imidazolium cation) in the interlayer spaces which inhibited the activity of the K-ImIL/Pd catalyst (Ngnie, Dedzo, & Detellier, 2016).

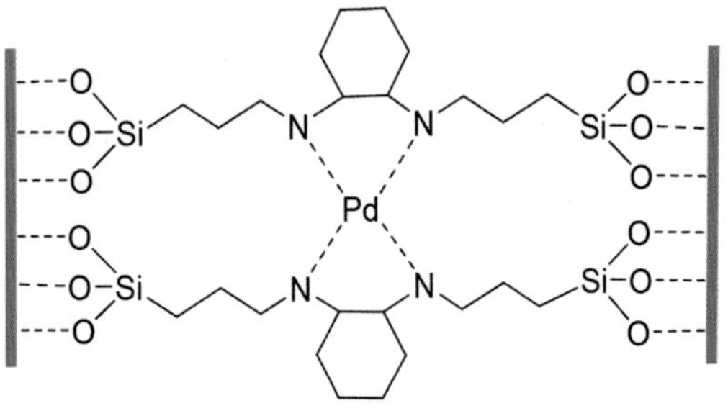

Figure 10.3 Structural representation of SBA-16 supported heterogeneous Pd complex.

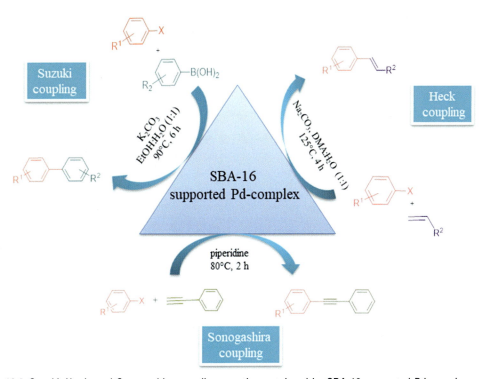

Figure 10.4 Suzuki, Heck, and Sonogashira coupling reaction catalyzed by SBA-16 supported Pd complex.

Hajipour et al. employed the use of "Click Chemistry" so as to synthesize a novel silica supported nanopalladium catalyst; Si-IL@Pd(0) NPs. The fabricated azide functionalized silica reacts with methylpropargylimidazolium bromide to produce the product of the click reaction. Further, Pd NPs were made to immobilize on to the click product which were further stabilized by coordination and electrostatic interactions (Fig. 10.5). The application of generated click ionic-silica supported Pd NPs were then explored in the Suzuki–Miyaura and Heck-Mizoroki cross-coupling reactions. Moreover, optimizations using various types of solvents, bases, and reaction temperatures were also carried out to obtain the best results (Hajipour, Abolfathi, & Mohammadsaleh, 2016). Similarly, silica-bonded N-propyl piperazine substrate was also utilized to immobilize the Pd NPs (Pd-SBNPP) as displayed in Fig. 10.6. The silica grafted piperazine species prevent the aggregation of Pd NPs, hence increasing the stability of active catalytic species. Both the aryl bromides and chlorides were used as substrate for the Heck and Suzuki reaction occurring in the presence of Pd-SBNPP species.

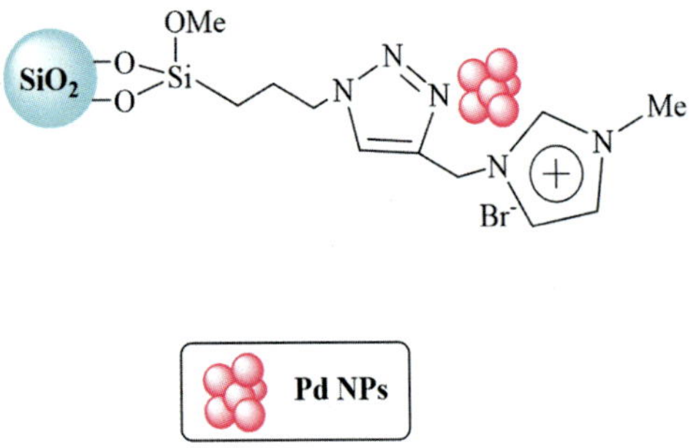

Figure 10.5 Structural illustration of Si-IL@Pd(0) NPs.

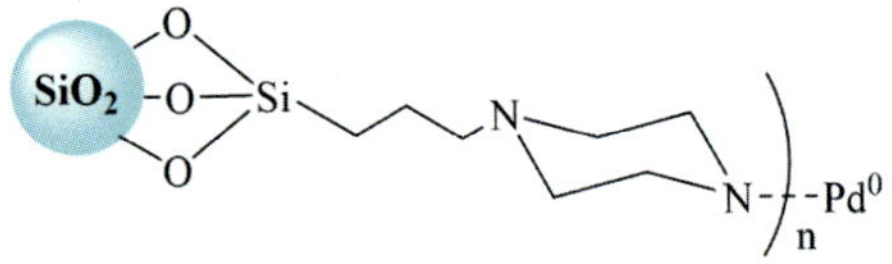

Figure 10.6 Structural representation of Pd-SBNPP.

The required products were figured out to be formed with good to high yield along with easy recoverability of the catalyst (Niknam, Habibabad, Deris, Panahi, & Reza Hormozi Nezhad, 2013).

Further, the Henry reaction or the nitro aldol reaction represents another type of C-C bond forming reaction. In this type of reaction, the nitroalkane reacts with the aldehyde in order to form a new C-C bond along with the formation of a new stereogenic center. The desire of the asymmetric synthesis raises the demand for the chiral catalytic system. For this, Sappino, Righi, and their coworkers immobilized homogeneous chiral β-amino alcohol ligand on to the superparamagnetic core–shell magnetite–silica NPs. Again, the click strategy was adopted in order to graft the chiral homogeneous catalyst onto the surface of silica-coated iron oxide (Fig. 10.7(A)). The fabricated chiral heterogeneous catalyst showed the comparable activities in terms of yield and enantioselectivity as that of the corresponding homogeneous one (Fig. 10.7(B)) (Sappino et al., 2019).

Sonogashira–Hagihara reactions comprise another class of C-C bond forming reactions in which the synthesis of substituted alkynes takes place by coupling between sp^2-hybridized carbon atoms of aryl, heteroaryl, or vinyl halides and sp-hybridized carbon atoms of terminal acetylenes. The traditional Sonogashira reaction generally took place in the existence of homogeneous Pd complex of phosphine and copper salts. It was shown that most of the time the side reaction, known as Glaser-type reaction, occurred by the homocoupling between the terminal alkynes, thus reducing the yield of the desired product. As a substitute, Moghadam et al. designed a copper-free, easily recoverable

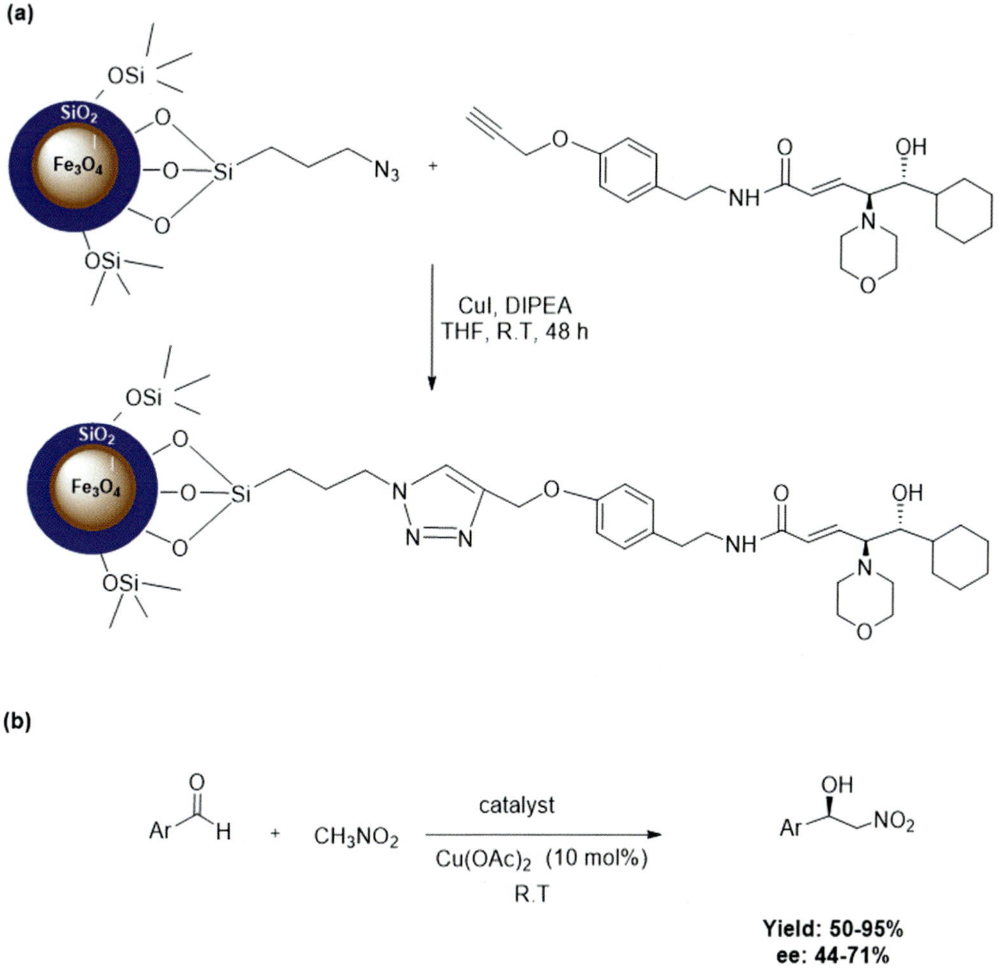

Figure 10.7 Schematic representation of (A) immobilization of chiral catalyst onto magnetite–silica via click reaction. (B) Henry reaction catalyzed by chiral nanocatalyst.

heterogeneous nanosilica supported palladium catalyst, denoted as Pd(II)Cl$_2$-BTP@TMSP-nSiO$_2$, as depicted in Fig. 10.8. The grafting of the Pd(II) complex of 3,5-bis(2-benzothiazolyl)pyridine onto the nanosilica functionalized with trimethoxysilylpropyl chloride was carried out. Thus the various aryl chlorides, aryl bromides, and aryl iodides were coupled with the phenylacetylene in the presence of Pd(II)Cl$_2$-BTP@TMSP-nSiO$_2$ and N,N-diisopropylethylamine (DIPEA) at room temperature to afford the required product in sufficient yields (Dehbanipour, Moghadam, Tangestaninejad, Mirkhani, & Mohammadpoor-Baltork, 2017). Likewise, the research group of

Hajipour explored the application of another copper- and phosphine-free palladium-based catalyst supported on acac-functionalized silica (SiO$_2$-acac-Pd) as a heterogeneous catalyst for a Sonogashira cross-coupling reaction occurring in aqueous media (Fig. 10.9) (Hajipour, Shirdashtzade, & Azizi, 2014)

The C-C bond forming aldol reaction generally requires the existence of a strong base for the production of β-hydroxy carbonyl compounds. Accordingly, Fattahi et al. loaded strong organic superbase, tetramethyl guanidine, on the magnetic Fe$_3$O$_4$ using 3-chloropropyltrimethoxysilane as a linker in order to synthesize a magnetically separable heterogeneous catalyst, represented as MNPs-TMG, for carbon–carbon bond formation

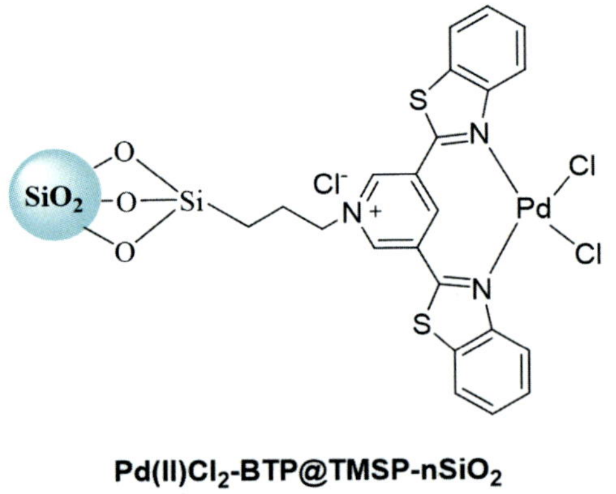

Figure 10.8 Structure of Pd(II) Cl2-BTP@TMSP-nSiO$_2$ nanohybrid.

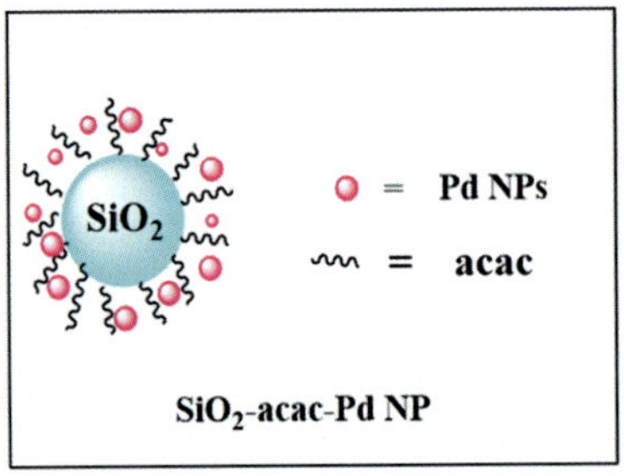

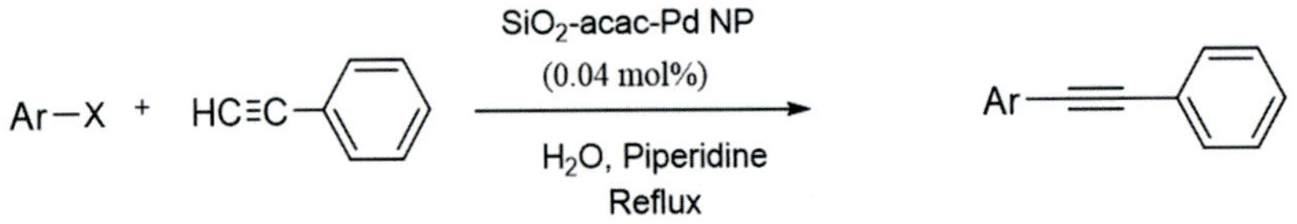

Pd(II)Cl$_2$-BTP@TMSP-nSiO$_2$

Figure 10.9 Sonogashira cross-coupling reaction of aryl halides with phenylacetylenecatalyzed by SiO$_2$-acac-Pd NPs.

between substituted aldehydes and ketones (Fig. 10.10). The lower reaction time, excellent yields, higher purity of isolated product, and recyclability were the tempting features, possessed by such kinds of catalysts (Fattahi, Ramazani, Ahankar, Asiabi, & Kinzhybalo, 2019).

10.3.1.2 C-N cross-coupling reactions

The significance of C-N coupling reactions lies in the synthesis of substituted amines which serves as essential building blocks of various biologically active molecules, pharmaceuticals, agrochemicals, dyes, drugs, and resins. Amongst the various available methods of C-N bond formation, the direct reductive amination of ketones seems to be the most effective approach for the generation of substituted amines. For this purpose, Sharma et al. had loaded a quite inexpensive metal onto the magnetic silica so to develop a potential and cost-effective hybrid nanocatalyst for the reaction. The nanocatalyst (Ni-ACF@Am-SiO$_2$@Fe$_3$O$_4$) was prepared by anchoring of 2-acetyl furan on the surface of amine functionalized silica-coated magnetic nanosupport proceeded by the complexation with nickel acetate. The prepared complex was then successfully employed as an effective catalyst for the one-pot reductive amination of ketones in the existence of NaBH$_4$ as reducing agent under mild and solvent free conditions (Fig. 10.11) (Kumar, Sriparna, & Shivani, 2016). The N-arylation reactions comprises the another category of C-N coupling reactions. The transition metal catalyzed preparation of substituted aryl

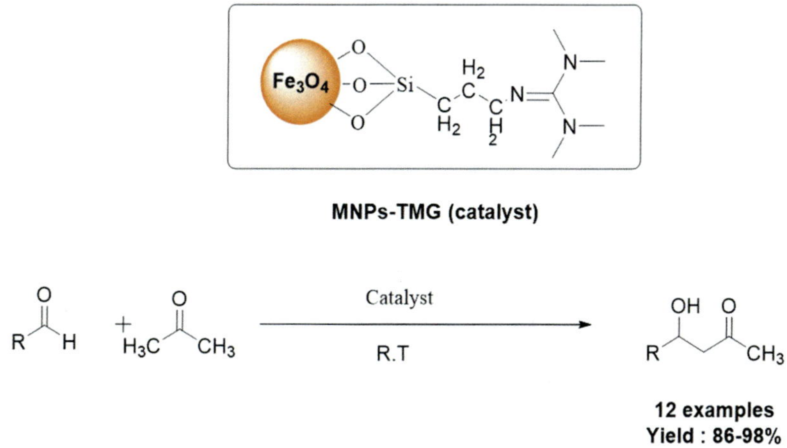

Figure 10.10 Aldol reaction catalyzed by MNPs-TMG.

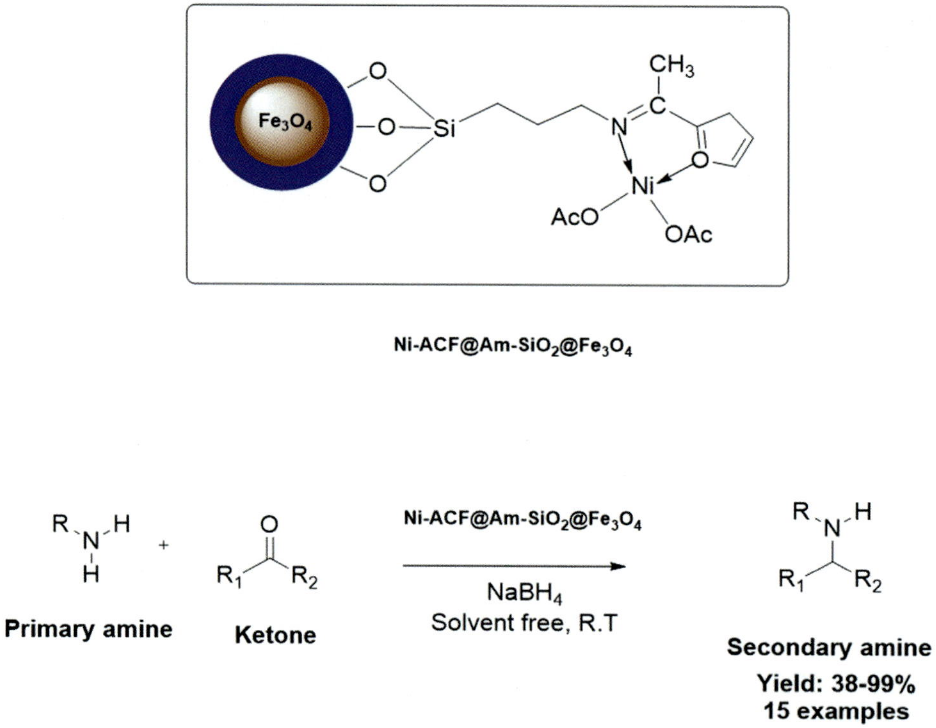

Figure 10.11 Schematic illustration of synthesis of secondary amines via Ni-ACF@Am-SiO2@Fe3O4 catalyzed reductive amination of ketones.

amines by construction of C-N bonds is generally represented as Buchwald-Hartwig reaction. Veisi et al. introduced the application of green tea extract-encapsulated silica gel nanobiocomposite as a catalyst. The inherent capability of the green tea extract to behave as reducing and stabilizing agent further facilitated the immobilization Pd NPs on its surface. The catalytic activity of fabricated SiO_2@green tea/Pd nanocomposite was further examined for Buchwald–Hartwig amination reaction between various aryl halides and amines (like, piperidine, imidazole, etc.). This method serves as a greener protocol for the production of C-N bonds because of its simple and cost-effective nature, easy purification of products, excellent yields, and multiple reuse of catalyst without significant decline in its activity (Veisi, Tamoradi, Karmakar, & Hemmati, 2020). However, Zahedi et al. employed the use of magnetically active imine supported copper complex as catalyst for the N-arylation reaction between aryl halides and benzimidazole or pyrazole to produce the required products in sufficient yields in less reaction time (Fig. 10.12). The desired catalyst (Cu-HB@AS-MNPs) was synthesized by anchoring of copper complex onto the surface of amine-functionalized MNPs with the aid of 2-hydroxybenzophenone (Zahedi, Asadi, & Firuzabadi, 2020).

Apart from the various nanohybrid structures, the bare NPs of silica can also behave as a catalytic surface owing to the presence of reactive hydroxyl groups on its surface. Hasaninejad and Zare developed a solvent-free strategy for the synthesis of quinoxalines using SiO_2 NPs by condensing 1,2-diamines with 1,2-diketones at room temperature (Hasaninejad, Shekouhy, & Zare, 2012). On the other hand, Javidi et al. formulated a magnetically active Fe_3O_4@SiO_2-imid-PMA^n nanocatalyst for the synthesis of quinoxaline derivatives at room temperature (Fig. 10.13). The active catalyst was prepared by the immobilization of NPs of phosphomolybdic acid on imidazole functionalized Fe_3O_4@SiO_2. Or we can say that the methodology to heterogenize the active catalytic species, that is phosphomolybdic acid onto the solid support was adopted in order to overcome the limitations of its homogeneous counterpart (Jaber & Mohsen, 2016)

10.3.1.3 C-S and C-O cross-coupling reactions

Although the carbon–sulfur bond formation reactions lead to the manufacturing of industrially and pharmaceutically important aryl-sulfur compounds, these kinds of coupling reactions are comparatively less studied than C-C and C-N bond formation reactions because of the inherent tendency of the sulfur-containing compounds to undergo parallel oxidative S-S coupling reactions. In addition to this, sometimes the deactivation of the metal catalyst

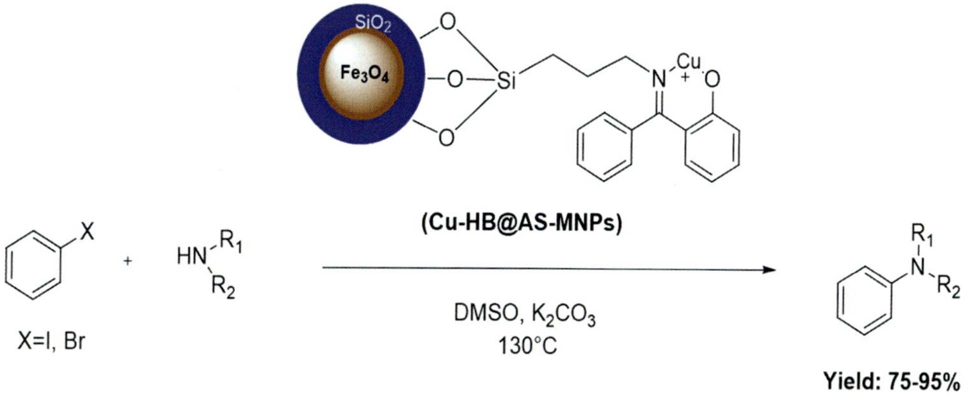

Figure 10.12 N-arylation of amines with aryl halides catalyzed by Cu-HB@AS-MNPs.

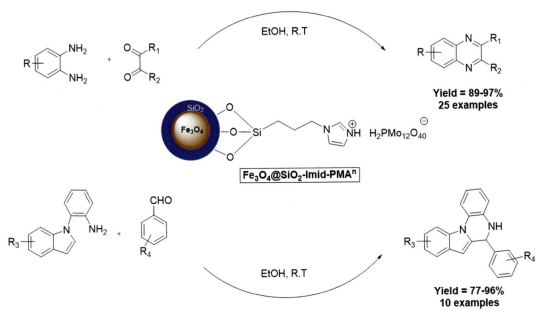

Figure 10.13 Preparation of quinoxaline derivatives in the presence of Fe_3O_4@SiO_2-imid-PMAn.

can also take place by binding with the sulfur heteroatom (Correa, Carril, & Bolm, 2008). In spite of these considerations, a wide variety of catalysts incorporating the chelated and thus stabilized metal atoms have been designed for the C-S cross-coupling reaction of aryl halides and thiols. Bhaumik et al. presented the application of Cu-grafted furfural functionalized mesoporous organosilica in the aryl-sulfur coupling reaction between various aryl halides and thiophenols for the synthesis of various symmetrical and unsymmetrical diaryl sulfides (Fig. 10.14). The metal-grafted functionalized mesoporous material was prepared by the hydrothermal cocondensation of TEOS with the Schiff-base generated by the condensation of furfural and 3-aminopropyltriethoxysilane (APTES) followed by its complexation with $Cu(OAC)_2$ (Mondal, Modak, Dutta, & Bhaumik, 2011). Similarly, in the same year Jun, Lee, and their research group designed another metal chelated complex comprising a backbone of magnetic silica for the C-S cross-coupling reactions, as demonstrated in Fig. 10.15. The synthesis of the active catalyst was carried out by immobilization of N-heterocyclic carbene-nickel (NHC-Ni) complex on magnetite/silica NPs. The fabricated catalyst exhibited excellent recoverability and recyclability up to several times without requiring any additional activation treatment (Yoon et al., 2010).

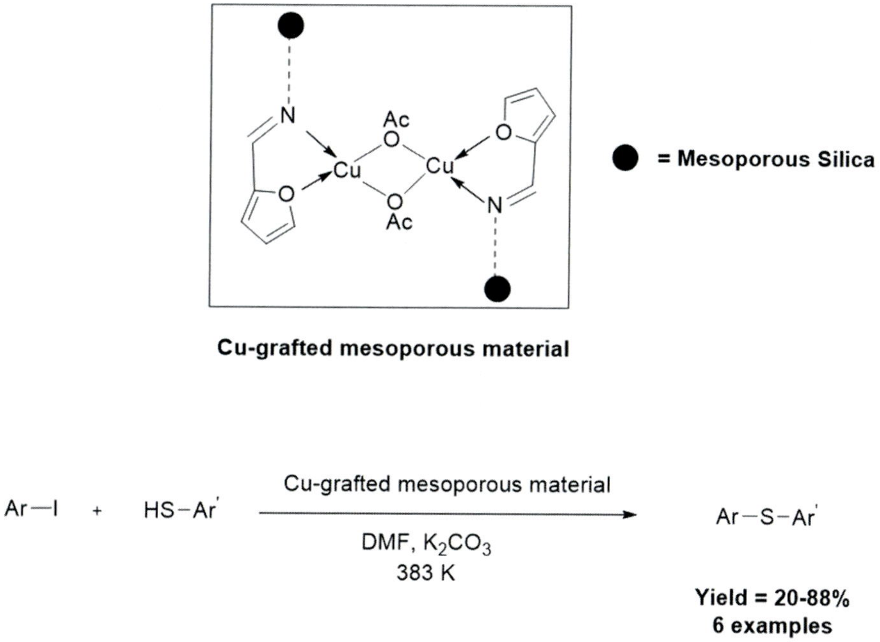

Cu-grafted mesoporous material

Ar—I + HS—Ar' $\xrightarrow[\text{DMF, K}_2\text{CO}_3]{\text{Cu-grafted mesoporous material}}$ Ar—S—Ar'

383 K

Yield = 20–88%
6 examples

Figure 10.14 Schematic illustration of C-S coupling reaction over Cu-grafted functionalized mesoporous material.

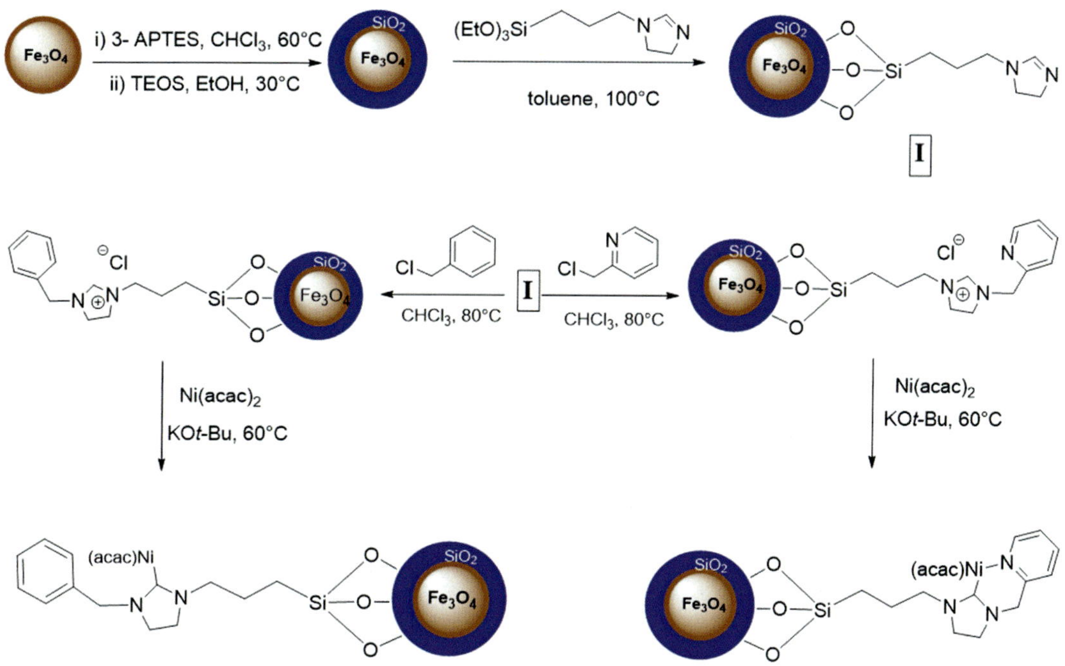

Figure 10.15 Synthetic route to the magnetite/silica NPs-supported NHC-Ni catalyst.

Ashraf et al. reported the immobilization of copper complex on the glycerol functionalized $Fe_3O_4@SiO_2$ so as to synthesize a highly efficient and reusable catalyst, represented as $Fe_3O_4@SiO_2$-Glycerol-Cu(II), for C-S and C-O cross-coupling reactions. The various aryl halides were tested to react with the thiourea and phenols in the presence of catalytic amount (1.25 mol%) of the synthesized complex in order to prepare diversely substituted diarylsulfides and ethers, respectively (Fig. 10.16). The significance of construction of C-O bond lies in the fact that despite of holding application in the manufacturing of fragrances, cosmetics, pharmaceuticals, and dye stuffs; the ethers can also acts as effective reaction medium for various organic reactions (Ashraf, Liu, Peng, & Zhou, 2020).

Usually, water is considered to be an ideal and greener reaction medium owing to its inexpensive, nontoxic, and eco-friendly nature. Thus due to its characteristics as well its natural abundance, water has been widely utilized as a reaction medium in several cross-coupling reactions. Zolfigol et al. introduced a highly stable and magnetically separable Pd-containing

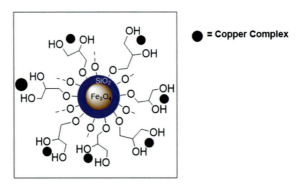

Figure 10.16 Representation of C-S and C-O cross-coupling reactions catalyzed by $Fe_3O_4@SiO_2$-Glycerol-Cu(II).

phosphorus silica magnetite [Fe_3O_4@SiO_2@PPh_2@Pd(0)] for the aqueous phase O-arylation of phenols with aryl halides. High efficiency, simplicity, cleaner reaction profile, and easy recoverability were some salient features of this methodology (Zolfigol et al., 2013).

10.3.2 Oxidation reactions

The oxidation reactions comprise another well-explored family of organic reactions. The conventional oxidation reactions generally take place in the presence of stoichiometric oxidants possessing toxic and expensive heavy metals, thus insignificant from an economic point of view. Alternatively, the modernized oxidation reactions generally require the presence of molecular oxygen or hydrogen peroxide as the fundamental oxygen source, along with the recyclable catalysts and comparatively nontoxic solvents, thus encouraging the greener processes. However, Hashemi et al. carried out the oxidant-free dehydrogenation of alcohols in the presence of magnetically retrievable silver NPs supported on silica-coated ferrite (Fe_3O_4@SiO_2-Ag). The catalyst was found to exhibit high chemoselectivity for the conversion of a wide variety of alcohols to the corresponding carbonyl compounds in the presence of a C-C double bond (Fig. 10.17). Apart from this, the catalyst also underwent recyclability up to eight times without any appreciable loss in its catalytic activity (Ahmad, Mehdi, Nona, & Mohammad, 2015).

Deng et al. took advantage of the green oxidant H_2O_2 for the selective oxidation of benzyl alcohol to benzoic acid (99% selectivity) in the presence of photoirradiation using aqueous media. The catalyst was designed in such a way to mimic the cytochrome P450 enzyme, the natural oxidation catalysts. Thus they had chosen cobalt octakis(butylthio) porphyrazine complex (CoPz(S-Bu)$_8$) as the active catalyst for this work since the

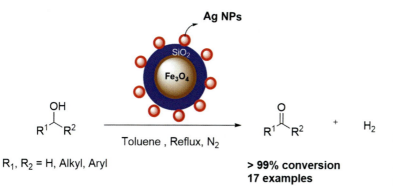

Figure 10.17 Schematic illustration of oxidant-free dehydrogenation of alcohols in the presence of Fe_3O_4@SiO_2-Ag.

chemical properties of this type of complex are quite similar to metalloporphyrins. Further, the immobilization of (CoPz(S-Bu)$_8$) on the silica-coated magnetic nanosphere was carried out in order to make the recovery of the catalyst more facile. It was demonstrated that the benzyl alcohol substituted with electron donating substituents showed the enhanced conversion to corresponding acid comparative to others (Li et al., 2017). Likewise, another greener protocol utilizing the aqueous H_2O_2 as oxidant was developed by Rostamnia et al. for the clean and selective oxidation of organic sulfides. The oxidation reaction was thus successfully carried out in the presence of a catalytic amount of silica-coated magnetite immobilized by NH_2-coordinated tris(8-quinolinolato)iron (Fe_3O_4@SiO_2-FeQ_3) to selectively produce sulfoxides in comparatively higher yields (Fig. 10.18). The reaction took place under mild conditions in the presence of fabricated catalyst, thus avoiding the overoxidation of sulfides to sulfones (Rostamnia, Gholipour, Liu, Wang, & Arandiyan, 2018).

Considering the importance of optically active sulfoxides as an important class of pharmaceutical compounds such as esomeprazole, modafinil, and sulindac, Shen et al. presented

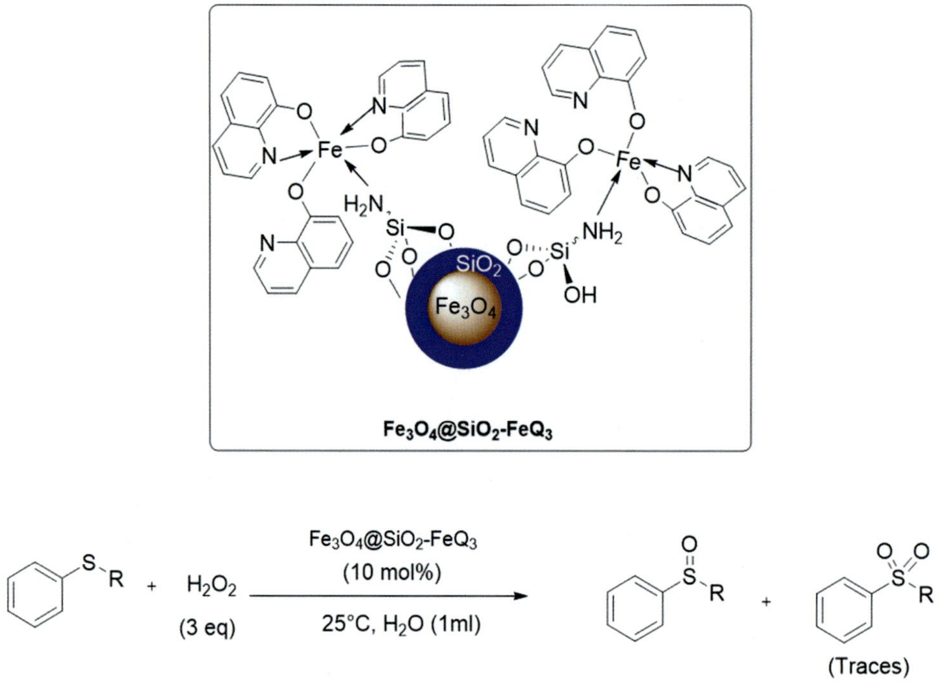

Figure 10.18 Schematic illustration of Fe_3O_4@SiO_2-$FeQ3$ catalyzed oxidation of sulfides.

the synthesis of a highly active and selective silica supported chitosan@vanadium nanocatalyst (SiO$_2$-CS@vanadium). The fundamental approach behind the construction of this catalyst was to immobilize chiral scaffold (chitosan) on to the solid support material in order to make a chiral heterogeneous catalyst for the asymmetric oxidation of aryl alkyl sulfides. The asymmetric sulfoxidation proceeded smoothly in the presence of aqueous H$_2$O$_2$ oxidant and the fabricated catalyst so to produce the desired chiral products in high yield (up to 95%) along with good enantioselectivity (up to 68% ee up to 95%) (Shen et al., 2017).

Freire et al. introduced another vanadium-incorporated silica-based complex for the selective oxidative catalysis. The hybrid nanocatalyst was prepared by immobilization of oxidovanadium^{IV} acetylacetonate [VO(acac)$_2$] onto the APTES functionalized silica NPs. The catalytic activity of synthesized complex, [VO(acac)$_2$APTES@SiO$_2$], was then explored in the epoxidation of geraniol. (Fig. 10.19). It was found that despite having high activity and regioselectivity for the epoxidation of geraniol, the synthesized catalyst also exhibited 100% substrate conversion and 99% selectivity toward the 2,3-expoxygeraniol product. This silica anchored vanadylacetylacetonate catalyst showed the higher catalytic performance comparative to the bulk heterogeneous vanadylacetylacetonate owing to the reduction in the dimensions of the support material which permitted

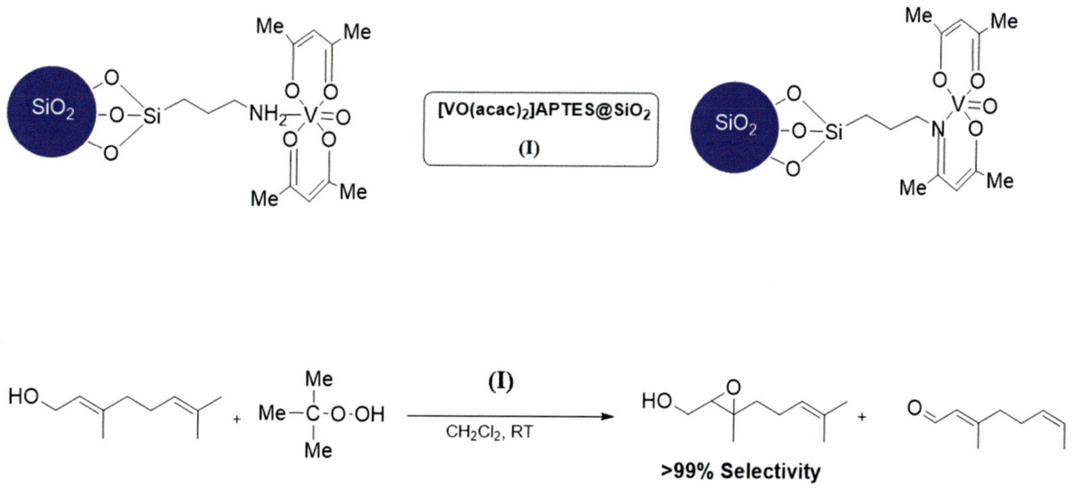

Figure 10.19 Schematic illustration of epoxidation of geraniol catalyzed by silica-anchored [VO(acac)2] hybrid nanocatalyst.

the immobilization of a higher concentration of active catalytic species on its surface. Moreover, the figure enhanced the rate of reaction, which can be ascribed to the higher dispersion of catalyst in the medium, as a consequence of which the contact between the substrate, oxidant species, and the catalyst gets intensified (Pereira et al., 2011).

10.3.3 Reduction reactions

The reduction reactions represent a class of industrially important chemical transformations which are widely utilized in the synthesis of various complex organic moieties and value-added products. The classical reduction reactions usually required the presence of limited and expensive metals like Ru, Pd, Pt, Bi, Pt/Ni, Pd/Ni, etc. Therefore the recovery and reusability of the metal catalysts always remained a central focus point for the scientific community. Recently, the diverse supported materials have been developed for production of diverse range of metal immobilized recoverable heterogeneous catalyst for the hydrogenation reactions. These reduction reactions usually take place in the presence of a mild reducing agent such as sodium borohydride or hydrazine hydrate (Naseem, Begum, & Farooqi, 2017; Zeynizadeh & Karami, 2019)

Beller et al. introduced an effective way for the selective semihydrogenation of alkynes to alkenes. They had carried out the synthesis of silica-supported N-graphitic modified cobalt NPs (Co/phen@SiO$_2$) for the selective catalysis. The prepared catalyst exhibited high activity and selectivity for the reduction of diverse internal alkynes to the Z-isomer of the corresponding alkene (Fig. 10.20(A)). In addition to this, numerous terminal alkynes including sensitive functionalized compounds were also readily reduced to terminal alkenes. The chemoselective nature of the fabricated catalyst was also examined in the hydrogenation of phenylacetylene in the presence of excess of styrene (Fig. 10.20(B)). Thus we can say that this method allowed the purification of alkenes by selective hydrogenation of corresponding alkynes in the presence of excess of olefins (Chen et al., 2017). On the other side, Pitchumani et al. efficiently synthesized the nickel NPs encapsulated K10-montmorillonite clay acquiring a nickel content of 2.84 wt.%. The resultant catalytic system was thus served as an economic and eco-friendly protocol for the reduction of alkenes and alkynes in the presence of hydrazine hydrate as a reducing agent (Dhakshinamoorthy & Pitchumani, 2008).

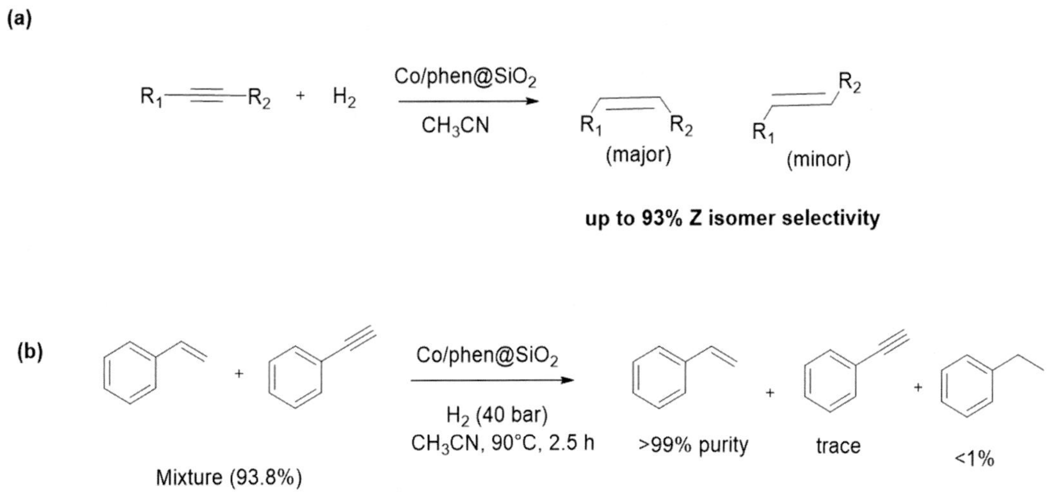

Figure 10.20 Schematic representation of (A) Co/phen@SiO$_2$ catalyzed selective hydrogenation of alkynes; (B) illustration of alkene purification using Co/phen@SiO$_2$.

However, exploiting the impact of magnetic recoverability, Xu et al. designed a magnetic aminoclay (AC)-based highly stable and efficient AC@Fe$_3$O$_4$@Pd nanocatalyst by depositing Pd NPs on the surface of magnetic aminoclay nanocomposite. The synthesized AC@Fe$_3$O$_4$@Pd nanocatalyst displayed excellent catalytic activity for the reduction of various nitrophenols and nitroanilines to corresponding aminobenzene utilizing NaBH$_4$ as reducing agent. High stability, dispersibility, and efficient magnetism were some of the significant features possessed by this protocol (Jia et al., 2018). Likewise, Wang et al. introduced another silica-based core–shell structured nanocomposite for the hydrogenation of nitrobenzene using hydrogen as reductant (Fig. 10.21). They had developed size-uniformed mesoporous Ag@SiO$_2$ nanocatalyst by a one-pot method for catalytic purposes. During its synthesis, AgNO$_3$ was reduced in situ by different types of aldehydes which adversely affected its size and hence catalytic activity. It was observed that the catalyst prepared by a strong reducing agent like CH$_2$O exhibited higher activity toward the hydrogenation reaction (Zhao et al., 2020). Further, Das et al. carried out the bioinspired fabrication of Ag NPs onto the surface of nanosilica. They demonstrated that the protein immobilized nanosilica served as an efficient template material for the facile and in situ growth of Ag NPs. Since the protein molecules adsorbed on the surface of nanosilica served as an efficient reducing, capping, and stabilizing agent, the

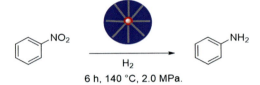

Figure 10.21 Ag@SiO$_2$ catalyzed hydrogenation of nitrobenzene.

reduction of silver ions was successfully achieved without the use of any external reducing agent. The synthesized Ag@nanosilica hybrid nanomaterial showed excellent activity, recovery, and reusability in the reduction of 4-nitrophenol (Das, Khan, Guha, & Naskar, 2013).

Hu et al. investigated the catalytic activity of the synthesized Ag-decorated silica NPs for the removal of methylene blue dye from wastewater. In this approach, the desired Ag/SiO$_2$ nanocatalyst was prepared by adopting a quite different technique to that previously mentioned, that is firstly the variety of silica NPs with rough surfaces was synthesized, followed by immobilization of highly dispersed silver NPs on their surface using a simple wet impregnation method. The significance of fabricated nanocomposite was then explored as an adsorbent and subsequently as a catalyst for the reduction of methylene blue in the presence of excess of NaBH$_4$. It was predicted that the Ag/SiO$_2$ nanocomposite with smaller particle size and higher Ag loading displayed high adsorption capacity and catalytic activity for the removal of methylene blue (Hu, Yan, Hu, Feng, & Zhou, 2019).

10.3.4 Multicomponent reactions

These types of reactions are highly atom economical as several bonds are formed in a single step with an efficient manner requiring less time and less energy. Since, three or more substrates react together in one pot to afford the highly selective product, these kinds of reactions are also known as convergent reactions. Numerous reports have been found in the literature regarding the utilization of silica-based nanohybrids as catalysts in these highly efficient multicomponent reactions (MCRs).

Recently, Sharma et al. fabricated titania–silica NPs assemblies by loading the TiO_2 NPs onto SiO_2 NPs using sol–gel methodology. The catalyst was further used as a heterogeneous catalyst for one-pot preparation of pharmacologically important 2,3-diaryl-3,4-dihydroimidazo[4,5-b] indole scaffolds by reacting isatin, various anilines, aldehydes, and ammonium acetate in the presence of methanol (Fig. 10.22) (Geedkar, Kumar, Reen, & Sharma, 2020).

Hamidian et al. introduced a quite simple, inexpensive, and easy preparative method for the production of silica NPs from rice husk. They took advantage of the fact that large volumes of rice husk are generated as a waste product in the rice milling industries. Refluxing of the rice husk ash with 1 M NaOH and subsequent pH adjustment resulted in nanosized silica. Further, the developed SiO_2 NPs were made to react with chlorosulfonic acid in order to generate nano SiO_2-SO_3H. The synthesized acid-functionalized NPs were then served as an effective catalyst for the synthesis of triazolo[1,2-a]indazole-1,3,8-trione derivatives by the reaction of arylaldehydes with urazole and dimedone under solventless conditions. (Fig. 10.23) (Hamidian, Fozooni, Hassankhani, & Mohammadi, 2011). On the other side, Niknam et al. explored the application of silica grafted N-propyl-imidazolium hydrogen sulfate ([Sipim]HSO_4) as recyclable heterogeneous catalyst for the production of 3,4-dihydropyrano[c]-chromenes and pyrano[2,3-c]pyrazoles derivatives under solventless conditions (Fig. 10.24) (Khodabakhsh & Abolhassan, 2013).

Sarrafi et al. introduced an efficient method for the production of 4H-Chromene derivatives by one-pot condensation of aldehydes or isatins with malononitrile and cyclic 1,3 diketones (Fig. 10.25). The reaction proceeded in the presence of MSNs which serves as a highly effective, neutral, biocompatible, and

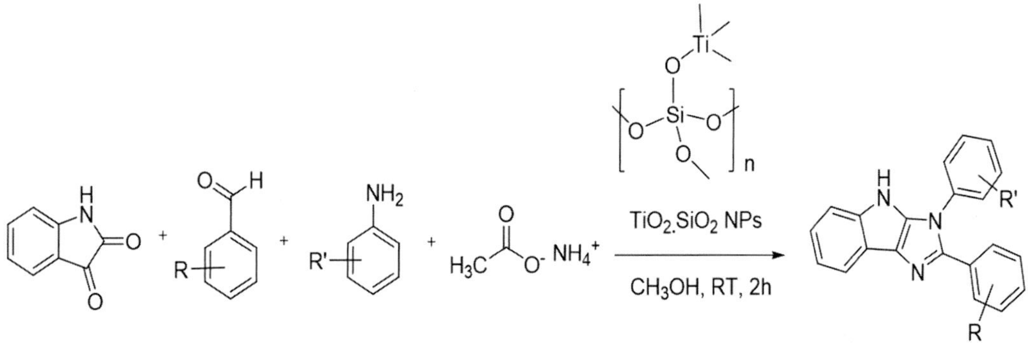

Figure 10.22 Synthetic route to the production of 2,3-diaryl-3,4-dihydroimidazo[4,5-b] indole derivatives in the presence of $TiO_2.SiO_2$ NPs.

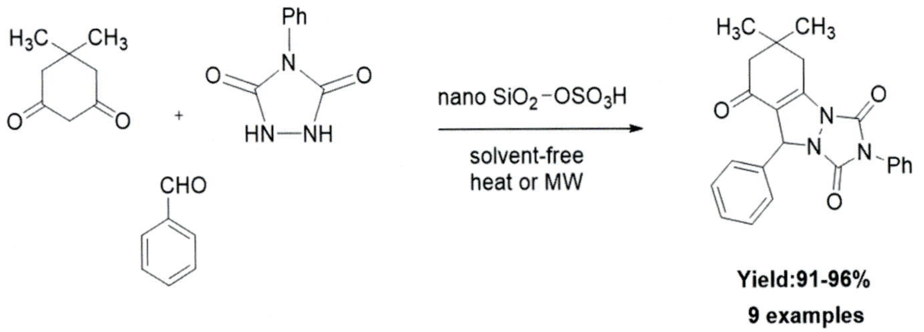

Figure 10.23 Nano SiO₂-SO₃H catalyzed the synthesis of triazolo[1,2-a]indazole-trione derivatives.

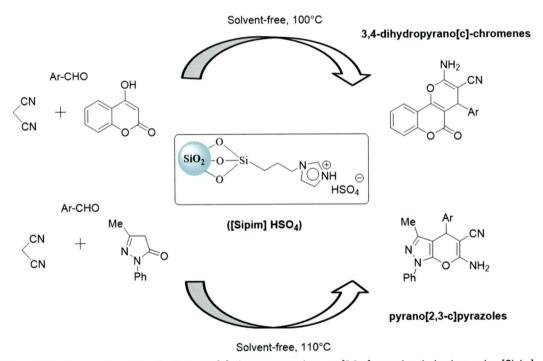

Figure 10.24 Preparation of 3,4-dihydropyrano[c]-chromenes and pyrano[2,3-c]pyrazoles derivatives using [Sipim] HSO₄.

recoverable catalyst for the generation of an excellent yield of chromene derivatives along with high reaction rates (Sarrafi, Mehrasbi, Vahid, & Tajbakhsh, 2012). Kohzadian et al. reported a highly effective and rapid method for the preparation of pyrido[2,3-d:6,5-d′]dipyrimidines via the one-pot multicomponent reaction of arylaldehydes, 2-thiobarbutaric acid, and NH$_4$OAc. The desired reaction proceeded smoothly under solventless conditions in the existence of recoverable mesoporous silica-based nanomaterial (nano-[SiO$_2$-R-NMe$_2$SO$_3$H][Cl]) (Fig. 10.26). High yield of products, short reaction time, and mild reaction temperature and other conditions are some of the significant advantages offered by the current protocol (Kohzadian & Zare, 2020).

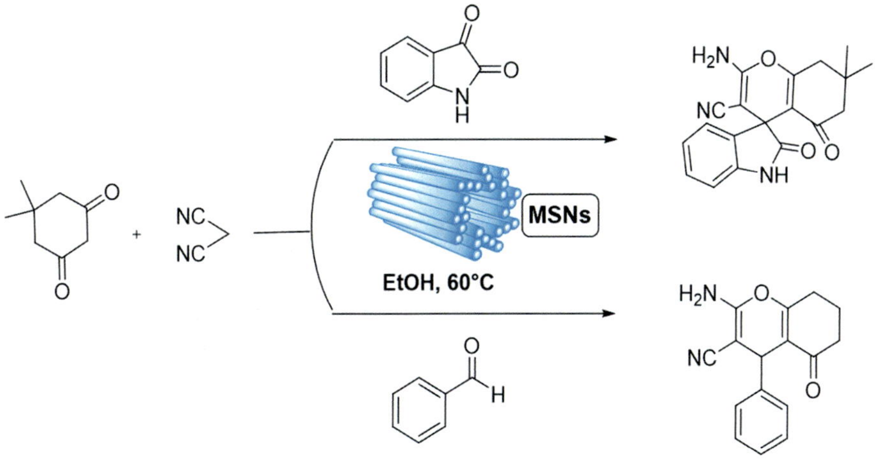

Figure 10.25 Preparation of functionalized 4H-Chromene derivatives in the presence of MSNs.

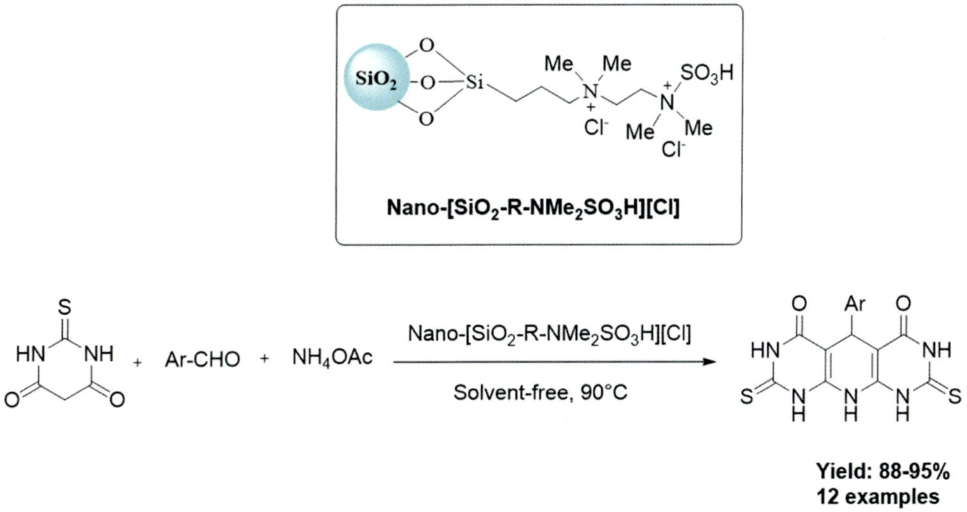

Figure 10.26 One-pot preparation of pyrido[2,3-d:6,5-d′]dipyrimidines using nano-[SiO$_2$-R-NMe$_2$SO$_3$H][Cl].

However, Nasresfahani et al. modified the mesoporous silica NPs by functionalization with homopiperazine sulfamic acid, denoted as MSNs-HPZ-SO$_3$H for the one-pot three-component Strecker reaction of aldehydes or ketones with amines and trimethylsilylcyanides (TMSCN) under solventless conditions, so as to prepare α-amino nitriles (Fig. 10.27(A)). High surface area, quite acidic nature, and recoverability were the key highlights of these types of nanocatalyst (Nasresfahani, Kassaee, & Eidi, 2019). On the other side, Rajput et al. presented the application of magnetically recoverable Fe$_3$O$_4$@SiO$_2$-Pr-NH-SO$_3$H NPs as an effective catalyst for the Strecker reaction in an aqueous medium (Fig. 10.27(B)). Two different catalysts were prepared by silanization of two different alkoxy groups, that is 3-APTES and 3-APTMS, on the surface of SiO$_2$-coated Fe$_3$O$_4$ NPs, followed by immobilization of chlorosulfonic acid on their surfaces. It was predicted that the sulfonic acid functionalized NPs,

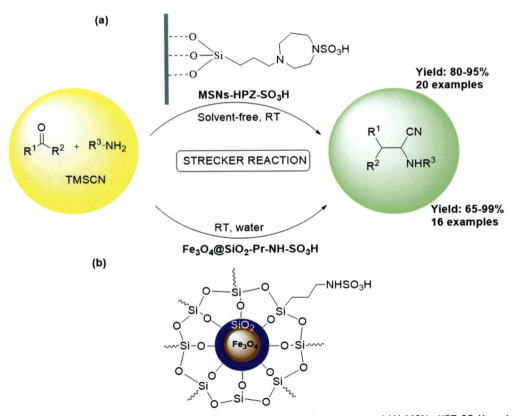

Figure 10.27 Preparation of α-amino nitriles via Strecker reaction in the presence of (A) MSNs-HPZ-SO$_3$H; and (B) Fe$_3$O$_4$@SiO$_2$-Pr-NH-SO$_3$H NPs.

which employed APTES as silanizing agent exhibited better activity, in terms of yield and reaction time, than APTMS. This was supposed to be because of the comparative higher surface area and pore volume offered by the APTES silanizing agent than APTMS (Singh, Rajput, Arora, & Jigyasa, 2016).

Similarly, Rajput et al. had carried out the synthesis of oxytyramine grafted SiO_2-coated Fe_3O_4 NPs for one-pot multicomponent synthesis of cyclohexane carbonitrile derivatives. The admirable features offered by this protocol were simpler catalyst preparation method, short reaction time, quite simple workup, high purification of products, easier recovery, and recyclability of catalyst (Arora, Rajput, & Singh, 2015). Apart from this, the heterogenization of silicotungstic acid on the amino-functionalized silica-coated MNPs were also accomplished by this research group (as depicted in Fig. 10.28) so as to develop an easily recoverable and recyclable nanocatalyst. It was thus deduced that comparative to corresponding homogenous counterpart, the prepared heterogeneous catalyst (STA-amine-Si-magnetite) showed the better activity for the preparations of 1H-Pyrazolo[1,2-b] phthalazinedione derivatives (Arora & Rajput, 2018). Moghanian et al. depicted the preparation of sulfanilic acid functionalized silica-coated nano-Fe_3O_4 particles (MNPs-$PhSO_3H$) using a simple, facile, and inexpensive method. The catalytic activity of this magnetic organic–inorganic hybrid nanocomposite was investigated in the solvent-free condensation reaction of aldehydes, 2-naphthols, and amides/urea/amine for the production of 1-amido and 1-aminoalkyl-2-naphthol derivatives (Fig. 10.29(A)). The supermagnetic nature of the developed catalyst allowed its facile recoverability and thus reusability for several consecutive runs without having any appreciable loss in its efficiency (Moghanian, Mobinikhaledi, Blackman, & Sarough-Farahani, 2014).

In addition to this, sulfonic acid-functionalized silica-coated $CuFe_2O_4$ NPs were synthesized and well characterized by

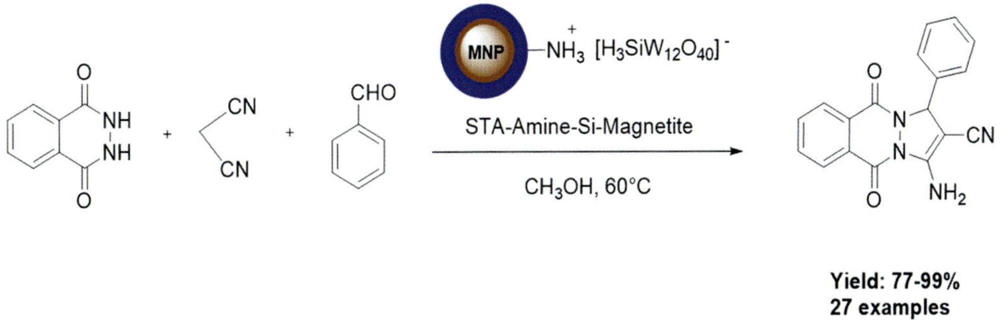

Yield: 77-99%
27 examples

Figure 10.28 STA-Amine-Si-magnetite NPs catalyzed synthesis of 1H-Pyrazolo[1,2-b] phthalazinedione derivatives.

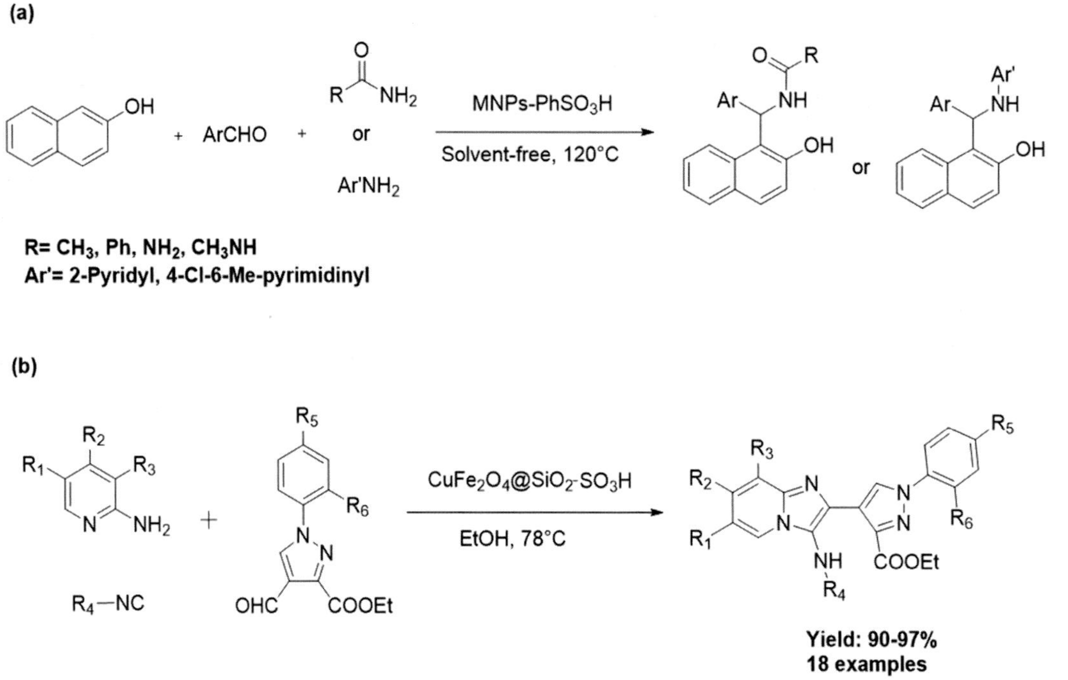

Figure 10.29 Schematic illustration of the synthesis of (A) 1-amido and 1-aminoalkyl-2-naphthol derivatives, and (B) 2-pyrazole-3-amino-imidazo-fused polyheterocycles catalyzed by MNPs-PhSO$_3$H and CuFe$_2$O$_4$@SiO$_2$-SO$_3$H NPs, respectively.

Shrivastava et al. As shown in Fig. 10.29(B), the catalytic activity of synthesized CuFe$_2$O$_4$@SiO$_2$-SO$_3$H NPs was then examined in the production of biologically and medicinally significant 2-pyrazole-3-amino-imidazo-fused polyheterocycles (Swami, Agarwala, & Shrivastava, 2016).

Mahmoodi et al. introduced the application of another functionalized silica-coated MNPs for MCRs. They had formulated the synthesis of piperidinium benzene-1,3 disulfonate functionalized SiO$_2$@Fe$_3$O$_4$ nanocatalyst (PBDS-SCMNPs). The fabricated nanocatalyst was further found to be effective for the green synthesis of pyran derivatives in sufficient yields (Fig. 10.30) (Ghorbani-Vaghei, Izadkhah, & Mahmoodi, 2017).

10.3.5 Carbon dioxide conversion

Carbon dioxide (CO$_2$) has been designated as the major anthropogenic greenhouse gas. Since, the concentration of CO$_2$ is constantly increasing in the atmosphere, there exists the

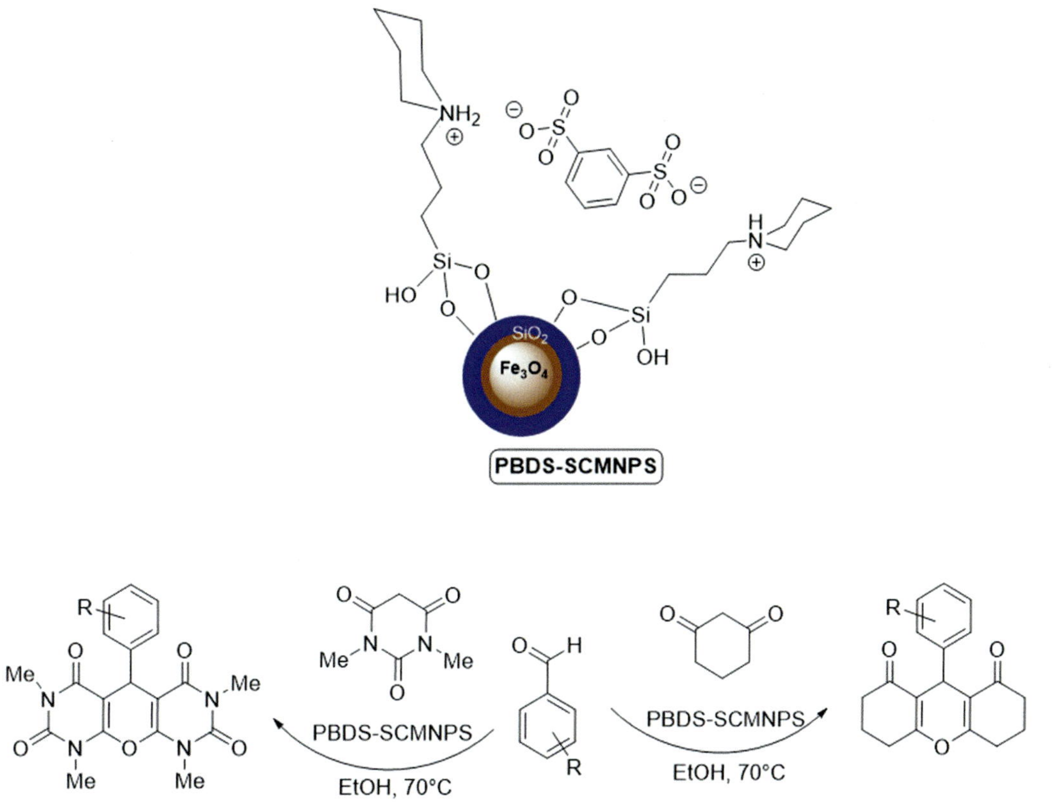

Figure 10.30 Schematic representation of one-pot production of pyran derivatives using PBDS-SCMNPs.

immediate need to design the new technologies which are capable of reducing the environmental impact of CO_2 (Khdary & Abdelsalam, 2020). Recently, the catalytic conversion of carbon dioxide into valuable chemicals has been considered to be an effective method for the fixation of atmospheric CO_2. The numerous homogeneous and heterogeneous catalysts have been found to be active for fixation of CO_2. The following section will cover some of the important CO_2 conversion reactions catalyzed by silicon-based nanohybrids.

The cyclic carbonates are known to possess diverse applications, such as an aprotic high-boiling polar solvent, electrolytes for batteries, precursors for polymeric materials, fuel additives, plastic materials, and significant intermediates for the synthesis of fine chemicals, such as dialkyl carbonates, pyrimidines, glycols, and carbamates. Thus the production of cyclic carbonates by the reaction of CO_2 with epoxides is considered to be one of the crucial pathways for CO_2 utilization (Fig. 10.31). The traditional method for the synthesis of cyclic carbonates relies on the use of highly toxic, hazardous, and corrosive phosgene. Apart from this, huge amounts of chlorinated solvents are produced as by-products, thus diminishing the industrial application of this procedure (Calabrese, Giacalone, & Aprile, 2019). Therefore Park et al. introduced a viable industrial-scale sulfonic acid tethered mesoporous silica (SBA-15-SO_3H) catalyst for the synthesis of styrene carbonate from CO_2 and styrene. The reaction took place efficiently under mild conditions in the presence of tetrabutyl ammonium bromide (TBAB) as a cocatalyst (Table 10.1, entry 1). The scope of the reaction was generalized with a variety of epoxides and the effect of various reaction parameters such as effect of temperature; CO_2 pressure, etc. was also investigated in order to select the optimum condition for the reaction. It was demonstrated that a turnover of 920 can be achieved at a mild reaction temperature of 80°C. Apart from this, the fabricated catalyst was found to be thermally stable and capable of being recycled a number of times without significant reduction in its activity (Ravi, Roshan, Tharun, Kathalikkattil, & Park, 2015).

The application of functionalized silica supported 1,4-diazabicyclo [2.2.2] octane (DABCO)-based ammonium salts as an effective recoverable heterogeneous catalyst for the preparation of cyclic carbonates from epoxides and $_2$ was reported by Hajipour et al. (Table 10.1, entry 2). The reaction followed the green chemistry protocols as it was carried out smoothly in the presence of synthesized acid-base bifunctional catalyst without the presence of any metal cocatalyst and organic dissolvable. It was demonstrated that the immobilized N-benzyl DABCO bromide having a bulkier group and more nucleophilic halide anion exhibited much better catalytic performance than others. The research group also illustrated that since Si-OH groups present on the silica exhibited a synergistic effect with the halide ion, hence the silica grafted heterogeneous catalyst showed higher catalytic activity than its homogeneous analog (Hajipour, Heidari, & Kozehgary, 2015). Park et al. had carried out the synthesis of series of carboxyl-group-functionalized

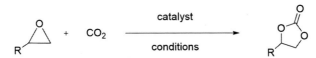

Figure 10.31 General reaction for the generation of cyclic carbonates via cycloaddition of epoxide with CO_2.

Table 10.1 Numerous catalysts utilized for CO_2 conversion.

Sr. No	Type of catalyst	Method of synthesis	Conditions	Yield	Selectivity
1.	SBA-15-SO$_3$H	Sol–gel	0.1 mol% catalyst, TBAB (0.4 mol%), 80°C, CO_2 (1 MPa), 8 h	11%–93%	98%–99%
2.	SiO$_2$-supported ammonium salt based on DABCO	–	0.6 g catalyst, 100°C, CO_2 (4 atm), solventless, 30 h	20%–95%	70%–99%
3.	Silica grafted carboxyl-group functionalized imidazolium-based ionic liquids (CILX-Si)	–	0.5 g catalyst, CO_2 (1.62 MPa), 110°C–115°C, 3–24 h	82%–99%	–
4.	Cu-ABF@ASMNPs	Coprecipitation followed by immobilization	50 mg catalyst, DBU (4 mol%), CO_2 (1 atm), 12 h, 80°C	89%–99%	–
5.	Co@Si0.95	–	0.2 g catalyst, P = 2.0 MPa, 320°C	3.0 mmol$_{methanol}$g$_{catalyst}^{-1}$h^{-1}	70.5%
6.	Porous Si NPs	Magnesiothermic reduction of silica NPs	100°C–150°C, P = 10 bars	0.203 mmol$_{methanol}$	–
7.	Cu/ZnO/ZrO$_2$@ SBA-15	Impregnation-sol–gel autocomustion-combined strategy	250°C, P = 3.0 MPa	376 mg$_{methanol}$ g$_{catalyst}^{-1}$h^{-1}	30.6%

imidazolium-based ionic liquids (CIL's) bearing different halides anions (Cl$^-$, Br$^-$, I$^-$) and further grafted it onto the silica in order to order to design a heterogeneous catalyst, represented as CLX-Si, for cycloaddition reaction of epoxide and CO_2 (Table 10.1, entry 3). The catalytic activity of this catalyst was also compared with the silica grafted hydroxyl-group-functionalized ionic liquid and silica grafted nonfunctionalized ionic liquid. It was observed that the presence of –COOH groups significantly accelerated the reaction owing to the existence of greater synergistic effect with the halide anions. Thus it was concluded that amongst the three fabricated catalysts, the CIL grafted silica showed the highest activity and selectivity (Han, Choi, Choi, Liu, & Park, 2011). However, Sharma et al. employed the use of metal-based functionalized silica-coated magnetic nanohybrid as a catalyst for this reaction. They had formulated the immobilization of 2-acetylbenzofuran onto the surface of amine modified silica-coated MNPs, followed by its coordination with copper acetate (Cu-ABF@ASMNPs) (Table 10.1, entry 4). In the presence of the synthesized nanocatalyst, the fixation of CO_2 was preceded effectively at atmospheric pressure. Easy magnetic recovery, short reaction times, relatively lower temperature, solvent-free and organic-halide free reaction conditions were the noteworthy highlights of the developed protocol (Sharma et al., 2018).

The conversion of CO_2 into the value-added chemicals such as ethanol, methanol and formic acid comprises another significant CO_2 fixation reaction. According to Wang et al. silica accelerates the selective hydrogenation of CO_2 to methanol on the cobalt catalysts. The catalyst (Co@Si$_x$) was synthesized by incorporating Cobalt NPs onto the amorphous silica, which acted as a support and ligand. The designed catalyst exhibited the Co-O-SiO$_n$ linkages which favors the formation and hence stabilization of intermediate methoxy species (*CH$_3$O) required for the methanol production rather than dissociation of C-O for hydrocarbon formation. Thus we can say that the fabricated silica based catalyst, represented as Co@Si0.95, (Table 10.1, entry 5) provided the optimum surface for methanol formation by avoiding the side reactions (Wang et al., 2020). However, Dasog et al. selectively produced methanol from the CO_2 using high surface area, hydride terminated porous silicon NPs (Table 10.1, entry 6) without the need of any cocatalyst. The synthesis of porous Si-NPs was based on the magnesiothermic reduction method, preceded by the etching with nitric acid/ hydrofluoric acid to obtain the hydride terminated Si-NPs. It was investigated that the stoichiometric reaction between the CO_2 and the hydride surface of Si-NPs efficiently took place at a minimum temperature and pressure of 100°C and 10 bars, respectively. The mechanistic insights showed that the physical trapping of CO_2 molecules within the porous Si framework allowed the efficient hydrogen transfer between the two. The recyclability test of the fabricated catalyst was also carried out and it was inferred that the Si-NPs was recyclable up to four cycles without any critical reduction in methanol yield (Dasog, Kraus, Sinelnikov, Veinot, & Rieger, 2017).

Mureddu et al. accomplished the hydrogenation reaction for the production of methanol from CO_2 using a mesoporous support material; SBA-15. They had carried out the synthesis of Cu/ZnO@SBA-15 and Cu/ZnO/ZrO$_2$@SBA-15 nanocomposites by an innovative impregnation-sol—gel auto combustion combined strategy (Table 10.1, entry 7). The large surface area of the support material allowed the high dispersion of active phase into/over the well-ordered mesoporous channels. The significant confinement of the active phase in the SBA-15 structure facilitated its ability to interact with H_2 and CO_2, thus enhances the catalytic performance too. It was found that the SBA-15 supported catalyst with the lowest Cu/Zn molar ratio exhibited much higher catalytic activity than the corresponding unsupported catalyst (Mureddu, Ferrara, & Pettinau, 2019). However, Wei et al. demonstrated that the particle size and the distribution of the active phase were not the only factors responsible for high activity of any catalyst. For this, they studied the effect of additives (metal oxides) on the catalytic performance of CuO-ZnO/SBA-15 catalyzed methanol production. The research group had synthesized three different metal oxide supported porous catalysts that were CuO-ZnO/SBA-15, CuO-ZnO-MnO_2/SBA-15 and CuO-ZnO-ZrO_2/SBA-15 by impregnation method. They observed that the introduction of metal-oxide in the catalyst induced the formation of complex multi-oxide layer on its surface which significantly enhanced the dispersion of CuO grains on SBA-15 surface. It was also deduced that comparative to Mn, the presence of Zr significantly increases the basic site of the catalyst, thus resulted into the quite good interaction between the metal and oxygen in the catalyst and hence responsible for better catalytic activity than other (Min, Wei, Hai-chuan, Hai-hui, & Wen-gui, 2019).

10.3.6 Addition reactions

Addition reactions constitute a group of organic reactions in which two or more moieties combine to give a large adduct. These types of reactions are limited to unsaturated compounds, such as alkenes, alkynes, carbonyl, or imine-containing moieties. Banerjee et al. demonstrated the application of neutral and stable SiO_2 NPs in the production of biologically and chemically important thio-adducts via addition reactions. They had observed that the silica NPs effectively catalyze 1,2-addition reaction of thiols to alkenes, alkynes, and alkyl/acyl halides and 1,4-addition of thiols to conjugated alkenes at room temperature, so as to produce thioethers, thioesters, vinyl thioethers, and thio-Michael adducts, respectively (Fig. 10.32). It was also indicated that under neutral pH conditions, the existence of SiO^- polarized the SH hydrogen of thiols with the Si-OH-promoted removal of halide ion or acetate ion. The significant features such protocol were gentle reaction condition, high yields of thioethers and thioesters, the utilization of inexpensive, nontoxic catalyst, and quite good recyclability (Banerjee, Das, Alvarez, & Santra, 2010).

The same research group also introduced a simple and straightforward route for the bis-Michael addition of active methylene compounds to α,β-unsaturated ketones, esters and nitriles in a single step using silica NPs (Fig. 10.33). The synthesis of silica NPs was based on the well-known Stöber method, which involved the basic hydrolysis and condensation reaction of TEOS in 1:1 water-ethanol mixture. Since under neutral pH conditions, the hydroxyl groups present on the surface of silica NPs exist in a partly deprotonated form, thus they predicted that the OH group stabilized the enol form of active methylene

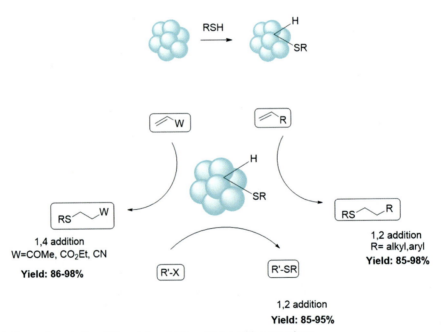

Figure 10.32 Illustration of silica NPs catalyzed 1,2-and 1,4-addition reactions.

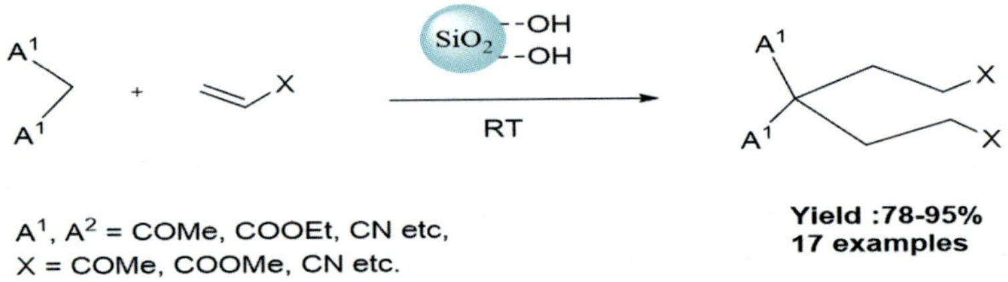

Figure 10.33 Silica NPs catalyzed bis-Michael addition of active methylene compounds to conjugated alkenes.

compounds and polarized the conjugated alkene via H-bonding with carbonyl oxygen, whereas SiO⁻ promoted the nucleophilic attack of the enolized active methylene compound. The catalyst silica NPs were also found to be reusable for up to seven runs without any detectable activity loss (Banerjee & Santra, 2009). On the other side, Guo et al. explored the application of acid-functionalized silica (SiO_2-SO_3H) as a catalyst in the Michael addition reactions. The research group had introduced a highly efficient, inexpensive, convenient, and green protocol for chemoselective aza-Michael addition reaction of amines/thiols to α,β-unsaturated compounds using silica sulfuric acid (SSA) (SiO_2-SO_3H), as represented in Fig. 10.34. Moreover, the reaction was proceeded efficaciously at room temperature in the absence of any organic solvent (Wang, Yuan, & Guo, 2009).

10.3.7 Miscellaneous reaction

Apart from the above discussed classification of various types of organic reactions, the silicon-based nanohybrids have also been used as catalyst in several other kinds of reactions. Since, formic acid is known to be a safe and convenient source for fuel production; Mielby et al. performed the decomposition of formic acid over silica-encapsulated and amine-functionalized gold NPs in

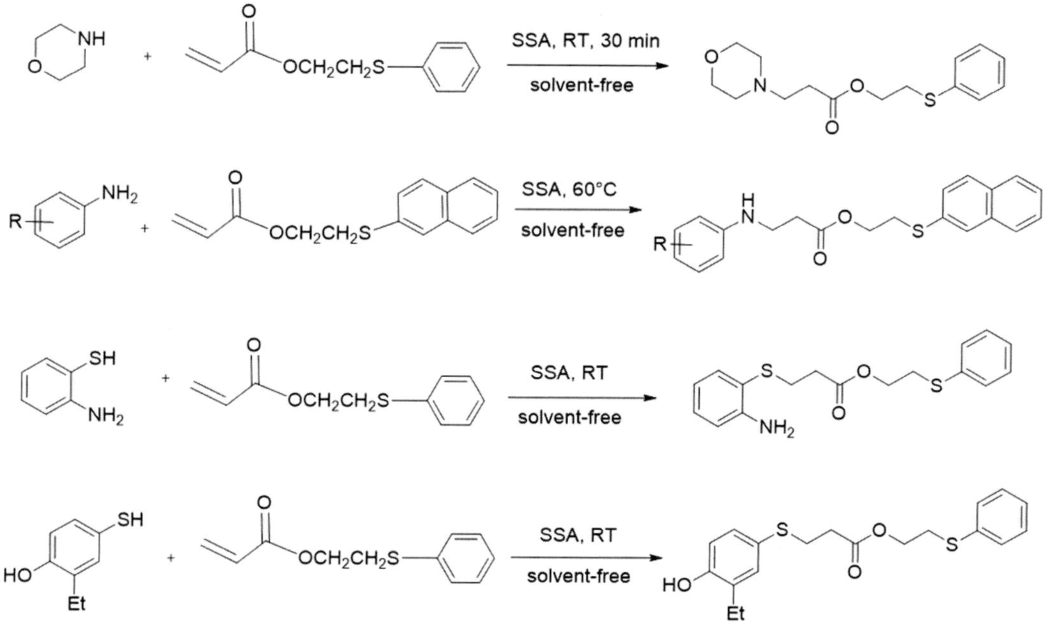

Figure 10.34 SSA catalyzed aza-Michael addition reactions of amine/thiols to α,β-unsaturated compounds.

order to produce hydrogen gas (Fig. 10.35). The Au@SiO$_2$ core–shell nanocatalyst was synthesized in a reverse micelle system which made it convenient to control the particle size distribution of Au NPs and shell thickness of SiO$_2$ (Mielby, Kunov-Kruse, & Kegnæs, 2017). However, Jin et al. utilized mesoporous silica-supported Pd-MnO$_X$ catalyst for the room temperature decomposition of formic acid (Jin et al., 2016).

An effective and environment-friendly approach for the preparation of 1,3,5-triarylbenzenes, which are significant intermediates for the production of dendrimers, conjugated star polyaromatics, pharmaceuticals, and buckminsterfullerenes, was introduced by Ghanbaripour et al. The synthesis of 1,3,5-triarylbenzenes was carried out via triple-self condensation of acetophenone using nanosilica sulfuric acid (SSA) as a catalyst under microwave irradiation and solventless conditions (Fig. 10.36(A)). High yields of products, simple workup, lower reaction times, green conditions, and reusable catalyst were the noteworthy features of this approach (Ghanbaripour et al., 2012).

Tayebee et al. modified the surface of pretreated natural zeolite nanoclinoptilolite (NCP) by hexadecyltrimethylammoniumcation (HDTMA) in order to synthesize a cost-effective and efficient catalyst (NCP/HDTMA) for the production of 1,3,5-triarylbenzene

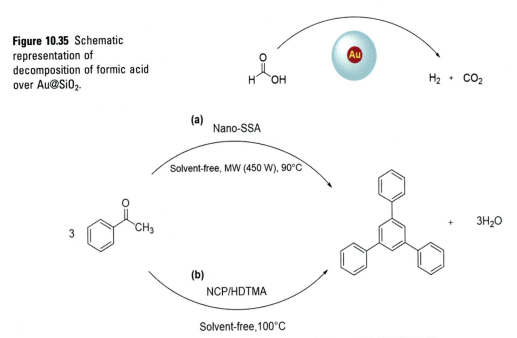

Figure 10.35 Schematic representation of decomposition of formic acid over Au@SiO$_2$.

Figure 10.36 Preparation of 1,3,5-triarylbenzene using (A) nano-SSA, and (B) NCP/HDTMA.

from acetophenone (Fig. 10.36(B)). Such types of cyclotrimerization reactions are highly recommended with respect to the green chemistry aspects as the only side product of the reaction was water (Tayebee et al., 2015).

Among the various biomass-derived chemicals, 5-hydroxymethylfurfural(5-HMF) serves as an important intermediate for various fine chemicals and useful fuels. In an attempt to synthesize 5-HMF, the methodology of dehydration of fructose using acid catalyst was adopted. However, in that case, 5-HMF was either again rehydrated to produce levulinic acid and formic acid or oligomerized/polymerized to produce some other by-products, thus lowering the yield of 5-HMF. Hence, as an alternative, various research groups adopted the method of utilization of ionic liquids as a novel reaction medium and catalyst for this reaction. Sidhpuria et al. carried out the immobilization of ionic liquid 1-(tri-ethoxysilyl-propyl)-3-methyl-imidazolium hydrogen sulfate on the surface of silica NPs in order to design an efficient heterogeneous catalyst, Si-IL-HSO$_4$, for the dehydration of fructose selectively to 5-HMF, represented in Fig. 10.37. In addition to this, they also demonstrated that the synthesized catalyst can be recycled and reused suitably for seven times without losing its catalytic activity for fructose conversion and 5-HMF production (Sidhpuria, Daniel-Da-silva, Trindade, & Coutinho, 2011).

Chen and research group introduced the effect of silica coating on the performance of MNPs-supported Grubbs–Hoveyda catalysts for ring-closing metathesis. In the consideration of this, they had immobilized the Grubbs–Hoveyda type ruthenium–carbene complex on both, silica-coated and uncoated MNPs having an imidazolium-based ionic liquid moiety as linker. It was illustrated that both these MNPs-supported catalysts exhibited comparable catalytic activity in the ring-closing metathesis of dienes as that of the corresponding homogenous

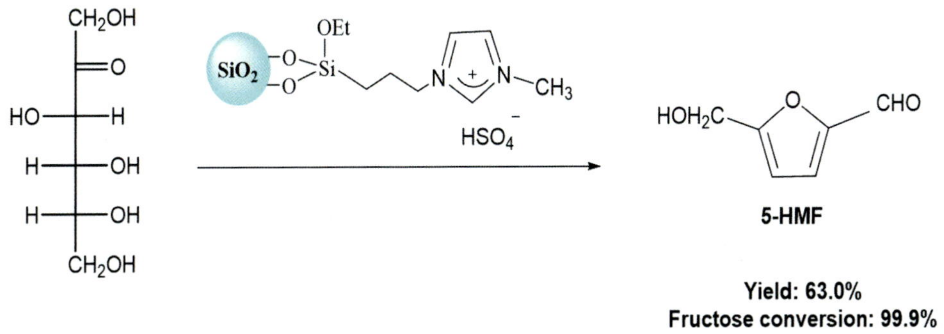

Figure 10.37 Selective dehydration of fructose to 5-HMF using Si-IL-HSO$_4$.

imidazolium tagged ruthemium catalyst (Fig. 10.38). However, in terms of recyclability the silica-coated-MNP-supported catalysts was observed to be superior to the silica-uncoated one, as it offered reusability for 14 times, while maintaining its catalytic activity. This was due to the minimal leaching of Ru and high compatibility of the ionic support material with the reaction medium in the case of silica-coated MNP-supported catalyst. The presence of a relatively larger amount of the hydroxyl functional groups on the surface of silica-coated MNP-supported catalyst allowed the grafting of an increased amount of the chelating ligand moiety on its surface, which ultimately led to the improvement in its recyclability by minimizing the leaching of active Ru species from its surface (Chen et al., 2015; Chen et al., 2017).

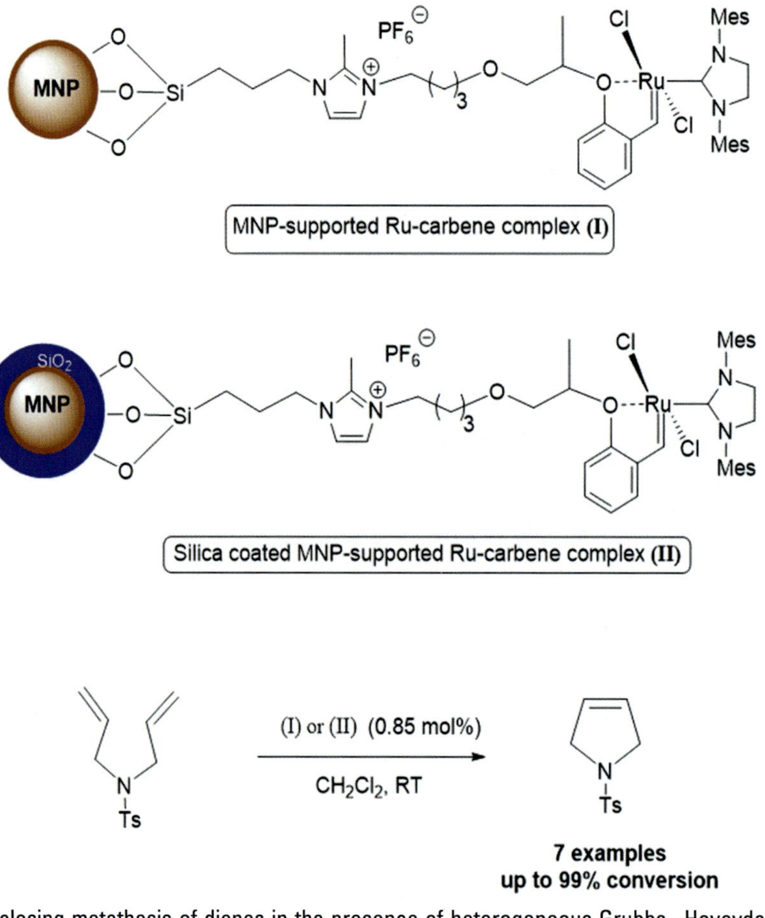

Figure 10.38 Ring closing metathesis of dienes in the presence of heterogeneous Grubbs–Hoveyda type Ru–carbene complex.

10.4 Conclusion and future scope

Due to their widespread application, much consideration has been devoted toward the manufacturing of the silicon-based hybrid nanomaterials. The high surface area, outstanding mechanical and thermal stability, nanometer range of silica, and the existence of silanol groups on its surface permits immobilization of various groups on its surface. Apart from this, silica can also be grafted on the surface of MNPs in order to prevent them from agglomeration or to provide chemical stability. Thus the main motive of this chapter was to briefly summarize the catalytic application of various known diversely structured silicon-incorporated hybrid NPs in various important organic reactions, such as formation of C-C, C-N, C-O bonds via coupling and cross-coupling reactions, oxidation, reduction, decomposition reactions, one-pot preparation of heterocyclic compounds, and addition reactions. In addition to high activity and selectivity, easy workup, efficient recoverability, and recyclability are the salient features of these hybrid nanomaterials. Although remarkable progress has been made in the field of nanocatalysis, still there is a need to discover some advanced, green, and sustainable catalysts for various leading organic transformations. The future demand for catalytic systems lies in the development of multifunctional, robust, recyclable, and recoverable nanohybrid materials which can effectively catalyze distinct organic reactions with high atom-efficiencies, yield, enantioselectivity, and chemoselectivity.

References

Ahmad, B., Mehdi, S.-F., Nona, E., & Mohammad, M. H. (2015). Silver nanoparticles supported on silica-coated ferrite as magnetic and reusable catalysts for oxidant-free alcohol dehydrogenation. *RSC Advances*, 22503–22509. Available from https://doi.org/10.1039/C4RA15498C.

Arora, P., & Rajput, J. K. (2018). Amelioration of $H_4[W_{12}SiO_{40}]$ by nanomagnetic heterogenization: For the synthesis of 1H−pyrazolo[1,2-b]phthalazinedione derivatives. *Applied Organometallic Chemistry*, *32*(2). Available from https://doi.org/10.1002/aoc.4001.

Arora, P., Rajput, J. K., & Singh, H. (2015). Nanostructured oxytyramine catalyst for the facile one-pot synthesis of cyclohexanecarbonitrile derivatives. *RSC Advances*, *5*(118), 97212–97223. Available from https://doi.org/10.1039/c5ra20150k.

Ashraf, M. A., Liu, Z., Peng, W. X., & Zhou, L. (2020). Glycerol Cu(II) complex supported on Fe_3O_4 magnetic nanoparticles: A new and highly efficient reusable catalyst for the formation of aryl-sulfur and aryl-oxygen bonds. *Catalysis Letters*, *150*(4), 1128–1141. Available from https://doi.org/10.1007/s10562-019-02973-7.

Banerjee, S., & Santra, S. (2009). Remarkable catalytic activity of silica nanoparticle in the bis-Michael addition of active methylene compounds to conjugated alkenes. *Tetrahedron Letters, 50*(18), 2037−2040. Available from https://doi.org/10.1016/j.tetlet.2009.01.154.

Banerjee, S., Das, J., Alvarez, R. P., & Santra, S. (2010). Silica nanoparticles as a reusable catalyst: A straightforward route for the synthesis of thioethers, thioesters, vinyl thioethers and thio-Michael adducts under neutral reaction conditions. *New Journal of Chemistry, 34*(2), 302−306. Available from https://doi.org/10.1039/b9nj00399a.

Cai, M., Xu, Q., & Huang, Y. (2007). Heterogeneous Suzuki reaction catalyzed by MCM-41-supported sulfur palladium(0) complex. *Journal of Molecular Catalysis A: Chemical, 271*(1−2), 93−97. Available from https://doi.org/10.1016/j.molcata.2007.02.040.

Calabrese, C., Giacalone, F., & Aprile, C. (2019). Hybrid catalysts for CO_2 conversion into cyclic carbonates. *Catalysts, 9*(4). Available from https://doi.org/10.3390/catal9040325.

Cavallaro, G., Lazzara, G., Rozhina, E., Konnova, S., Kryuchkova, M., Khaertdinov, N., & Fakhrullin, R. (2019). Organic-nanoclay composite materials as removal agents for environmental decontamination. *RSC Advances, 9*(69), 40553−40564. Available from https://doi.org/10.1039/c9ra08230a.

Chaturvedi, S., Dave, P. N., & Shah, N. K. (2012). Applications of nano-catalyst in new era. *Journal of Saudi Chemical Society, 16*(3), 307−325. Available from https://doi.org/10.1016/j.jscs.2011.01.015.

Chen, S. W., Zhang, Z. C., Zhai, N. N., Zhong, C. M., & Lee, S. G. (2015). The effect of silica-coating on catalyst recyclability in ionic magnetic nanoparticle-supported Grubbs-Hoveyda catalysts for ring-closing metathesis. *Tetrahedron, 71*(4), 648−653. Available from https://doi.org/10.1016/j.tet.2014.12.024.

Chen, F., Kreyenschulte, C., Radnik, J., Lund, H., Surkus, A. E., Junge, K., & Beller, M. (2017). Selective semihydrogenation of alkynes with N-graphitic-modified cobalt nanoparticles supported on silica. *ACS Catalysis, 7*(3), 1526−1532. Available from https://doi.org/10.1021/acscatal.6b03140.

Correa, A., Carril, M., & Bolm, C. (2008). Iron-catalyzed S-arylation of thiols with aryl iodides. *Angewandte Chemie - International Edition, 47*(15), 2880−2883. Available from https://doi.org/10.1002/anie.200705668.

Cuong, D.-V., Housseinou, B., Zora, E.-B., Jean-Mario, N., Marc, M. J., Yuefeng, L., & Cuong, P.-H. (2016). Silicon carbide foam as a porous support platform for catalytic applications. *New Journal of Chemistry*, 4285−4299. Available from https://doi.org/10.1039/C5NJ02847G.

Das, S. K., Khan, M. M. R., Guha, A. K., & Naskar, N. (2013). Bio-inspired fabrication of silver nanoparticles on nanostructured silica: Characterization and application as a highly efficient hydrogenation catalyst. *Green Chemistry, 15*(9), 2548−2557. Available from https://doi.org/10.1039/c3gc40310f.

Dasog, M., Kraus, S., Sinelnikov, R., Veinot, J. G. C., & Rieger, B. (2017). CO_2 to methanol conversion using hydride terminated porous silicon nanoparticles. *Chemical Communications*, 3114−3117. Available from https://doi.org/10.1039/C7CC00125H.

Davidson, M., Ji, Y., Leong, G. J., Kovach, N. C., Trewyn, B. G., & Richards, R. M. (2018). Hybrid mesoporous silica/noble-metal nanoparticle materials − Synthesis and catalytic applications. *ACS Applied Nano Materials, 1*(9), 4386−4400. Available from https://doi.org/10.1021/acsanm.8b00967.

Dehbanipour, Z., Moghadam, M., Tangestaninejad, S., Mirkhani, V., & Mohammadpoor-Baltork, I. (2017). Nano−silica supported palladium catalyst: Synthesis, characterization and application of its activity in Sonogashira cross−coupling reactions. *Journal of Organometallic Chemistry, 853*, 5−12. Available from https://doi.org/10.1016/j.jorganchem.2017.10.006.

Dhakshinamoorthy, A., & Pitchumani, K. (2008). Clay entrapped nickel nanoparticles as efficient and recyclable catalysts for hydrogenation of olefins. *Tetrahedron Letters, 49*(11), 1818−1823. Available from https://doi.org/10.1016/j.tetlet.2008.01.061.

Faheem, U. (2008). Clays, nanoclays, and montmorillonite minerals. *Metallurgical and Materials Transactions A*, 2804−2814. Available from https://doi.org/10.1007/s11661-008-9603-5.

Fattahi, N., Ramazani, A., Ahankar, H., Asiabi, P. A., & Kinzhybalo, V. (2019). Tetramethylguanidine-functionalized Fe_3O_4/Chloro-Silane core-shell nanoparticles: an efficient heterogeneous and reusable organocatalyst for Aldol reaction. *Silicon, 11*(3), 1441−1450. Available from https://doi.org/10.1007/s12633-018-9954-5.

Gawande, M. B., Monga, Y., Zboril, R., & Sharma, R. K. (2015). Silica-decorated magnetic nanocomposites for catalytic applications. *Coordination Chemistry Reviews, 288*, 118−143. Available from https://doi.org/10.1016/j.ccr.2015.01.001.

Geedkar, D., Kumar, A., Reen, G. K., & Sharma, P. (2020). Titania-silica nanoparticles ensembles assisted heterogeneous catalytic strategy for the synthesis of pharmacologically significant 2,3-diaryl-3,4-dihydroimidazo[4,5-b]indole scaffolds. *Journal of Heterocyclic Chemistry, 57*(4), 1963−1973. Available from https://doi.org/10.1002/jhet.3925.

Ghanbaripour, R., Mohammadpoor-Baltork, I., Moghadam, M., Khosropour, A. R., Tangestaninejad, S., & Mirkhani, V. (2012). Nano-silica sulfuric acid catalyzed the efficient synthesis of 1,3,5-triarylbenzenes under microwave irradiation. *Journal of the Iranian Chemical Society, 9*(5), 791−798. Available from https://doi.org/10.1007/s13738-012-0089-0.

Ghorbani-Vaghei, R., Izadkhah, V., & Mahmoodi, J. (2017). One-pot synthesis of pyran derivatives using silica-coated magnetic nanoparticles functionalized with piperidinium benzene-1,3-disulfonate as a new efficient reusable catalyst. *Research on Chemical Intermediates, 43*(4), 2299−2314. Available from https://doi.org/10.1007/s11164-016-2762-x.

Ghosh, S. (2019). Mesoporous silica-based nano drug-delivery system synthesis. *Characterization, and Applications*, 285−317. Available from https://doi.org/10.1016/B978-0-12-814033-8.00009-6.

Gogoi, S., Barua, S., Khan, R., & Karak, N. (2019). *Silicon-Based Nanomaterials and Their Polymer Nanocomposites. Nanomaterials and Polymer Nanocomposites* (pp. 261−305). Elsevier.

Hajipour, A. R., Abolfathi, P., & Mohammadsaleh, F. (2016). A click strategy for the immobilization of palladium nanoparticles onto silica: Efficient and recyclable catalysts for carbon-carbon bond formation under mild reaction conditions. *RSC Advances, 6*(81), 78080−78089. Available from https://doi.org/10.1039/c6ra11734a.

Hajipour, A. R., Heidari, Y., & Kozehgary, G. (2015). Silica grafted ammonium salts based on DABCO as heterogeneous catalysts for cyclic carbonate synthesis from carbon dioxide and epoxides. *RSC Advances, 5*(29), 22373−22379. Available from https://doi.org/10.1039/c4ra16083e.

Hajipour, A. R., Shirdashtzade, Z., & Azizi, G. (2014). Copper- and phosphine-free Sonogashira coupling reaction catalyzed by silica-(acac)-supported palladium nanoparticles in water. *Applied Organometallic Chemistry, 28*(9), 696–698. Available from https://doi.org/10.1002/aoc.3184.

Hamidian, H., Fozooni, S., Hassankhani, A., & Mohammadi, S. Z. (2011). One-pot and efficient synthesis of triazolo[1,2-a]indazoletriones via reaction of arylaldehydes with urazole and dimedone catalyzed by silica nanoparticles prepared from rice husk. *Molecules, 16*(11), 9041–9048. Available from https://doi.org/10.3390/molecules16119041.

Han, L., Choi, H. J., Choi, S. J., Liu, B., & Park, D. W. (2011). Ionic liquids containing carboxyl acid moieties grafted onto silica: Synthesis and application as heterogeneous catalysts for cycloaddition reactions of epoxide and carbon dioxide. *Green Chemistry, 13*(4), 1023–1028. Available from https://doi.org/10.1039/c0gc00612b.

Hasaninejad, A., Shekouhy, M., & Zare, A. (2012). Silica nanoparticles efficiently catalyzed synthesis of quinolines and quinoxalines. *Catalysis Science and Technology, 2*(1), 201–214. Available from https://doi.org/10.1039/c1cy00332a.

Hu, M., Yan, X., Hu, X., Feng, R., & Zhou, M. (2019). Synthesis of silver decorated silica nanoparticles with rough surfaces as adsorbent and catalyst for methylene blue removal. *Journal of Sol-Gel Science and Technology, 89*(3), 754–763. Available from https://doi.org/10.1007/s10971-018-4871-z.

Jaber, J., & Mohsen, E. (2016). $Fe_3O_4@SiO_2$ –imid–PMA n magnetic porous nanosphere as recyclable catalyst for the green synthesis of quinoxaline derivatives at room temperature and study of their antifungal activities. *Materials Research Bulletin*, 409–422. Available from https://doi.org/10.1016/j.materresbull.2015.10.002.

Jia, L., Zhang, W., Xu, J., Cao, J., Xu, Z., & Wang, Y. (2018). Facile fabrication of highly active magnetic aminoclay supported palladium nanoparticles for the room temperature catalytic reduction of nitrophenol and nitroanilines. *Nanomaterials, 8*(6). Available from https://doi.org/10.3390/nano8060409.

Jin, M. H., Oh, D., Park, J. H., Lee, C. B., Lee, S. W., Park, J. S., ... Lee, D. W. (2016). Mesoporous silica supported Pd-MnOx catalysts with excellent catalytic activity in room-temperature formic acid decomposition. *Scientific Reports, 6*. Available from https://doi.org/10.1038/srep33502.

Khdary, N. H., & Abdelsalam, M. E. (2020). Polymer-silica nanocomposite membranes for CO_2 capturing. *Arabian Journal of Chemistry, 13*(1), 557–567. Available from https://doi.org/10.1016/j.arabjc.2017.06.001.

Khodabakhsh, N., & Abolhassan, P. (2013). Silica-grafted ionic liquids as recyclable catalysts for the synthesis of 3,4-dihydropyrano[c]chromenes and pyra-no[2,3-c]pyrazoles. *Green and Sustainable Chemistry*, 1–8. Available from https://doi.org/10.4236/gsc.2013.32A001.

Kohzadian, A., & Zare, A. (2020). Effective and rapid synthesis of pyrido[2,3-d:6,5-d′]dipyrimidines catalyzed by a mesoporous recoverable silica-based nanomaterial. *Silicon, 12*(6), 1407–1415. Available from https://doi.org/10.1007/s12633-019-00235-0.

Kumar, S. R., Sriparna, D., & Shivani, S. (2016). Nickel(ii) complex covalently anchored on core shell structured $SiO_2@Fe_3O_4$ nanoparticles: A robust and magnetically retrievable catalyst for direct one-pot reductive amination of ketones. *New Journal of Chemistry*, 2089–2101. Available from https://doi.org/10.1039/C5NJ02495A.

Li, H., Cao, L., Yang, C., Zhang, Z., Zhang, B., & Deng, K. (2017). Selective oxidation of benzyl alcohols to benzoic acid catalyzed by eco-friendly cobalt thioporphyrazine catalyst supported on silica-coated magnetic nanospheres.

Journal of Environmental Sciences, 60, 84–90. Available from https://doi.org/10.1016/j.jes.2017.05.044.

Martínez, S. S., Volzone, C., & Huck, L. (2015). Nanoclay as adsorbent: Evaluation for removing dyes used in the textile industry. *Procedia Materials Science*, 586–591. Available from https://doi.org/10.1016/j.mspro.2015.04.112.

McInnes, S. J. P., & Voelcker, N. H. (2009). Silicon-polymer hybrid materials for drug delivery. *Future Medicinal Chemistry, 1*(6), 1051–1074. Available from https://doi.org/10.4155/fmc.09.90.

Mielby, J., Kunov-Kruse, A. J., & Kegnæs, S. (2017). Decomposition of formic acid over silica encapsulated and amine functionalised gold nanoparticles. *Journal of Catalysis, 345*, 149–156. Available from https://doi.org/10.1016/j.jcat.2016.11.020.

Min, L., Wei, N., Hai-chuan, Y., Hai-hui, H., & Wen-gui, G. (2019). Effect of additive on CuO-ZnO/SBA-15 catalytic performance of CO_2 hydrogenation to methanol. *Journal of Fuel Chemistry and Technology*, 1214–1225. Available from https://doi.org/10.1016/s1872-5813(19)30048-9.

Moghanian, H., Mobinikhaledi, A., Blackman, A. G., & Sarough-Farahani, E. (2014). Sulfanilic acid-functionalized silica-coated magnetite nanoparticles as an efficient, reusable and magnetically separable catalyst for the solvent-free synthesis of 1-amido- and 1-aminoalkyl-2-naphthols. *RSC Advances, 4*(54), 28176–28185. Available from https://doi.org/10.1039/c4ra03676j.

Mondal, J., Modak, A., Dutta, A., & Bhaumik, A. (2011). Facile C-S coupling reaction of aryl iodide and thiophenol catalyzed by Cu-grafted furfural functionalized mesoporous organosilica. *Dalton Transactions, 40*(19), 5228–5235. Available from https://doi.org/10.1039/c0dt01771j.

Mureddu, M., Ferrara, F., & Pettinau, A. (2019). Highly efficient CuO/ZnO/ZrO$_2$@SBA-15 nanocatalysts for methanol synthesis from the catalytic hydrogenation of CO_2. *Applied Catalysis B: Environmental, 258*. Available from https://doi.org/10.1016/j.apcatb.2019.117941.

Narayan, R., Nayak, U. Y., Raichur, A. M., & Garg, S. (2018). Mesoporous silica nanoparticles: A comprehensive review on synthesis and recent advances. *Pharmaceutics, 10*(3). Available from https://doi.org/10.3390/pharmaceutics10030118.

Naseem, K., Begum, R., & Farooqi, Z. H. (2017). Catalytic reduction of 2-nitroaniline: A review. *Environmental Science and Pollution Research, 24*(7), 6446–6460. Available from https://doi.org/10.1007/s11356-016-8317-2.

Nasresfahani, Z., Kassaee, M. Z., & Eidi, E. (2019). Efficient synthesis of α-aminonitriles over homopiperazine sulfamic acid functionalized mesoporous silica nanoparticles (MSNs-HPZ-SO3H), as a reusable acid catalyst. *Journal of the Iranian Chemical Society, 16*(9), 1819–1825. Available from https://doi.org/10.1007/s13738-019-01654-x.

Ngnie, G., Dedzo, G. K., & Detellier, C. (2016). Synthesis and catalytic application of palladium nanoparticles supported on kaolinite-based nanohybrid materials. *Dalton Transactions, 45*(22), 9065–9072. Available from https://doi.org/10.1039/c6dt00982d.

Niknam, K., Habibabad, M. S., Deris, A., Panahi, F., & Reza Hormozi Nezhad, M. (2013). Modification of silica using piperazine for immobilization of palladium nanoparticles: A study of its catalytic activity as an efficient heterogeneous catalyst for Heck and Suzuki reactions. *Journal of the Iranian Chemical Society, 10*(3), 527–534. Available from https://doi.org/10.1007/s13738-012-0188-y.

Pereira, C., Silva, J. F., Pereira, A. M., Araújo, J. P., Blanco, G., Pintado, J. M., & Freire, C. (2011). Hybrid catalyst: From complex immobilization onto silica nanoparticles to catalytic application in the epoxidation of geraniol. *Catalysis Science and Technology, 1*(5), 784–793. Available from https://doi.org/10.1039/c1cy00090j.

Ravi, S., Roshan, R., Tharun, J., Kathalikkattil, A. C., & Park, D. W. (2015). Sulfonic acid functionalized mesoporous SBA-15 as catalyst for styrene carbonate synthesis from CO_2 and styrene oxide at moderate reaction conditions. *Journal of CO_2 Utilization, 10*, 88–94. Available from https://doi.org/10.1016/j.jcou.2015.01.003.

Rostamnia, S., Gholipour, B., Liu, X., Wang, Y., & Arandiyan, H. (2018). NH_2-coordinately immobilized tris(8-quinolinolato)iron onto the silica coated magnetite nanoparticle: Fe_3O_4@SiO_2-FeQ_3 as a selective Fenton-like catalyst for clean oxidation of sulfides. *Journal of Colloid and Interface Science, 511*, 447–455. Available from https://doi.org/10.1016/j.jcis.2017.10.028.

Sappino, C., Primitivo, L., De Angelis, M., Domenici, M. O., Mastrodonato, A., Romdan, I. B., ... Righi, G. (2019). Functionalized magnetic nanoparticles as catalysts for enantioselective Henry reaction. *ACS Omega, 4*(26), 21809–21817. Available from https://doi.org/10.1021/acsomega.9b02683.

Sarkar, S. M., Rahman, M. L., & Yusoff, M. M. (2015). Heck, Suzuki and Sonogashira cross-coupling reactions using ppm level of SBA-16 supported Pd-complex. *New Journal of Chemistry, 39*(5), 3564–3570. Available from https://doi.org/10.1039/c4nj02319f.

Sarrafi, Y., Mehrasbi, E., Vahid, A., & Tajbakhsh, M. (2012). Well-ordered mesoporous silica nanoparticles as a recoverable catalyst for one-pot multicomponent synthesis of 4H-chromene derivatives. *Cuihua Xuebao/Chinese Journal of Catalysis, 33*(9), 1486–1494. Available from https://doi.org/10.1016/s1872-2067(11)60423-3.

Sharma, R. K., Gaur, R., Yadav, M., Goswami, A., Zbořil, R., & Gawande, M. B. (2018). An efficient copper-based magnetic nanocatalyst for the fixation of carbon dioxide at atmospheric pressure. *Scientific Reports, 8*(1). Available from https://doi.org/10.1038/s41598-018-19551-3.

Sharma, R. K., Sharma, S., Dutta, S., Zboril, R., & Gawande, M. B. (2015). Silica-nanosphere-based organic-inorganic hybrid nanomaterials: Synthesis, functionalization and applications in catalysis. *Green Chemistry, 17*(6), 3207–3230. Available from https://doi.org/10.1039/c5gc00381d.

Shen, C., Qiao, J., Zhao, L., Zheng, K., Jin, J., & Zhang, P. (2017). An efficient silica supported Chitosan@vanadium catalyst for asymmetric sulfoxidation and its application in the synthesis of esomeprazole. *Catalysis Communications, 92*, 114–118. Available from https://doi.org/10.1016/j.catcom.2017.01.018.

Shylesh, S., Wang, L., Demeshko, S., & Thiel, W. R. (2010). Facile synthesis of mesoporous magnetic nanocomposites and their catalytic application in carbon-carbon coupling reactions. *ChemCatChem, 2*(12), 1543–1547. Available from https://doi.org/10.1002/cctc.201000215.

Sidhpuria, K. B., Daniel-Da-silva, A. L., Trindade, T., & Coutinho, J. A. P. (2011). Supported ionic liquid silica nanoparticles (SILnPs) as an efficient and recyclable heterogeneous catalyst for the dehydration of fructose to 5-hydroxymethylfurfural. *Green Chemistry, 13*(2), 340–349. Available from https://doi.org/10.1039/c0gc00690d.

Singh, H., & Rajput, J. K. (2017). Sustainable catalysis. *Encyclopedia of Physical Organic Chemistry, Major Reference Works*.

Singh, H., Rajput, J. K., Arora, P., & Jigyasa. (2016). Role of (3-aminopropyl)tri alkoxysilanes in grafting of chlorosulphonic acid immobilized magnetic nanoparticles and their application as heterogeneous catalysts for the green synthesis of α-aminonitriles. *RSC Advances, 6*(88), 84658−84671. Available from https://doi.org/10.1039/c6ra20095h.

Stöber, W., Fink, A., & Bohn, E. (1968). Controlled growth of monodisperse silica spheres in the micron size range. *Journal of Colloid And Interface Science, 26*(1), 62−69. Available from https://doi.org/10.1016/0021-9797(68)90272-5.

Sugimoto, H., Fujii, M., & Imakita, K. (2016). Silicon nanocrystal-noble metal hybrid nanoparticles. *Nanoscale, 8*(21), 10956−10962. Available from https://doi.org/10.1039/c6nr01747a.

Swami, S., Agarwala, A., & Shrivastava, R. (2016). Sulfonic acid functionalized silica-coated $CuFe_2O_4$ core-shell nanoparticles: An efficient and magnetically separable heterogeneous catalyst for the synthesis of 2-pyrazole-3-amino-imidazo-fused polyheterocycles. *New Journal of Chemistry, 40*(11), 9788−9794. Available from https://doi.org/10.1039/c6nj02264b.

Tayebee, R., Jarrahi, M., Maleki, B., Kargar Razi, M., Mokhtari, Z. B., & Baghbanian, S. M. (2015). A new method for the preparation of 1,3,5-triarylbenzenes catalyzed by nanoclinoptilolite/HDTMA. *RSC Advances, 5*(15), 10869−10877. Available from https://doi.org/10.1039/c4ra11216d.

Tolinski, M. (2009). *Overview of fillers and fibers*, 93−119. Available from https://doi.org/10.1016/B978-0-8155-2051-1.00007-8.

Veisi, H., Tamoradi, T., Karmakar, B., & Hemmati, S. (2020). Green tea extract − modified silica gel decorated with palladium nanoparticles as a heterogeneous and recyclable nanocatalyst for Buchwald-Hartwig C−N cross-coupling reactions. *Journal of Physics and Chemistry of Solids, 138*. Available from https://doi.org/10.1016/j.jpcs.2019.109256.

Wang, W., & Kumta, P. N. (2010). Nanostructured hybrid silicon/carbon nanotube heterostructures: Reversible high-capacity lithium-ion anodes. *ACS Nano, 4*(4), 2233−2241. Available from https://doi.org/10.1021/nn901632g.

Wang, Y., Yuan, Y. Q., & Guo, S. R. (2009). Silica sulfuric acid promotes aza-Michael addition reactions under solvent-free condition as a heterogeneous and reusable catalyst. *Molecules, 14*(11), 4779−4789. Available from https://doi.org/10.3390/molecules14114779.

Wang, L., Guan, E., Wang, L., Wang, L., Gong, Z., Cui, Y., ... Xiao, F. S. (2020). Silica accelerates the selective hydrogenation of CO_2 to methanol on cobalt catalysts. *Nature Communications, 11*(1). Available from https://doi.org/10.1038/s41467-020-14817-9.

Weinhold, F., & West, R. (2011). The nature of the silicon-oxygen bond. *Organometallics, 30*(21), 5815−5824. Available from https://doi.org/10.1021/om200675d.

Xu, Q., Hao, W., & Cai, M. (2007). Mercapto-functionalized MCM-41 anchored palladium(0) complex as an efficient catalyst for the heterogeneous Suzuki reaction. *Catalysis Letters, 118*(1−2), 98−102. Available from https://doi.org/10.1007/s10562-007-9157-y.

Yoon, H. J., Choi, J. W., Kang, H., Kang, T., Lee, S. M., Jun, B. H., & Lee, Y. S. (2010). Recyclable NHC-Ni complex immobilized on magnetite/silica nanoparticles for C-S cross-coupling of aryl halides with thiols. *Synlett, 16*, 2518−2522. Available from https://doi.org/10.1055/s-0030-1258545.

Zahedi, R., Asadi, Z., & Firuzabadi, F. D. (2020). A highly active, recyclable and cost-effective magnetic nanoparticles supported copper catalyst for N-arylation reaction. *Catalysis Letters, 150*(1), 65−73. Available from https://doi.org/10.1007/s10562-019-02929-x.

Zeynizadeh, B., & Karami, S. (2019). Synthesis of Ni nanoparticles anchored on cellulose using different reducing agents and their applications toward reduction of 4-nitrophenol. *Polyhedron, 166*, 196–202. Available from https://doi.org/10.1016/j.poly.2019.03.056.

Zhao, B., Dong, Z., Wang, Q., Xu, Y., Zhang, N., Liu, W., ... Wang, Y. (2020). Highly efficient mesoporous core-shell structured ag@sio$_2$ nanosphere as an environmentally friendly catalyst for hydrogenation of nitrobenzene. *Nanomaterials, 10*(5). Available from https://doi.org/10.3390/nano10050883.

Zolfigol, M. A., Khakyzadeh, V., Moosavi-Zare, A. R., Rostami, A., Zare, A., Iranpoor, N., ... Luque, R. (2013). A highly stable and active magnetically separable Pd nanocatalyst in aqueous phase heterogeneously catalyzed couplings. *Green Chemistry, 15*(8), 2132–2140. Available from https://doi.org/10.1039/c3gc40421h.

11

Silicon-based biosensor

Sandeep Arya[1], Anoop Singh[1], Asha Sharma[1] and Vinay Gupta[2]
[1]Department of Physics, University of Jammu, Jammu, India [2]Department of Mechanical Engineering, Khalifa University of Science and Technology, Abu Dhabi, United Arab Emirates

11.1 Introduction

Nanomaterials have gained much attention from researchers because of their remarkable properties (Papaefthymiou, 2009). Among the various metal-based hybrid nanoparticles, silicon-based hybrid nanoparticles are preferred because of their diverse applications in biosensing, energy storage device, catalysis, etc. Hybrid nanomaterials are the chemical conjugates of inorganic and/or organic materials (Karak, 2009). The superiority of the silicon-based hybrid nanoparticles over the other hybrid nanoparticles lies in their excellent barrier properties, light weight, and reinforcement of mechanical performance (Das, Mandal, Upadhyay, Chattopadhyay, & Karak, 2013). Silicon-based nanomaterials like polyhedral oligomeric silsesquioxanes, silica nanoparticles, nanoclay, etc. are familiar with their reinforcement effect for the configuration of polymeric nanocrystals (Liu, Hsu, & Wu, 2020). In comparison with pristine polymer, polymer silicate have manifold enhancement in their material properties. In terms of thermal and mechanical stability, biodegradability, gas/liquid permeability, flammability, etc. manyfold enhancement has been observed (Ray, Yamada, Okamoto, & Ueda, 2002). Silicon-based hybrid nanoparticles have potential applications in biosensing. A biosensor is an integrated device that provides specific semiquantitative or quantitative analytical information with the help of a bioreceptor, which is in close contact with the transducer. Thus the biosensor generally comprises two main components (McNaught & Wilkinson, 1997). The first component is a biological recognition element which is also known as a capture probe that can be an antigen (or antibody), enzyme (or substrate), receptor protein, or a complementary DNA. Due to the highly selective nature of the recognition element toward the target analyte, the biosensor shows better results compared with the chemical sensors. A transduction element, that is electrochemical, electrical, mass, or optical, comprises the second component of the biosensor, which helps to convert the concentration of the analyte into a measurable electrical signal. The high specificity, label-free, reusability, sensitivity, and compact size of the biosensor are the enhanced properties of the sensing device, which make the biosensor an attractive alternative to conventionally used techniques (Thevenot, Toth, Durst, & Wilson, 2001). There are various applications of the biosensor but the major one in the medical field is for the diagnosis of various chronic diseases, for example drug analysis, drug discovery, monitoring blood glucose in diabetic patients, cancer detection, and analysis of blood. Furthermore, biosensors have also found applications in homeland security, food technology, industrial process control, and environmental monitoring (Jane, Dronov, Hodges, & Voelcker, 2009). On the basis of the transduction principle, the biosensors are categorized as acoustic, mechanical, electrochemical, optical, and calorimetric, etc. Among these, the optical biosensors are very advantageous to us because of their high sensitivity, label-free operation, and multiplexed capabilities (Fan & White, 2011). The nanostructured porous silicon attracts much attention toward itself in comparison with various promising platforms because of its unique optical and physical properties. Thereafter, various researches were carried out for the synthesis of the single and multilayer structure of porous silicon for a large amount of applications in biomedicine, such as bioimaging (Gu et al., 2013), chemical and biosensing (Ensafi, Ahmadi, & Rezaei, 2017), drug delivery (Maniya, Patel, & Murthy, 2015), tissue engineering (Whitehead et al., 2008), and biomolecular screening (Tong et al., 2016). Porous silicon biosensors are responsible for the real-time monitoring of a target analyte by determining the optical properties. The porous silicon is used as an implantable device in the human body because porous silicon is biocompatible and produces nontoxic orthosilicic acid after its degradation, and these acids can be excreted through urine (Cheng et al., 2008).

11.2 Fabrication of silicon-based hybrid nanoparticles

There are various ways to fabricate a biosensor based on hybrid nanomaterials. Some of the methods to synthesize these materials are discussed here. To study the drug effects on 3D cultures in real time, a particularly designed sensor chip was fabricated for nondestructive, fast, and extracellular recording. In between the four electrodes of gold, the precultured spheroids were trapped, as shown in Fig. 11.1.

For the fabrication of the biosensor a piece of silicon wafer was selected to form a test system and the dimension of each chip was 20 × 20 × 0.5 mm. A technology such as microelectromechanical systems (MEMS) was used for the fabrication of silicon chips (see Fig. 11.1D). Firstly, the wet chemical anisotropic etching was utilized to structure the silicon wafer at 90₀C in a 40% KOH solution. The transferring of cavity structure into silicon dioxide (masking material) can be done through a SiO_2-etching process and lithography step. The lithography on the surfaces of structured wafer was done to develop the thick

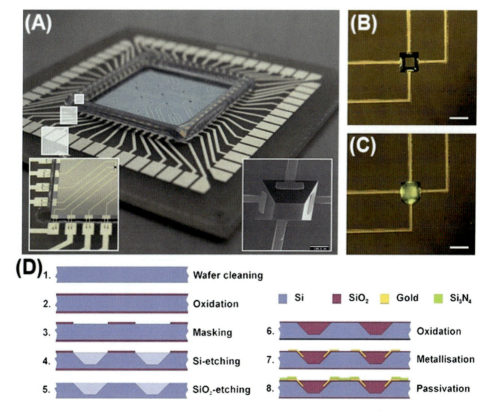

Figure 11.1 Image and fabrication scheme of the silicon-based multielectrode biosensor. (A) Complete biosensor chip. (B) and (C) Light microscopy images of a single microcavity with and without a positioned spheroid. (D) Manufacturing steps: A 4 in. silicon wafer is structured by standard etching processes. Reprinted with permission from Klo, D. et al. (2008). Biosensors and Bioelectronics, *23*, 1473. Copyright @ Elsevier B.V. (2008).

resist layer. Thereafter, sputtering was carried out to form a combined layer of metals (gold 500 nm, titanium 10 nm). A 700-nm silicon nitride layer was deposited on the surface of a wafer for upward passivation by a PECVD (plasma enhanced chemical vapor deposition) process on a Surface Technology Systems 310 at 350₀C with NH_3 and SiH_4 (Klo et al., 2008). The electrode elements were micromachined using a combination of both deep reactive ion etching (DRIE) and standard photolithography on both sides of polished silicon wafers. All the main steps that were involved in the process of microfabrication are summarized in Fig. 11.2A (Dávila, Esquivel, Sabat, & Mas, 2011), while Fig. 11.2B demonstrates the components of a sensor for applications in biosensing by porous silicon using various techniques (Dhanekar & Jain, 2013).

The field effect glucose biosensor based on silicon nanochannels can be fabricated with the help of glucose oxidase. This biosensor comprises silicon nanowires as nanochannels, 6 μm long, 100 nm high, and 50–100 nm wide, fabricated from a SOI wafer (silicon-on-insulator). The scanning electron

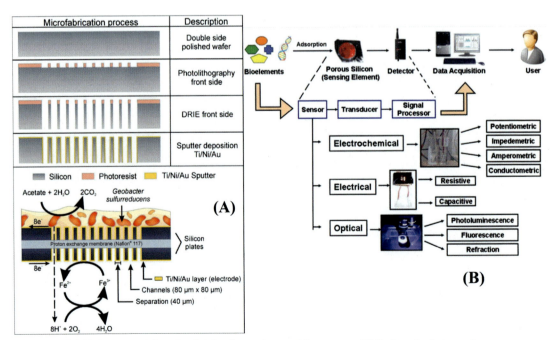

Figure 11.2 (A) Main steps of the microfabrication and assembly process. (B) Various techniques for porous silicon based biosensing. (A) Reprinted with permission from Dávila, D. et al.(2011) Biosensors and Bioelectronics, 26, 2426. Copyright @ Elsevier B.V. (2011). (B) Reprinted with permission from Dhanekar, S. & Jain, S. (2013). Biosensors and Bioelectronics, 41, 54. Copyright @ Elsevier B.V. (2013).

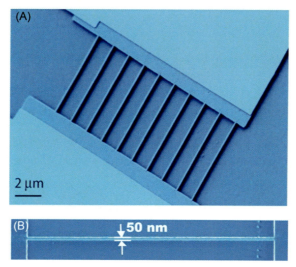

Figure 11.3 (A) Scanning electron micrograph of silicon nanowires (B) Top view of a 50 nm wide nanowire shows the controlled linear geometry. Reprinted with permission from Wang, X., et al. (2008). Applied Physics Letters, 92, 013903. Copyright @ American Institute of Physics (2008).

micrograph of the fabricated device is shown in Fig. 11.3 (Wang, Chen, Gibney, Erramilli, & Mohanty, 2008).

As demonstrated in Fig. 11.4A (i–v), the fabrication of each reconstituted wafer was done by utilizing two glass slides with 75×50 mm dimensions. On the carrier silicone O-ring was successfully deposited. After that a die was placed at the center of the O-ring and pressed gently. For improving the adhesion, it was kept inside the oven for 10 min at 65°C (Laplatine et al., 2018). The fabrication of the single-layer porous silicon samples can be done by electrochemical etching of silicon wafer doped with boron that has 0.01–0.02 Ω cm of resistivity. The fabrication of porous silicon was carried out in 1:3 volume to volume ratios of hydrofluoric acid (40%) and ethanol (99.9%). Firstly, the fabrication of the sacrificial layer can occur by applying 20 mA/cm^2 current density for 60 s. In the last step, 25 mA/cm^2 of constant current density was applied to complete the etching process in 180 s. On the porous silicon platform the immobilization of anti-HSP70 antibody is outlined in Fig. 11.4B (Maniyaa & Srivastavaa, 2020).

The label free and high sensitive detection can be achieved by the silicon nanowires when it is configured as field effect transistor. The electrical readout in silicon nanowires is based on the detection of charges on its surface passivated with a dielectric material, such as silicon dioxide, which is clearly demonstrated in

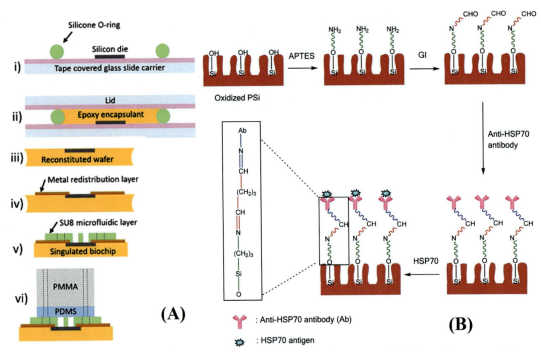

Figure 11.4 (A) Schematic of the lab-scale fan-out wafer-level-packaging process. (B) Schematic representation displaying development of label-free silicon based optical biosensor. (A) Reprinted with permission from Laplatine, L. et al. (2018). Sensors & Actuators B: Chemical, 273, 1610. Copyright @ Elsevier B.V. (2018). (B) Reprinted with permission from Maniyaa, N.H., & Srivastavaa, D. N. (2020). Materials Science in Semiconductor Processing, 115, 105126. Copyright @ Elsevier B.V. (2020).

Fig. 11.5. The novel scalable pixel-based biosensor was developed after the integration of silicon nanowires with the complementary metal-oxide-semiconductor (Jayakumar & Ostling, 2019).

Double-sided polished wafers of silicon were used for the preparation of samples. The thickness of the silicon wafer was 440 μm. The photolithography technique was used for the patterning of the substrates. The Bosch process was implemented to complete the silicon etches in a deep silicon reactive ion etcher. The Bosch process is a sequence of etches which firstly passivates by utilizing C_4F_8 as the source gas and after that etches are done by utilizing O_2 and SF_6 as the source gases. In the first step 150-μm thickness of the film is obtained with the help of a deep silicon etcher, as shown in Fig. 11.8(A). Fig. 11.8B demonstrates the backside lithography was done using an EV Group 620 aligner to define an aperture in the center of the thinned region which was also etched in the STS deep silicon etcher. After etching, the electrical insulating layer was formed as represented in Fig. 11.8C. A 75 μm coating layer was created

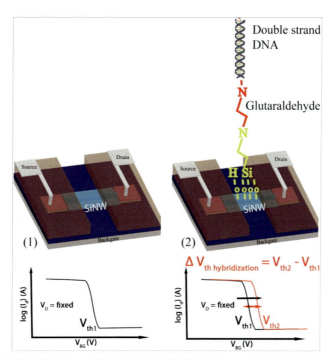

Figure 11.5 A schematic depicting the working principle of a SiNW sensor. (1) In the first step, the threshold voltage (Vth1) of the SiNW sensor is measured before bio molecule addition. (2) In the next step, the surface of the SiNW is functionalized with the target molecule such as double strand DNA where it undergoes hybridization process. Reprinted with permission from Jayakumar, G., & Ostling, M. (2019). Nanotechnology, 30, 225502. Copyright @ IOP Publications (2019).

on the surface of the device and also photolithography was conducted to pattern this layer on the device (Fig. 11.8D). A silver layer were formed on both sides of the surface of the wafer and then etched with 20:1:1 mixture of water, hydrogen peroxide, and sodium hydroxide (Fig. 11.8E). A PTFE layer (30 nm) was deposited chemically (CVD) using C_4F_8 as the gas source and the deep reactive ion etcher (Fig. 11.8F) (Wilk et al., 2007) (Fig. 11.6).

The detection of human blood groups can be done with the help of a silicon chip-based biosensor, which is shown in Fig. 11.7A (Jha & Sharma, 2010). For the production of the semiconducting nanowires biosensor, two nanofabrication processes were used such as bottom-up or top-down approaches. There are various factors, such as surface chemistry, mobilities, carrier densities, and diameters that alter the performance of the silicon nanowires-based biosensor. In Fig. 11.7B, the flow chart of the fabrication process is well demonstrated and in Fig. 11.7C

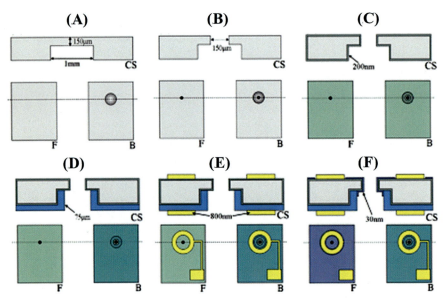

Figure 11.6 (A) A 1 mm recess is etched into the backside of a silicon substrate. (B) Using backside alignment techniques and deep silicon etch process, a 150 μm diameter aperture is etched through the center of the recessed region. (C) The surface is thermally oxidized with 300 nm of silicon dioxide. (D) A noise reducing 75 μm thick SU-8 layer is patterned on the backside of the device. (E) Silver is evaporated and wet etched to form electrodes on both sides of the device. (F) PTFE is chemically vapor deposited on the surface of the device using lift-off techniques to remove the material from over the active area of integrated electrodes. Reprinted with permission from Wilk, S. J., et al. (2007). Biosensors and Bioelectronics, 23, 183. Copyright @ Elsevier B.V. (2007).

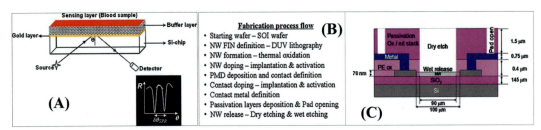

Figure 11.7 (A) Proposed Si-chip-based SPR biosensor set-up for the detection of human blood groups. (B) Schematic diagram of the fabrication process flow of a SiNW biosensor. (C) Schematic diagram of a cross-sectional view of a SiNW sensor. Reprinted with permission from Jha, R., & Sharma, A. K. (2010). Sensors and Actuators B, 145, 200. Copyright @ Elsevier B.V. (2010). Reprinted with permission from Zhang, G. J., & Ning, Y. (2012). Analytica Chimica Acta, 749, 1. Copyright @ Elsevier B.V. (2012).

the cross-sectional view of the silicon nanowires-based sensor is clearly shown. The fabrication of silicon nanowires was done through conventional lithography, oxidation, and etching (Zhang & Ning, 2012).

11.3 Attachment of bioreceptor on silicon-based materials

The attachment of the bioreceptor on the silicon-based materials is very important to enhance the chemical properties of the material. The specific detection of the analytes totally depends on the interaction of the bioreceptor and the target sample. There are various characteristics that can be improved through the immobilization of the bioreceptor. The enhanced performances of the biosensor due to the immobilization of the bioreceptor are the change in the hydrophilicity or hydrophobicity of the surface, preventing surface fouling and inducing selectivity in the sample. The most used process is bioreceptor physiosorption that can be mainly preferred for the designing of the new optical sensors (Freeman et al., 2011).

11.3.1 Chemical surface modification

The self-assembly of organofunctional alkoxysilanes is employed to functionalize the surface of silicon. The surface chemistry of the functionalized group plays an important role in the modification of the chemical surface of the silane-based layers. This surface chemistry brings very good results in the functionalization of silica-on-silicon optical devices. In the case of materials of silicon nitride and silicon, the layer of silicon oxide generates a hydroxyl group (Massad-Ivanir, Shtenberg, Tzur, Krepker, & Segal, 2011). The formation of the assembly on the gold surfaces by the thiol molecules is not much more complex than the silane self-assembled monolayers. The monolayers of alkylsiloxane are mainly generated through a process of chemisorptions of various self-assembling molecules like trimethoxy, triethoxy, or trichlorosilanes, onto the substrate (Ulman, 1996). Despite the thorough investigation of the formation mechanism, still there are a few controversies (Wouters, Hoeppener, Sturms, & Schubert, 2006). It is widely recognized that the formation mechanism depends upon certain characteristics such as age of the solution, solvent used, water content, temperature, and deposition time (Aswal, Lenfant, Guerin, Yakhami, & Villaume, 2006). Epoxy chemistry is considered as an alternative coupling system for the immobilization of biomolecules, proving its reactivity and stability to various nucleophiles, like sulfhydryl and amine groups under aqueous conditions (Thierry, Jasienak, de Smet, Vasilev, & Griesser, 2008). Chemicals such as 3-glycidoxypropyltrimethoxysilane can be preferred for covalently coating the surface. The resultant coating layer can be utilized to conjugate hydroxyl-, amine-, or thiol-containing ligands. Ramachandran et al. successfully reported that the 3-glycidoxypropyltrimethoxysilane has been employed to attach aminated oligonucleotides and antibody covalently by an epoxide ring (Ramachandran et al., 2008). Schneider, Dickinson, Vach, Hoijer, and Howard (2000) fabricated an integrated optical biosensor which depends upon the Hartman interferometer, that carried the oxidation of various epoxy moieties into the aldehyde groups by utilizing sodium periodate, After 3-glycidoxypropyltrimethoxysilane silanization, De Vos et al. (2009) introduced an alternative approach by using a poly ethylene glycol thin layer in order to prevent a microring resonator biosensor from nonspecific binding. Finally, biotinylation was carried out by 1-ethyl-3-(3-dimethylaminopropyl) carbodiimide/N-hydroxysuccinimide (EDC/NHS), in the first case NH_2—biotin and in second case NHS—biotin is used (De Vos, Bartolozzi, Schacht, & Bienstman, 2007). In another approach, the employment of thiol-ended organosilane helps to leave nucleophilic functionality on the surface. Sepúlveda et al. (2006) successfully reported the employment of 3-mercaptopropyltriethoxysilane (MPTS) for the functionalization of an integrated microsystem (Mach—Zehnder interferometer). This causes the attachment of a thiolated oligonucleotide to the surface through disulfide bond linkage. Xu, Suarez, and Gottfried (2007) demonstrated in their study that m-maleimidobenzoyl-N-hydroxysuccinimide ester (MBS), used as a hetero-bi-functional cross-linker, helps to attach probes through their amine groups, permitted by thiol functionality. It is also reported that by using this principle the biotinylation of silicon oxide surfaces can be carried easily, which helps to achieve good performance, and which is currently used in the ring resonator-based biosensor (Escorihuela, Bañuls, Puchades, & Maquieira, 2012). After giving their reactivity to epoxy, carboxylic acid, and aldehyde functionalities, 3-aminopropyltrime-thoxysilane (Biggs, Hunt, & Armani, 2012), and 3-aminopropyltriethoxysilane (Xu et al., 2010) (APTMS and APTES, respectively) have become the most used linker agents for biofunctionalization. There is a possibility of hydrogen bond formation between the amine of APTES and the SiOx surface which results in a disordered layer (Vanderberg et al., 1991). Furthermore, when cross-linking occurs between the units of alkoxysilane, it helps to obtain the silane structures (oligomerized) which result in producing the thicker and rough layers in comparison with a monolayer. The silanization which is solvent based by utilizing APTES can be investigated on a planar surface under optimal conditions (Howarter & Youngblood, 2006). The experiments with 1% concentration of APTES generates good quality of films with a less reaction time, that is less than one hour, and with the increase in the reaction time these obtained films becomes thicker and thicker. An aminated surface is used for the immobilization of antibodies through adsorption (Densmore et al., 2009) or to strongly fix an N-hydroxysuccinimide biotin (García-Rupérez et al., 2010; Scheler et al., 2012). N-Hydroxysuccinimide ester successfully combined with the primary amines to form a stable amide bond. Silane coupling agents containing carboxylate groups have also been utilized to functionalize integrated optical devices with carboxylic acids for the subsequent conjugation with amine-containing molecules. Duval et al. (2012) employed carboxyethylsilanetriol sodium salt on silicon nitride bimodal waveguide interferometers, and proteins were conjugated to the surface by 1-ethyl-3-(3-dimethylaminopropyl) carbodiimide/ N-hydroxysuccinimide (EDC/NHS). A microcavity sensor was carboxylated by Zlatanovic et al. (2009) using an effective chelator of metal ions, such as N-(trimethoxysilylpropyl) ethylene-diamine triacetic acid, while proteins were further conjugated by EDC/NHS. After organosilane layer formation, the remaining steps were performed online, and no data on yields or surface characterization are provided, except for the limit of detection for the biorecognition event. In addition, carboxylic acid-ended dimethyl monomethoxy organosilane has been used as a horizontal spacer to form mixed monolayers on a Mach–Zehnder interferometer by the coadsorption of binary solutions containing both a carboxyl-ended organosilane and another bearing a biotin moiety (Weisser et al., 1999). Isocyanatepropyltriethoxysilane (ICPTS) has been used to link proteins onto silicon photonic crystals without cross-linkers or activation steps being needed. The isocyanate moiety reacts with amines to form isourea bonds, and with hydroxyl groups to form urethanes. Oligonucleotides are also attached to the ICPTS surface using biotinylated probes which are affinity-captured by streptavidin covalently linked to the isocyanate modified surface (Escorihuela, Banuls, Castelló, et al., 2012). This is a simple one-step approach, but the experimental conditions must be well controlled to achieve an acceptable degree of reproducibility. Thus at a certain basic pH, there is a risk of decarboxylation, which provides an amine-ended surface instead of an isocyanate-ended one.

11.4 Applications of biosensor

The silicon-based hybrid nanoparticles have mostly been used in biomedical applications. For this purpose, mainly optical biosensing is preferred. Optical biosensing can be carried out on silicon-based hybrid nanoparticles for the detection of various biomolecules, such as enzymes, viruses, antibody, short DNA oligonucleotides, and other molecules, through immobilizing its capture probe on the surface of porous silicon.

11.4.1 DNA detection

The target DNA can bind with the capture probe of the DNA biosensor through DNA–DNA hybridization. In 1997 a very exciting work on porous silicon optical biosensor was done by Sailor and colleagues that developed much interest for the application of porous silicon for optical biosensing (Lin, Motesharei, Dancil, Sailor, & Ghadiri, 1997). More recently, the ring resonator structure of silicon was developed for the successful and selective detection of DNA (Rodriguez, Hu, & Weiss, 2015). The porous silicon slab waveguide was used for patterning the ring resonator structure. After that it was functionalized with succinimidyl 3-(2-pyridyldithio) propionate and aminopropyltriethoxysilane for the immobilization of 16mer probe DNA (thiol modified). Similarly other biosensors that are based on porous silicon grating coupled with Bloch subsurface wave (BSSW) and Bloch surface wave (BSW) propagating mode can be utilized for the detection of small molecules and DNA (Rodriguez, Ryckman, Jiao, & Weiss, 2014). The comparison of experimental and stimulated performance of the BSW/BSSW sensor are shown in Fig. 11.8. In the recent study, it was demonstrated that the high specific and sensitive optical biosensor for the detection of DNA shows 1 nM limit of detection (Vilensky, Bercovici, & Segal, 2015).

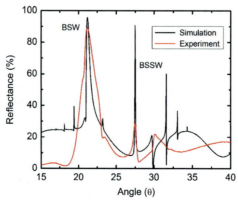

Figure 11.8 Comparison of simulated and experimental performance of the BSW/BSSW sensor. Reprinted with permission from Rodriguez, G. A. et al. (2014). Nanoscale Research Letters, 9, 383. Copyright @ Elsevier B.V. (2014).

The resonant waveguide biosensor is also helpful for the detection of DNA (Rong, Najmaie, Sipe, & Weiss, 2008). The different current densities of porosity layer are applied for the preparation of a waveguide structure of a biosensor. Glutaraldehyde and aminopropyltriethoxysilane are used to link the probe of DNA with an oxidized porous silicon waveguide.

11.4.2 Enzyme detection

Enzymes are biomarkers for various diseases such as neurodegeneration, cardiovascular, chronic wounds, and cancer. The idea of abnormal or normal cellular function can be obtained from the activities of the enzymes which play a crucial role in the detection of infections. One of the important properties of the enzyme is that under particular reaction conditions, that is ionic strength, temperature, and pH, it reacts highly with the corresponding substrate. By calculating the rate of formation of product it becomes easier to utilize an optical biosensor for the detection of enzymes. Based on porous silicon microcavity, a biosensor was fabricated for the multiplexed detection of enzyme from wound fluids (Krismastuti, Cavallaro, Prieto Simon, & Voelcker, 2016). In recent decades, selective, sensitive, and real-time detection of metalloproteinases found in exudates was confirmed by utilizing a fluorescence-based specific porous silicon biosensor (resonant microcavity) (Krismastuti, Pace, & Voelcker, 2014). The fluorophore embedded in the porous silicon matrix has higher fluorescence intensity in comparison with the fluorophore embedded in the buffer solution. This shows that the enhancement in the fluorescence intensity is due to the photonic structure of porous silicon. L-lactate dehydrogenase is an intracellular enzyme mainly found in the tissues of human. It plays an effective role in the diagnosis of disease. It is generated from dead cells, and therefore it is observed that the level of enzyme increases in the blood when there is injury or disease. In addition, the increase in the level of L-lactate dehydrogenase has been associated with leukemia, chronic wounds, melanoma, and pulmonary cancer. Therefore, for the detection of L-lactate dehydrogenase the biosensor based on luminescence-enhancing porous silicon microcavity was successfully prepared (Jenie, Prieto-Simon, & Voelcker, 2015). The enhancement in the intensity of fluorescence inside the porous silicon microcavity biosensor is due to the phenomena of immobilization of resazurin dye (nonfluorescent in nature) on the surface of porous silicon via acylation, thermal hydrosilylation, and thermal hydrocarbonization.

11.4.3 Antibody detection

A porous silicon biosensor can be manufactured by utilizing antigen as a capture molecule for the detection of antibody in the sample. Immunoglobulin G can be detected with the help of an optical interference immunosensor. This type of immunosensor was fabricated by coating TiO_2 on the porous silicon film (Li & Sailor, 2014). The chemical stability of the porous silicon surface was enhanced by coating with a layer of oxide. The layer of TiO_2 was deposited through a sol–gel technique by using a spin coater followed by strong heating in furnace at 500°C in air. The sensor based on the coating of biosensor showed greater stability in the pH range of 2–12. Because of the high stability, this sensor can be utilized in immunosensor application. Porous silicon with protein A adsorbed on it has an ability to bind with rabbit antisheep immunoglobulin G. Because of the lack of affinity of protein A toward chicken immunoglobulin G, this showed the specificity of protein A. A specific porous silicon biosensor was fabricated for the detection of immunoglobulin G in the serum as well as in whole blood samples (Bonanno & DeLouise, 2007a). The pore size of the porous silicon microcavity structure was designed in a specific manner that enhances the filtering capabilities of the biosensor. The designed porous silicon microcavity was oxidized thermally and after that silanized by using 3- aminopropyltrimethoxysilane. An amine-terminated surface of porous silicon was functionalized with chemicals like biotin–streptavidin to immobilize antirabbit immunoglobulin G. A porous silicon microcavity biosensor showed minimal cross-reactivity, high target binding specificity, and 2–10 mgmL^{-1} detection range for rabbit immunoglobulin G.

11.4.4 Cell detection

The cell detection of yeast, bacteria, and microorganisms is very important in water and food safety studies. Porous silicon on modification with antibody can be used for the detection of cells by optical biosensors. Recently, a new technique for the detection of bacteria due to the effect of the blockage of a porous silicon nanopore arrangement has been successfully reported (Tang, Li, Luo, Liu, & Wu, 2016). The *Escherichia coli* bacteria were rapidly and selectively captured on the arrangements of nanopore, resulting in pore blockages that can be measured by utilizing indirect mode Fourier-transformed spectroscopy (reflectometric interference) (FT-RIS). In a direct method, the shift in effective optical thickness is directly

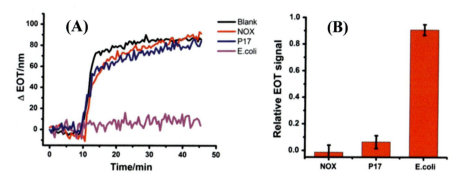

Figure 11.9 (A) EOT shift in respond to the same density of NOX, P17, and *E. coli*. (B) The histogram of normalized relative EOT shift in response to the same density of NOX, P17, and *E. coli*. Reprinted with permission from Tang, Y. et al. (2016). Biosensors & Bioelectronics, 79, 715. Copyright @ Elsevier B.V. (2016).

proportional to the concentration of analyte. However, cells, virus, and large bacteria with the size in micrometers are not able to enter the porous matrix because of a very small diameter of pores. But the large size of bioanalytes can be detected by utilizing an indirect FT-RIS method. The shift in effective optical thickness was linearly decreased with an increase in the bacterial density from 10^3-10^7 cfu mL^{-1} range because of the pore blockage. The selectivity of the antibody toward *Escherichia coli* was further confirmed by treating with P17, Nox, and nontarget bacteria, which did not show any variation in the shift in effective optical thickness because it not reacts with the antibody (Fig. 11.9). This approach of sensing is particularly very important for the label-free and rapid detection of bacteria. The *Lactobacillus acidophilus* bacteria are directly detected by fabricating aptamer-based porous silicon biosensor (Urmann, Arshavsky-Graham, Walter, Scheper, & Segal, 2016).

11.4.5 Virus detection

Porous silicon biosensors have been successfully fabricated for the detection of viruses by encapsulating an antibody on the surface of the biosensor. Various chemical substances such as latex nanospheres, M13KO7 bacteriophage, glutaraldehyde, and aminopropyltriethoxysilane (APTES) can be detected by a biosensor which is fabricating from the combination of porous silicon Bloch surface and subsurface waves (Rodriguez, Lonai, Mernaugh, & Weiss, 2014). On the surface of porous silicon the bacteriophage was attached through glutaraldehyde and APTES linker molecules. Both the glutaraldehyde and APTES were penetrated through the porous matrix because of their small size with a shift in the resonance for the first Bloch subsurface wave and Bloch surface wave mode of 1.97°; 2.66°, and 1.6°; 2.18°, respectively. But bacteriophage did not penetrate through the porous matrix because of its large size. In this case the resonance shift of 0.31° was observed only in the Bloch surface wave mode. The single-layer porous silicon biosensor was used for the detection of the bacteriophage Ms2 (Rossi, Wang, Reipa, & Murphy, 2007). The two different techniques comprising aryldiazirine and carbodiimide cross-linker were utilized for attaching AMN antibody, such as rabbit anti-Ms2, on the porous silicon surface. Because of the hydrophilic surface, the carbodiimide-functionalized porous silicon showed strong binding efficiency and displayed good pore penetration toward the antibody, whereas porous silicon functionalized with diazirine displayed less binding efficiency due to the hydrophobic surface. The biosensor based on the porous silicon can successfully detect the bacteriophage Ms2 by calculating the fluorescence data within a 1×10^6 to 1×10^{12} plaque-forming units (pfu)mL^{-1} viral concentration range and 2×10^7 pfu mL^{-1} limit of detection.

11.4.6 Protein detection

Mariani, Strambini, and Barillaro (2016) fabricated a biosensor on the nanostructured porous silicon interferometer for the detection of proteins in very small concentration, that is femtomole. In this case a much lower detection limit is obtained (20 p.m.). This technique is very sensitive and based on the calculated average value of the spectral interferogram. These values are obtained by subtracting the reflectance spectra from the adsorption of protein commonly called bovine serumalbumin. Additionally, good signal-to-noise ratio is also estimated with greater accuracy.

11.4.7 Small analyte detection

The optical biosensor based on the porous silicon can be successfully used for the detection of heavy metals, urea, and glucose (Syshchyk, Skryshevsky, Soldatkin, & Soldatkin, 2015). A conception of luminescent biosensors for heavy metals, urea, and glucose detection is demonstrated in Fig. 11.10.

In the medium, the chemical reaction between substrate and enzyme was take place and this causes the change in the pH of the medium. The variation in the pH of the medium is responsible for the change in photoluminescence of the porous silicon biosensor. In the porous silicon biosensor the enzymes such as glucose oxidase and urease were utilized as a bioselective material.

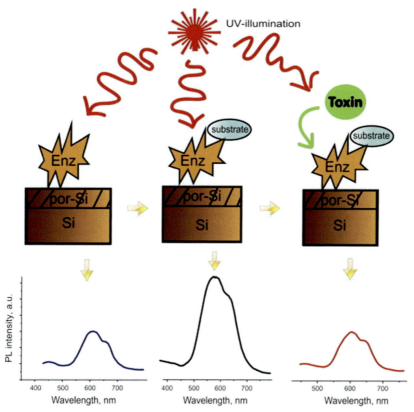

Figure 11.10 Conception of luminescent biosensors based on urease for heavy metals, urea, and glucose detection. Reprinted with permission from Syshchyk, O. et al. (2015). Biosensors & Bioelectronics, 66, 89. Copyright @ Elsevier B.V. (2015).

Glucose oxidase helps to convert the glucose into gluconic acid and the pH of the medium shifts to acidic, whereas urease helps to converts the urea into ammonia and shifts the pH of the medium into alkaline. With the increase in the concentration of urea of glucose from 0 to 3 mM in the respective urease- and glucose oxidase-containing medium, the photoluminescence intensity of porous silicon-based biosensor was increased by 1.45 and 1.7 times, respectively. But the photoluminescence quantum yield of porous silicon was restored in the presence of Cd^{2+}, Pb^{2+}, and Cu^{2+} in the solution containing urease and glucose oxidase enzyme-catalyzed reactions. Pacholski, Perelman, VanNieuwenhze, and Sailor (2009) has synthesized the double-layer biosensor based on the porous silicon for the detection of vancomycin by utilizing spectroscopy, that is reflective interferometric Fourier-transform spectroscopy. In the double-layer biosensor based on porous silicon, the lower layer acts as a bottom layer, whereas the

Table 11.1 Summary of detection of various biomolecules using porous silicon optical structures.

Analyte family	Analyte	Transducer	Detection range	Response time	References
DNA	24 oligonucleotides	Bragg reflector	200 μM	2 h	Rea et al. (2010)
	40 oligonucleotides	Bloch surface wave	50 μM	60 min	Rodriguez, Ryckman, et al. (2014)
	15 oligonucleotides	Single layer	1×10^{-9}–10×10^{-9} M	20 min	Vilensky et al. (2015)
Enzyme	Subtilisin	Bloch surface wave	0.01 mg mL^{-1}	20 min	Qiao et al. (2010)
	Sortase A	Microcavity	4.6×10^{-12}–4.6×10^{-8} M	30 min	Krismastuti et al. (2016)
	L-lactate dehydrogenase	Microcavity	0.16–6.5 U mL^{-1}	10 min	Jenie et al. (2015)
Antibody	Rabbit IgG	Microcavity	0.07–3 mg mL^{-1}	60 min	Bonanno and DeLouise (2007)
	Human IgG	Single layer	0.2–100 μg mL^{-1}	20 min	Szili et al. (2011)
	Sheep IgG	Single layer	10–500 μg mL^{-1}	90 min	Li and Sailor (2014)
Virus	Bacteriophage Ms2	Single layer	1×10^{6}–1×10^{12} pfu mL^{-1}	45 min	Rossi et al. (2007)
	M13K07 Bacteriophage	Bloch surface wave	32 μg mL^{-1}	20 min	Rodriguez, Lonai, et al. (2014)
Protein	Streptavidin	Microcavity	0.5–5 μM	20 min	Zhao et al. (2016)
	Bovine serum albumin	Single layer	150 pm–15 μM	60 min	Mariani et al. (2016)

top layer acts as a sensing channel. Furthermore, porous silicon double layer displayed both quantitative information about binding events and qualitative specific detection of vancomycin, for example equilibrium binding constants. Porous silicon biosensor is also fabricated for the sensing of target compounds for drug discovery and screening. The detection of various biomolecules by utilizing porous silicon optical structures is summarized in Table 11.1 (Bonanno & DeLouise, 2007b; Jenie et al., 2015; Krismastuti et al., 2016; Li & Sailor, 2014; Mariani et al., 2016; Qiao, Guan, Gooding, & Reece, 2010; Rea et al., 2010; Rodriguez, Lonai, et al., 2014; Rodriguez, Ryckman, et al., 2014; Rossi et al., 2007; Szili et al., 2011; Vilensky et al., 2015; Zhao, Gaur, Retterer, Laibinis, & Weiss, 2016).

11.5 Conclusion

The diverse techniques for the synthesis of silicon-based materials, fabrication of silicon-based biosensor, and biomedical applications of hybrid nanoparticles of silicon are discussed in this chapter. The surface chemistry approaches for the attachment of bioreceptors on silicon-based materials are also mentioned after the fabrication processes. The specific detection of the analytes totally depends on the interaction of the bioreceptor and the target sample. Also in this chapter, the optical biosensing by utilizing silicon-based hybrid nanoparticles is productively discussed. Optical biosensing can be carried out on porous silicon for the detection of various biomolecules such as enzymes, viruses, antibody, short DNA oligonucleotides, and other molecules through immobilizing the capture probe on the surface of porous silicon. On the surface of the porous silicon optical biosensor the binding of the target analyte with the capture probe is very important. This binding of the target analyte with a capture probe can replace the water present in the porous structure of silicon.

Reference

Aswal, D. K., Lenfant, S., Guerin, D., Yakhami, J. V., & Villaume, D. (2006). *Analytica Chimica Acta, 568*, 84.

Biggs, B. W., Hunt, H. K., & Armani, A. M. (2012). *Journal of Colloid and Interface Science, 369*, 477.

Bonanno, L. M., & DeLouise, L. A. (2007a). *Biosensors & Bioelectronics, 23*, 444.

Bonanno, L. M., & DeLouise, L. A. (2007b). *Langmuir: The ACS Journal of Surfaces and Colloids, 23*, 5817.

Cheng, L., Anglin, E., Cunin, F., Kim, D., Sailor, M. J., Falkenstein, I., ... Freeman, W. R. (2008). *The British Journal of Ophthalmology, 92*, 705.

Das, B., Mandal, M., Upadhyay, A., Chattopadhyay, P., & Karak, N. (2013). *Biomedical Materials (Bristol, England), 8*, 035003.

Dávila, D., Esquivel, J. P., Sabat, N., & Mas, J. (2011). *Biosensors and Bioelectronics, 26*, 2426.

De Vos, K., Bartolozzi, I., Schacht, E., & Bienstman, P. (2007). *Optics Express, 15*, 7610.

De Vos, K., Girones, J., Popelka, S., Schacht, E., Baets, R., & Bienstman, P. (2009). *Biosensors & Bioelectronics, 24*, 2528.

Densmore, A., Vachon, M., Xu, D. X., Janz, S., Ma, R., Li, Y. H., ... Schmid, J. H. (2009). *Optics Letters, 34*, 3598.

Dhanekar, S., & Jain, S. (2013). *Biosensors and Bioelectronics, 41*, 54.

Duval, D., González-Guerrero, A. B., Dante, S., Osmond, J., Monge, R., Fernández, L. J., ... Lechuga, L. M. (2012). *Lab on a Chip, 12*, 1987.

Ensafi, A. A., Ahmadi, N., & Rezaei, B. (2017). *Sensors Actuators B: Chemical, 239*, 807.

Escorihuela, J., Bañuls, M. J., Puchades, R., & Maquieira, A. (2012). *Chemical Communications, 48*, 2116.

Escorihuela, J., Banuls, M. J., Castelló, J. G., Toccafondo, V., García-Rupérez, J., Puchades, R., & Maquieira, Á. (2012). *Analytical and Bioanalytical Chemistry, 404*, 2831.

Fan, X., & White, I. M. (2011). *Nature Photonics, 5*, 591.

Freeman, L. M., Li, S., Dayani, Y., Choi, H. S., Malmstadt, N., & Armani, A. M. (2011). *Applied Physics Letters, 98*, 143703.

García-Rupérez, J., Toccafondo, V., Bāṇuls, M. J., Castelló, J. G., Griol, A., Peransí-Llopis, S., & Maquieira, A. (2010). *Optics Express, 18*, 24276.

Gu, L., Hall, D. J., Qin, Z., Anglin, E., Joo, J., Mooney, D. J., ... Sailor, M. J. (2013). *Nature Communications, 4*.

Howarter, J. A., & Youngblood, J. P. (2006). *Langmuir: the ACS Journal of Surfaces and Colloids, 22*, 11142.

Jane, A., Dronov, R., Hodges, A., & Voelcker, N. H. (2009). *Trends in Biotechnology, 27*, 230.

Jayakumar, G., & Ostling, M. (2019). *Nanotechnology, 30*, 225502.

Jenie, S. N., Prieto-Simon, B., & Voelcker, N. H. (2015). *Biosensors & Bioelectronics, 74*, 637.

Jha, R., & Sharma, A. K. (2010). *Sensors and Actuators B, 145*, 200.

Karak, N. (2009). *Fundamentals of Polymers: Raw Materials to Finish Products*. New Delhi: PHI Learning Pvt. Ltd.

Klo, D., Kurz, R., Jahnke, H. G., Fischer, M., Rothermel, A., Anderegg, U., ... Robitzki, A. A. (2008). *Biosensors and Bioelectronics, 23*, 1473.

Krismastuti, F. S. H., Pace, S., & Voelcker, N. H. (2014). *Advanced Functional Materials, 24*, 3639.

Krismastuti, F. S. H., Cavallaro, A., Simon, B. P., & Voelcker, N. H. (2016). *Advancement of Science, 3*, 1500383.

Laplatine, L., Luan, E., Cheung, K., Ratner, D. M., Dattner, Y., & Chrostowski, L. (2018). *Sensors & Actuators B: Chemical, 273*, 1610.

Li, J., & Sailor, M. J. (2014). *Biosensors & Bioelectronics, 55*, 372.

Lin, V. S. Y., Motesharei, K., Dancil, K.-P. S., Sailor, M. J., & Ghadiri, M. R. (1997). *Science (New York, N.Y.), 278*, 840.

Liu, Y. C., Hsu, W. F., & Wu, T. M. (2020). *Journal of Applied Electrochemistry, 50*, 311.

Maniya, N. H., Patel, S. R., & Murthy, Z. V. P. (2015). *Applied Surface Science, 330*, 358.

Maniyaa, N. H., & Srivastavaa, D. N. (2020). *Materials Science in Semiconductor Processing, 115*, 105126.

Mariani, S., Strambini, L. M., & Barillaro, G. (2016). *Analytical Chemistry, 88*, 8502.

Massad-Ivanir, N., Shtenberg, G., Tzur, A., Krepker, M. A., & Segal, E. (2011). *Analytical Chemistry, 83*, 3282.

McNaught, A. D., & Wilkinson, A. (1997). *IUPAC. Compendium of Chemical Terminology, Gold Book*. Oxford: Blackwell Scientific Publications.

Pacholski, C., Perelman, L. A., VanNieuwenhze, M. S., & Sailor, M. J. (2009). *Physica Status Solidi (A), 206*, 1318.

Papaefthymiou, G. C. (2009). *Nano Today, 4*, 438.

Qiao, H., Guan, B., Gooding, J. J., & Reece, P. J. (2010). *Optics Express, 18*, 15174.

Ramachandran, A., Wang, S., Clarke, J., Ja, S. J., Goad, D., Wald, L., ... Little, B. E. (2008). *Biosensors & Bioelectronics, 23*, 939.

Ray, S. S., Yamada, K., Okamoto, M., & Ueda, K. (2002). *Nano Letters, 2*, 1093.

Rea, I., Lamberti, A., Rendina, I., Coppola, G., Gioffre, M., Iodice, M., ... De Stefano, L. (2010). *Journal of Applied Physics, 107*, 014513.

Rodriguez, G. A., Lonai, J. D., Mernaugh, R. L., & Weiss, S. M. (2014). *Nanoscale Research Letters, 9*, 383.

Rodriguez, G. A., Ryckman, J. D., Jiao, Y., & Weiss, S. M. (2014). *Biosensors & Bioelectronics, 53*, 486.

Rodriguez, G. A., Hu, S., & Weiss, S. M. (2015). *Optics Express, 23*, 7111.

Rong, G., Najmaie, A., Sipe, J. E., & Weiss, S. M. (2008). *Biosensors & Bioelectronics, 23*, 1572.

Rossi, A. M., Wang, L., Reipa, V., & Murphy, T. E. (2007). *Biosensors & Bioelectronics, 23*, 741.

Scheler, O., Kindt, J. T., Qavi, A. J., Kaplinski, L., Glynn, B., Barry, T., ... Bailey, R. C. (2012). *Biosensors & Bioelectronics, 36*, 56.

Schneider, B. H., Dickinson, E. L., Vach, M. D., Hoijer, J. V., & Howard, L. V. (2000). *Biosensors & Bioelectronics, 15*, 13.

Sepúlveda, B., Sánchez del Río, J., Moreno, M., Blanco, F. J., Mayora, K., & Lechuga, L. M. (2006). *Journal of Optics A: Pure and Applied Optics, 8*, S561.

Syshchyk, O., Skryshevsky, V. A., Soldatkin, O. O., & Soldatkin, A. P. (2015). *Biosensors & Bioelectronics, 66*, 89.

Szili, E. J., Jane, A., Low, S. P., Sweetman, M., Macardle, P., Kumar, S., ... Voelcker, N. H. (2011). *Sensors and Actuators B: Chemical, 160*, 341.

Tang, Y., Li, Z., Luo, Q., Liu, J., & Wu, J. (2016). *Biosensors & Bioelectronics, 79*, 715.

Thevenot, D. R., Toth, K., Durst, R. A., & Wilson, G. S. (2001). *Biosensors & Bioelectronics, 16*, 121.

Thierry, B., Jasienak, M., de Smet, L. C. P., Vasilev, K., & Griesser, H. J. (2008). *Langmuir: The ACS Journal of Surfaces and Colloids, 24*, 10187.

Tong, W. Y., Sweetman, M. J., Marzouk, E. R., Fraser, C., Kuchel, T., & Voelcker, N. H. (2016). *Biomaterials, 74*, 217.

Ulman, A. (1996). *Chemical Reviews, 96*, 1533.

Urmann, K., Arshavsky-Graham, S., Walter, J. G., Scheper, T., & Segal, E. (2016). *Analyst, 141*, 5432.

Vanderberg, E. T., Bertilsson, L., Liedberg, B., Uvdal, K., Erlandson, R., Elwing, H., & Lundstrom, I. (1991). *Journal of Colloid and Interface Science, 147*, 103.

Vilensky, R., Bercovici, M., & Segal, E. (2015). *Advanced Functional Materials, 25*, 6725.

Wang, X., Chen, Y., Gibney, K. A., Erramilli, S., & Mohanty, P. (2008). *Applied Physics Letters, 92*, 013903.

Weisser, M., Tovar, G., Mittler-Neher, S., Knoll, W., Brosinger, F., Freimut, H., ... Ehrfeld, W. (1999). *Biosensors & Bioelectronics, 14*, 405.

Whitehead, M. A., Fan, D., Mukherjee, P., Akkaraju, G. R., Canham, L. T., & Coffer, J. L. (2008). *Tissue Engineering Part A, 14*, 195.

Wilk, S. J., Petrossian, L., Goryll, M., Thornton, T. J., Goodnick, S. M., Tang, J. M., & Eisenberg, R. S. (2007). *Biosensors and Bioelectronics, 23*, 183.

Wouters, D., Hoeppener, S., Sturms, J. P. E., & Schubert, U. S. (2006). *Journal of Scanning Probe Microscopy, 1*, 45–50.

Xu, D. X., Vachon, M., Densmore, A., Ma, R., Delâge, A., Janz, S., ... Schmid, J. H. (2010). *Optics Letters, 35*, 2771.

Xu, J., Suarez, D., & Gottfried, D. S. (2007). *Analytical and Bioanalytical Chemistry, 389*, 1193.

Zhang, G. J., & Ning, Y. (2012). *Analytica Chimica Acta, 749*, 1.

Zhao, Y., Gaur, G., Retterer, S. T., Laibinis, P. E., & Weiss, S. M. (2016). *Analytical Chemistry, 88*, 10940.

Zlatanovic, S., Mirkarimi, L. W., Sigalas, M. M., Bynum, M. A., Chow, E., Robotti, K. M., ... Grot, A. (2009). *Sensors and Actuators B, 141*, 13.

Graphene-based field effect transistor (GFET) as nanobiosensors

Homa Farmani[1], Ali Farmani[1] and Tuan Anh Nguyen[2]
[1]School of Electrical Engineering, Lorestan University, Khoramabad, Iran
[2]Institute for Tropical Technology, Vietnam Academy of Science and Technology, Hanoi, Vietnam

12.1 Introduction

In the first stage, we provide a minireview about FET-based biosensors and models with graphene. Also, considering the extraordinary electrical, chemical, and optical properties of graphene, this material is introduced as a good candidate for transistor-based biosensors.

In this ever-changing world the new kind of coronavirus (CoV) has undoubtedly become a serious issue due to its rapid transmission from one human to another in a matter of minutes. COVID-19 has faster transfer capability compared to the other coronaviruses like SARS and MERS. The name corona is derived from the Latin word omeaning crown due to its similarity in shape under the electron microscope. CoVs have high adjustability in the host in terms of their mutation rate (Sheikhzadeh et al., 2020). The genetic structure of the coronavirus includes three parts. Two-thirds of it includes RNA synthesis materials for encoding polymerase and the open reading frame which are nonstructural polyprotein. The remaining part consists of four protein parts: nucleocapsid, spike, membrane, envelope, and other proteins. It has been found that the coding part of COVID-19 has a 92.67% and 96.92% similarity to nucleotides in pangolin and has a 97.82% and 98.67% similarity to amino acid in bat CoV genome. There is a wide range of experiments that have been done that have showed the most validated identification of this worldwide pandemic belongs to beta CoVs. In December 2019 the first signs of coronaviruses were discovered in Hubei province. The World Health Organization

(WHO) announced COVID-19 was of international concern. Early symptoms of this disease commonly include fever, shortness of breath, fatigue, and cough (Akinwande et al., 2019; Sheikhzadeh et al., 2020). In some patients symptoms like diarrhea, headache, and dyspnea have been observed. It is believed that males are more at risk than females, perhaps this because of their hormones and immune system which are more resistant. Most importantly this infection can spread quickly via respiratory droplets during talking, sneezing, and coughing. The recovery period of this disease is estimated to be 1 to 14 days. On the other hand the transmission of the disease is dependent on R0 (basic reproduction) and the range of R0 estimated for COVID-19 is between 3.3 and 5.5. This range of R leads to fast transmission from animals to human and humans to humans. To note the biggest concern is that patients may have the disease but with no symptoms but they can transmit virus to other people. Moreover, the number of infected and deaths have been dramatically increasing daily. Hence early diagnosis is very important to prevent it from spreading from one person to another and to cut the transmission chain and isolate the infected person. The WHO proposed a guideline as ASSURED (Affordable, Sensitive, Specific, User-friendly, Rapid and robust, Equipment free, and Deliverable to end-users) for the diagnosis of patients There are several current methods shown below, such as immunological assays, amplification method, and biosensors (Fig. 12.1).

12.1.1 Immunological assays

This method involves immunological assays [e.g., lateral flow immunoassay (LFIA), enzyme-linked immunosorbent assay (ELISA)].

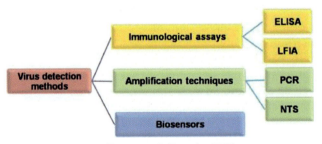

Figure 12.1 Different virus detection methods (Sengupta & Hussain, 2021). Reproduced with permission from Joydip Sengupta, Chaudhery Mustansar Hussain, Graphene-based field-effect transistor biosensors for the rapid detection and analysis of viruses: A perspective in view of COVID-19, Carbon Trends, Volume 2, January 2021, 100011. Copyright Elsevier 2021.

This method is based on the detection of antigen/antibody interactions. Some improved examples of immunoassays include peptide-based luminescent, automated chemiluminescent immunoassay, enzyme-linked immunosorbent assay, and immunochromatographic assay. This method requires the complex production routes of recombinant proteins and antibodies.

12.1.2 Amplification method

This method involves amplification-based techniques [e.g., reverse transcription polymerase chain reaction (RT-PCR), nanopore target sequencing (NTS)]. This technique was classified into groups: firstly, reverse transcription polymerase reaction which is based on amplification of RNA and genes extracted from biological samples; secondly, amplifying of nucleic acids at a steady temperature, called the isothermal nucleic acid technique. This method is divided into four advanced techniques including (1) loop-mediated isothermal amplification, (2) clustered regularly interspaced short palindromic, and (3) rolling circle amplification (Hess, Seifert, & Garrido, 2013).

All of these techniques are expensive, require skilled personnel, are difficult to work, and offer slow detection. This is the reason why biosensors have received great attention in the early detection of different samples with various concentrations.

12.1.3 Nanobiosensors

Biosensors can offer fast, inexpensive, time-efficient, highly stable, quick response, simultaneous detection, and multisample detection measurements. Localized surface plasmon resonance is one of the promising candidates for the early detection of real samples of coronavirus. LSPR biosensors have exponential sensitivity to detect any surrounding changes like the variation in refractive index RI with low limit of detection.

Applying nanoparticles in biosensors brought a revolution in biological detection due to their unique optical and physical properties. Due to their remarkable properties like high surface to area ratio, high conductivity, high biocompatibility with sample, and high bioaffinity, they are widely used in biosensors. One of the most significant nanoparticles is graphene. In 2004 graphene flakes were isolated by Andre Giem and Konstantin Novoselove (Farmani & Mir, 2020; Farmani et al., 2020; Farmani, Farmani, & Biglari, 2020; Ghodrati, Mir, & Farmani, 2020; Hamzavi-Zarghani et al., 2019; Han et al., 2020). The lightest known material with a two-dimensional structure has improved biosensors application in terms of its outstanding properties like high mobility, strong structure, high conductivity, transparency, and high capability to drug delivery and loading.

12.2 Graphene-based field effect transistor (FET) as biosensors

A field effect transistor (FET) is based on applying an electronic field to control the flow of modulated carrier (Amoosoltani, Zarifkar, & Farmani, 2019; Baqir et al., 2019; Farmani & Mir, 2019; Farmani, 2019a, 2019b; Sadeghi et al., 2019). In a graphene-based FET, graphene is employed as a channel of the structure that leads to a quick response and detection of different kind of viruses and bacteria. It was reported that graphene was suitable when used in the Si MOS FETs for highly sensitive biosensors. In graphene, its two-dimensional electron gas was bare and exposed directly to the liquid, whereas in the Si MOS FETs, its two-dimensional electron gas was covered by a thick SiO_2 layer (Matsumoto, Kenzo, Ohno, & Inoue, 2014).

Some GFET application have been reported for the detection of both human and avian influenza. Additionally, a graphene-based FET device was used in identifying Ebolavirus with a low detection limit. Furthermore, a graphene oxide nanogrid in a FET structure showed high ability to detect Hepatitis B virus.

Fig. 12.2 presents the optical microscope image of a GFET and a schematic image of biosensing by the GFET (Matsumoto et al., 2014).

Much work has been done to detect HIV virus by applying amine-functionalized graphene-based FET devices. Research groups has found that using a flexible kapton substrate improved the detection performance of norovirus. Likewise, a small amount of zika virus was detectable with the FET device. Recently scientists proposed graphene-FET in the detection of SARS-CoV-2 (COVID-19). In this case, 1-pyrenebuyanoic acid succinimidyl ester was applied as a reporter in functionalization of GFET with a COVID-19 spike antibody for identification (Farmani, 2019a; Farmani et al., 2017; Mozaffari and Farmani, 2019).

Table 12.1 shows some types of virus that could be detected using graphene-FET.

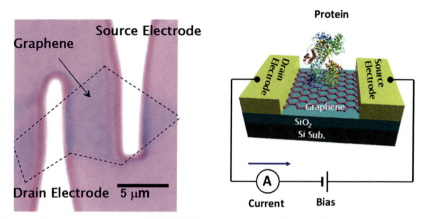

Figure 12.2 Optical microscope image of a GFET (left) and a schematic image of biosensing by the GFET (right) (Matsumoto et al., 2014). Reproduced with permission of Kazuhiko Matsumoto, et al., Recent advances in functional graphene biosensors, J. Phys. D Appl. Phys., 47, (9) (2014), Article 094005. Licensed under creative commons Attribution 4.0 International (CC BY 4.0).

Table 12.1 Graphene-based FET for biosensors

Virus sample	Basic material	Limit of detection
Human influenza	Graphene	1 nm/mL
Hepatitis B virus	Graphene	0.1 fm
Papilloma virus	Graphene oxide	1.75 nm
Rotavirus	Graphene oxide	100 pfu
Rotavirus	Graphene oxide	Na
Covid-19 virus	Graphene	1 fg/mL
Covid-19 virus	Graphene	0.2 pm

12.3 Conclusion

For the next-generation of biosensors, graphene plays an important role, especially in the field of COVID-19. Several recent advances in graphene-based materials for biosensors have been carefully reviewed, and advanced applications have been highlighted. The GFET biosensor is introduced as a reliable and selective method which shows a low limit of detection for the detection of COVID-19. GFET can play an important role in identifying SARS-CoV-2 at an early-stage and aid in protecting other people with no symptoms of COVID-19. Although GFET has paved the was for the detection of COVID-19, there are still problem like the preparation of samples and sometimes it is difficult to control immobilization and binding between virus and antibody.

References

Akinwande, D., et al. (2019). Graphene and two-dimensional materials for silicon technology. *Nature, 573*(7775), 507–518.

Amoosoltani, N., Zarifkar, A., & Farmani, A. (2019). Particle swarm optimization and finite-difference time-domain (PSO/FDTD) algorithms for a surface plasmon resonance-based gas sensor. *Journal of Computational Electronics, 18*(4), 1354–1364.

Baqir, M. A., et al. (2019). Tunable plasmon induced transparency in graphene and hyperbolic metamaterial-based structure. *IEEE Photonics Journal, 11*(4), 1–10.

Farmani, A. (2019a). Graphene plasmonic: Switching applications. *Handbook of Graphene: Physics, Chemistry, and Biology,* 455.

Farmani, A. (2019b). Three-dimensional FDTD analysis of a nanostructured plasmonic sensor in the near-infrared range. *JOSA B, 36*(2), 401–407.

Farmani, A., & Mir, A. (2019). Graphene sensor based on surface plasmon resonance for optical scanning. *IEEE Photonics Technology Letters, 31*(8), 643–646.

Farmani, A., & Mir, A. (2020). Nanosensors for street-lighting system. *Nanosensors for Smart Cities* (pp. 209–225). Elsevier.

Farmani, A., et al. (2017). Design of a tunable graphene plasmonic-on-white graphene switch at infrared range. *Superlattices and Microstructures, 112,* 404–414.

Farmani, A., et al. (2020). Optical nanosensors for cancer and virus detections. *Nanosensors for Smart Cities* (pp. 419–432). Elsevier.

Farmani, H., Farmani, A., & Biglari, Z. (2020). A label-free graphene-based nanosensor using surface plasmon resonance for biomaterials detection. *Physica E: Low-dimensional Systems and Nanostructures, 116,* 113730.

Ghodrati, M., Mir, A., & Farmani, A. (2020). Carbon nanotube field effect transistors–based gas sensors. *Nanosensors for Smart Cities* (pp. 171–183). Elsevier.

Hamzavi-Zarghani, Z., et al. (2019). Tunable mantle cloaking utilizing graphene metasurface for terahertz sensing applications. *Optics Express, 27*(24), 34824–34837.

Han, B., et al. (Eds.), (2020). *Nanosensors for Smart Cities.* Elsevier.

Hess, L. H., Seifert, M., & Garrido, J. A. (2013). Graphene transistors for bioelectronics. *Proceedings of the IEEE, 101*(7), 1780–1792.

Matsumoto, K., Kenzo, M., Ohno, Y., & Inoue, K. (2014). Recent advances in functional graphene biosensors. *Journal of Physics D: Applied Physics, 47*(9), Article 094005.

Mozaffari, M. H., & Farmani, A. (2019). On-chip single-mode optofluidic microresonator dye laser sensor. *IEEE Sensors Journal.*

Sadeghi, T., et al. (2019). Improving the performance of nanostructure multifunctional graphene plasmonic logic gates utilizing coupled-mode theory. *Applied Physics B, 125*(10), 189.

Sengupta, J., & Hussain, C. M. (2021). Graphene-based field-effect transistor biosensors for the rapid detection and analysis of viruses: A perspective in view of COVID-19. *Carbon Trends, 2,* 100011.

Sheikhzadeh, E., et al. (2020). Diagnostic techniques for COVID-19 and new developments. *Talanta,* 121392.

13

Biomedical applications

Jih-Hsing Chang[1], Narendhar Chandrasekar[2],
Shan-Yi Shen[1], Mohd. Shkir[3] and Mohanraj Kumar[1]
[1]Department of Environmental Engineering and Management, Chaoyang University of Technology, Taichung City, Taiwan [2]Department of Nanoscience and Technology, Sri Ramakrishna Engineering College, Coimbatore, India [3]Advanced Functional Materials & Optoelectronics Laboratory (AFMOL), Department of Physics, Faculty of Science, King Khalid University, Abha, Saudi Arabia

13.1 Introduction

Nanoparticles are defined as particles of less than 1000 nm in size by the US Food and Drug Administration. Nonetheless, nanoparticles are also classified in the medical field as less than 500 nm, as cells may be endocytotic to nanoparticles less than 500 nm. However, particles from 500 nm to 1000 nm can be used for medical purposes because, although they cannot enter cells, they can have very useful bulk properties, such as optical or magnetic properties. The nanoparticles' clinical activity is mainly based on drug administration, as some types of nanoparticles may act as stable vectors in vivo and have low therapeutic stability. The promising field as a potential drug delivery sector is seeing new work into compact nanoparticles, which contain both biological constituents, such as peptides or lipids, and typical chemical components, such as silica or iron oxide. The silica nanoparticles' surface may have an ordered internal mesoporous structure (of diameter between 2 nm to 10 nm) with a sizeable porous volume of $0.6 - 1$ cm^3/g and a large surface area of $700 - 1,000$ m^2/g. Their size, ranging from nano- to submicroscale (50–500 nm), shape, and charge can be customized, offering several possibilities for loading anticancer drugs (Fig. 13.1). One of the most important aspects of a drug delivery system is the EPR (enhanced permeation and retention) effect (Fig. 13.2) (Khosravian et al., 2016; Napierska et al., 2009; Vogt et al., 2010). The nanoparticles enhance the drugs' permeation into the tissue or cell used as a target. So this chapter focuses on silica nanoparticles that are particularly promising for biomedical applications.

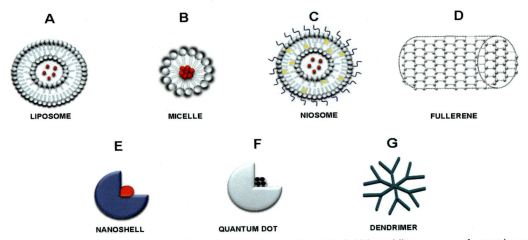

Figure 13.1 Schematic of several drug delivery vessels. From left to right; lipid-based liposomes, surfactant-based micelles, hexagonal carbon architectured carbon nanotubes, dendrimers, and nanoparticles. (Elisa, 2013).

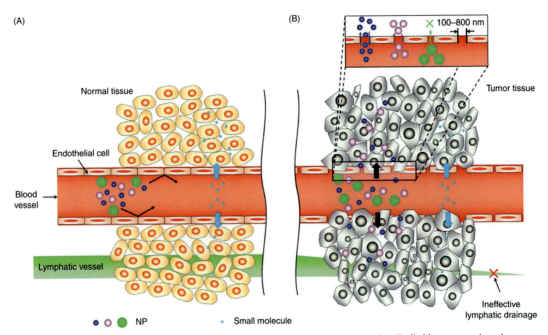

Figure 13.2 Blood transport represents nanomaterials or molecules from normal cells (left) compared to the tumors' EPR effect. (Neda & Aleksandra, 2017).

The normal tissues have a proper cell packing ratio which cuts off replication once they come into contact with other cells. The phenomenon is called contact inhibition. Cancer cells lack this property and keep replicating without contact inhibition and are densely packed.

Silica-based nanomaterials are a variety of nanoparticles that have recently emerged as promising vectors for biomedical applications. Furthermore, silicon-based materials have been widely investigated for applications in microelectronics, optics, and chemical/biological sensors because of the simple fabrication process for the semiconductor property of silicon and large-scale production. However, silicon-based hybrid nanomaterials have significant biomedical applications because particle size, shape, porosity, and monoscattering can be easily controlled when produced. There are two standard methods based on use for the production of silica nanoparticles. The first of these is to combine functional materials involved in biomedical applications, such as killing cancer cells or increasing the imaging variant, while providing other charged nanoparticles that effectively interact with the environment by increasing the specificity and reducing the specificity. The developed material should create an appropriate in vivo biointerface to be a successful transmission method and it not just about how robust it is in vitro. When the nanoparticle is taken into the body, it faces a complex environment, where it is recognized as a foreign object and the body attempts to eliminate it. For example, different proteins and cellular components in the bloodstream are rapidly reduced in effectiveness. The rapid uptake of nanoparticles by the reticuloendothelial system (RES) reduces nanoparticles' blood flow time and prevents them from reaching targets. Therefore scientists have been working to improve nanoparticle circulation time and to overcome bioadhesion in the bloodstream. However, Table 13.1 gives the main critique of the complete view of silica nanoparticles' biomedical applications in the twentieth century.

Moreover, for biosensing applications, the nanostructured porous silicon is incredibly appealing. The microstructure relies on the refractive index's modifications to its permeability or absorption into the living organisms' microstructure. The design is used for many modern drug delivery systems, such as biomedical platforms or electronic biosensors. Yet many biomedical applications of silica nanoparticles are given in Fig. 13.3. The scope of this chapter is to present innovations in the current biomedical role of silica nanomaterials and to conclude with a review of the possible directions for this technology and MSNs biosafety.

Table 13.1 An exclusive look at the biomedical uses of silica nanoparticles in the 20th century.

NPs type	Remarkable topics in the biomedical application era	Remarkable application in biomedical era	Reference with year
DDSNs	Functional dye-doped silica nanoparticles for bioimaging, diagnostics, and therapeutics	Bioimaging, diagnostics, and therapeutics	(Santra, Dutta, & Moudgil, 2005)
MHSNC	Drug-loaded, magnetic, hollow silica nanocomposites for nanomedicine	Nanomedicine	(Zhou, Gao, & Shao, 2005)
Pure silica, dye-doped silica, and silica-coated magnetic nanoparticles	Bioconjugated silica-coated nanoparticles for bioseparation and bioanalysis	Bioseparation and molecular imaging	(Smith, Wang, & Tan, 2006)
SiO_2	In vitro toxicity of silica nanoparticles in human lung cancer cells	Human lung cancer therapy	(Lin, Huang, Zhou, & Ma, 2006)
MSNs	Synthesis and functionalization of a mesoporous silica nanoparticle based on the sol–gel process and applications in controlled release	Drug delivery	(Trewyn, Slowing, Giri, Chen, & Lin, 2007)
MSN	Mesoporous silica nanoparticles for drug delivery and biosensing applications	Drug delivery and biosensing	(Slowing, Trewyn, Giri, & Lin, 2007)
MSNs	Mesoporous silica nanoparticles as controlled release drug delivery and gene transfection carriers	Drug delivery, gene delivery, and other therapeutic agents	(Slowing, Vivero-Escoto, Wu, & Lin, 2008)
MSNs	Mechanized nanoparticles for drug delivery	Drug delivery	(Cotí et al., 2009)
MSNs	Mesoporous silica nanoparticles for intracellular controlled drug delivery	Drug delivery	(Vivero-Escoto, Slowing, Trewyn, & Lin, 2010)
SiO_2 NPs MSNs	Silica-based nanoparticles for photodynamic therapy applications	Photodynamic therapy	(Couleaud et al., 2010)
MSNs	Toward biocompatible nanovalves based on mesoporous silica nanoparticles	Targeted drug delivery	(Yang, 2011)
MSNs	Multifunctional mesoporous silica nanocomposite nanoparticles for theranostic applications	Theranostic applications	(Lee, Lee, Kim, Kim, & Hyeon, 2011)
MSNs	Mesoporous silica nanoparticle based nanodrug delivery systems: synthesis, controlled drug release and delivery, pharmacokinetics and biocompatibility	Drug delivery	(He & Shi, 2011)
MSNs	Critical considerations in the biomedical use of mesoporous silica nanoparticles	Tumor therapy and biodistribution	(Lin, Hurley, & Haynes, 2012)
MSNs	Mesoporous silica nanoparticles: synthesis, biocompatibility and drug delivery	Drug delivery	(Tang, Li, & Chen, 2012)

(Continued)

Table 13.1 (Continued)

NPs type	Remarkable topics in the biomedical application era	Remarkable application in biomedical era	Reference with year
SiO$_2$ NPs MSNs	Silica-based nanoprobes for biomedical imaging and theranostic applications	Bioimaging and theranostic	(Vivero-Escoto, Huxford-Phillips, & Lin, 2012)
MSNs	Functionalized mesoporous silica materials for controlled drug delivery	Controlled drug delivery	(Yang, Gai, & Lin, 2012)
MSNs	Mesoporous silica nanoparticle nanocarriers: biofunctionality and biocompatibility	Targeted drug delivery	(Tarn, Ashley, Xue, Carnes, & Jeffrey Brinker, 2013)
MSNs	Mesoporous silica nanoparticles in medicine	Cancer therapy	(Mamaeva, Sahlgren, & Lindén, 2013)
X@MSNs	Magnetic mesoporous silica-based core/shell nanoparticles for biomedical applications	Hyperthermia treatment and targeted drug delivery	(Knezevic, Ruiz-Hernandez, Henninkc, & Vallet-Regíde, 2013)
MSNs	Mesoporous silica nanoparticles as antigen carriers and adjuvants for vaccine delivery	Vaccine delivery	(Mody et al., 2013)
MSNs	Multifunctional mesoporous silica nanoparticles as a universal platform for drug delivery	Drug delivery	(Argyo, Weiss, Bräuchle, & Bein, 2014)
MSNs	MSN anticancer nanomedicines: chemotherapy enhancement, overcoming of drug resistance, and metastasis inhibition	Anticancer nanomedicines and targeted drug delivery	(He & Shi, 2014)
MSNs	pH-responsive mesoporous silica nanoparticles employed in controlled drug delivery systems for cancer treatment	Controlled drug delivery	(Yang et al., 2014)
MSNs	Cell microenvironment stimuli-responsive controlled-release delivery systems based on mesoporous silica nanoparticles	Controlled drug delivery	(Zhu, Wang, Lin, Xie, & Cell, 2014)
MSNs	Smart multifunctional drug delivery toward anticancer therapy harmonized in mesoporous nanoparticles	Multifunctional drug delivery	(Baek et al., 2015)
MSNs	Large pore mesoporous silica nanomaterials for application in delivery of biomolecules	Drug and biomolecule delivery	(Knezevic & Durand, 2015)
MSNs	Mesoporous silica nanoparticles in drug delivery and biomedical applications	Controlled/targeted drug delivery systems	(Wang et al., 2015)
MSNs	Molecular and supramolecular switches on mesoporous silica nanoparticles	Disease therapy and cell imaging	(Songa & Yang, 2015)
PMO NPs	Syntheses and applications of periodic mesoporous organosilica nanoparticles	Catalysis and nanomedicine	(Croissant, Cattoen, Man, Durand, & Khashab, 2015)

(Continued)

Table 13.1 (Continued)

NPs type	Remarkable topics in the biomedical application era	Remarkable application in biomedical era	Reference with year
MSNs	Mesoporous-silica-functionalized nanoparticles for drug delivery	Drug delivery	(Giret, Man, & Carcel, 2015)
MSNs	Smart mesoporous silica nanocarriers for antitumoral therapy	Antitumoral therapy	(Baeza, Vallet-Regí, & Mesoporous, 2015)
MONs, PMO NPs, BS NPs	Chemistry of mesoporous organosilica in nanotechnology: molecularly organic–inorganic hybridization into frameworks	Nanomedicine	(Chen & Shi, 2016)
MONs, PMO NPs, BS NPs	Mesoporous silica nanoparticles with organobridged silsesquioxane framework as innovative platforms for bioimaging and therapeutic agent delivery	Bioimaging and therapeutic agent delivery	(Du et al., 2016)
MSNs	Shape matters when engineering mesoporous silica-based nanomedicines	Carriers for drug delivery	(Hao, Lia, & Tang, 2016)
MSNs	Mesoporous silica nanoparticles in tissue engineering — A perspective	Controlled drug delivery and stem cell tracking	(Rosenholm, Zhang, Linden, & Sahlgren, 2016)
MSNs	The application of mesoporous silica nanoparticle family in cancer theranostics	Cancer theranostics	(Feng et al., 2016)
MSN@Lipids	Protocells: modular mesoporous silica nanoparticle-supported lipid bilayers for drug delivery	Drug delivery	(Butler et al., 2016)
MSNs	Externally controlled nanomachines on mesoporous silica nanoparticles for biomedical applications	Drug delivery	(Ruhle, Saint-Cricq, & Zink, 2016)
MSNs	Synthesis, functionalization, and applications of morphology-controllable silica-based nanostructures: a review	Biological catalysis and drug delivery	(Sun, Zhou, & Zhang, 2016)
MSNs	Recent applications of the combination of mesoporous silica nanoparticles with nucleic acids: development of bioresponsive devices, carriers, and sensors	Biological sensors and nanomedicine	(Castillo, Baeza, & Vallet-Regi, 2017)
MSNs	Stimuli-responsive delivery vehicles based on mesoporous silica nanoparticles: recent advances and challenges	Drug delivery systems	(Zhu et al., 2017)
MSNs	pH-responsive mesoporous silica and carbon nanoparticles for drug delivery	Drug delivery	(Gisbert-Garzaran, Manzano, & Vallet-Regi, 2017)
MSNs	Progress in nanotheranostics based on mesoporous silica nanomaterial platforms	Multimodal imaging and cancer therapy	(Singh, Patel, Leong, & Kim, 2017)

(*Continued*)

Table 13.1 (Continued)

NPs type	Remarkable topics in the biomedical application era	Remarkable application in biomedical era	Reference with year
MSNs	Advances in mesoporous silica-based nanocarriers for codelivery and combination therapy against cancer	Cancer therapy	(Castillo, Colilla, & Vallet-Regi, 2017)
MSNs	Mesoporous silica materials for controlled delivery based on enzymes	Controlled drug delivery	(Llopis-Lorente, Lozano-Torres, Bernardos, Martinez-Manez, & Sancenon, 2017)
MSNs	Mesoporous silica molecular sieve based nanocarriers: transpiring drug dissolution research	Drug delivery	(Pattnaik & Pathak, 2017)
SiO_2 NPs	Bioactive effects of silica nanoparticles on bone cells are size, surface, and composition dependent	Nanomedicine	(Ha, Viggeswarapu, Habib, & Beck Jr, 2018)
MSNs, FMSNs	Ultrasound-mediated cavitation-enhanced extravasation of mesoporous silica nanoparticles for controlled-release drug delivery	Controlled drug delivery	(Paris, Mannaris, & Cabanas, 2017)
OD-SNPs, ID-SNPs, and DD-SNPs	Impact of the strategy adopted for drug loading in nonporous silica nanoparticles on the drug release and cytotoxic activity	Drug release and cytotoxic activity	(Riva et al., 2018)
MSNs	Optoacoustic imaging identifies ovarian cancer using a microenvironment targeted theranostic wormhole mesoporous silica nanoparticle	Optoacoustic imaging and theranostic nanomedicine	(Samykutty et al., 2018)
MSN, L-MSN	Luminescent mesoporous silica nanoparticles for biomedical applications: synthesis and characterization	Biomedical imaging	(Jain, Reeja, Mondal, & Sinha, 2018)
MSNs	Advances in mesoporous silica nanoparticles for targeted stimuli-responsive drug delivery	Drug delivery	(Castillo et al., 2019)
MMSN	A robust method for fabrication of monodisperse magnetic mesoporous silica nanoparticles with core–shell structure as anticancer drug carriers	Anticancer drug carriers	(Asgari, Soleymani, Miri, & Barati, 2019)
MSNs	Functionalized mesoporous silica nanoparticles with excellent cytotoxicity against various cancer cells for pH-responsive and controlled drug delivery	Controlled drug delivery	(Park et al., 2019)
Ag NC-MSN	Boosting antibacterial activity with mesoporous silica nanoparticles supported silver nanoclusters	Antibacterial activity	(Liu, Li, Fang, & Zhu, 2019)

(Continued)

Table 13.1 (Continued)

NPs type	Remarkable topics in the biomedical application era	Remarkable application in biomedical era	Reference with year
Au@MSNs	Novel design of NIR-triggered plasmonic nanodots capped mesoporous silica nanoparticles loaded with natural capsaicin to inhibition of metastasis of human papillary thyroid carcinoma B-CPAP cells in thyroid cancer chemophotothermal therapy	Cancer chemophotothermal therapy	(Yu et al., 2019)
GO/MSN	Constructing mesoporous silica-grown reduced graphene oxide nanoparticles for photothermal-chemotherapy	Photothermal and chemotherapy	(Guo, Yang, Zhang, & Chen, 2019)
Fe_3O_4@SiO_2	Antibacterial and anticancer activities of asymmetric lollipop-like mesoporous silica nanoparticles loaded with curcumin and gentamicin sulfate	Antibacterial and anticancer	(Cheng, Zhang, Deng, & Hu, 2020)
MSN@SiNPs	Multifunctional mesoporous silica nanoplatform based on silicon nanoparticles for targeted two-photon-excited fluorescence imaging-guided chemo/photodynamic synergetic therapy in vitro	Chemo/photodynamic therapy	(Li, Zhang, He, Li, & Zhang, 2020)
HMSNs	A pH/ROS-responsive, tumor-targeted drug delivery system based on carboxymethyl chitin gated hollow mesoporous silica nanoparticles for antitumor chemotherapy	Antitumor chemotherapy	(Ding et al., 2020)
SN-g-PCAAMC-b-PDMAEMA	UV-light cross-linked and pH de-cross-linked coumarin-decorated cationic copolymer grafted mesoporous silica nanoparticles for drug and gene co-delivery in vitro	Drug and gene codelivery	(Zhou, Ding, Wang, & Fu, 2020)
CNT@MS	Near infrared light responsive carbon nanotubes@mesoporous silica for photothermia and drug delivery to cancer cells	Drug delivery and photothermia	(Li et al., 2020)
MSN-AuNPs	Gold nanoparticle-capped mesoporous silica-based H_2O_2-responsive controlled release system for Alzheimer's disease treatment	Alzheimer's disease treatment	(Yang et al., 2016)

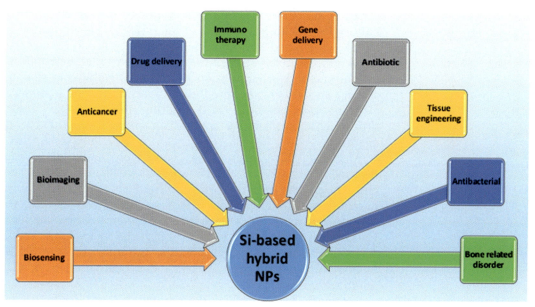

Figure 13.3 Biomedical applications of silicon-based hybrid nanoparticles.

13.2 Why silica nanoparticles are most suited for biomedical applications?

Silica nanoparticles have multiple therapeutic benefits in the medicine era. First, the chemistry is convenient to produce nanomaterials with specified properties of sizes, shapes, and surface area. In chemistry silica is easily adjustable and is important for optimum biocompatibility and biodistribution. Silica nanoparticles have been synthesized for over 60 years and have a history. In 1956 the synthetizations of monodispersed amorphous silica nanoparticles were initially performed by Kolbe, then upgraded by Fink and Stöber in 1968 (Kneuer et al., 2000). Stöber's method is used to measure silica nanoparticles' size from tens of nanometers to several micrometers in diameter by changing the catalyst and precursor ratio. In the early 1990s, scientists from Japan and the Mobil Corporation reported "Template"-guided synthesis methods for mesoporous silica nanoparticles (MSNs) (Kolbe, 1956; Stöber, Fink, & Bohn, 1968). Tunable silica nanomaterials, for example include solid spheres (Almeida, Chen, Foster, & Drezek, 2011), mesoporous hollow spheres (Wan & Zhao, 2007), mesoporous particles (Kresge, Leonowicz, Roth, Vartuli, & Beck, 1992), rattle-typed spheres

(Chen et al., 2014), foam-like nanoparticles (Zhang et al., 2015), nanotubes (Schmidt-Winkel et al., 1999), mesoporous red blood cell-shaped nanoparticles (Chen et al., 2017), etc. The silica nanoparticles' surface may also be easily altered/tailored by functional molecules such as colloidal stabilizing spacers, activity antibodies, imaging fluorophores, magnetic nanoparticles for manipulation (MRI), gating molecules to induce drug release, surface-free interaction groups, and therapeutic agents.

Moreover, silica nanoparticles can be biocompatible as they degrade to silica in vivo, which naturally exists in various tissues and can also be excreted from the body via urine. Despite the dosage, crystallinity, particle height, surface characteristics, and the administration route have effects on their toxicity. The higher dose raises a toxicity risk. Nanoparticles made from crystal silica are toxic and cause silicosis. But due to its excellent biocompatibility, amorphous silica has been used as a food additive for decades.

Moreover, silica, an essential mineral, is found in the human body in keratinized pieces, such as in nails and hair, and is vital to bones and connective tissues' growth. The FDA has approved it to be safe in adequate quantities. Silica is used as silica acid, a contaminated derivative of silica. In addition to nonporous Si-based nanoparticles, several MSNs are available, categorized by synthesis routes, degree of structural order, morphologies, scale, load, and pore size. MSNs are also synthesized as organic—inorganic hybrids. Organic partitions in their wall structure, known as MONs (metal—organic nanostructures) and functionalized MSNs, synthesized by in situ or postsynthesis modified phases. The saturation concentration of solidified amorphous silica or equivalent silica gels was found to be ~115 ppm at neutral pH and 25°C (Morey, Fournier, & Rowe, 1964; Vogelsberger, Seidel, & Rudakoff, 1992); it differed with pH and reaction conditions. Thus each MSN form's solubility depends on the respective structural parameters and is thought to be different for these various types of nano MSNs. Therefore not just in terms of stabilization and resorption, but also in terms of the biomedical properties of MSNs, the parameters of the synthesis, including unexpected side effects, will differ. Toxicity measurements of MSNs are typically established by viability experiments on particular cell types that show their biocompatibility; relatively little evidence is available on the degradability factors and other parameters of MSNs. Worldwide researchers are tracking and researching the degradation mechanisms of silica hybrids for MSNs in various applications.

13.3 Pharmaceutical applications of mesoporous silica nanoparticles

Silica-based nanomaterials, such as porous silica, have long been recognized in potential biomedical applications, especially drug delivery. Researchers worldwide have identified mesoporous silica drug delivery systems with unusual pore characteristics such as scale, form, accessibility, and morphologies range. An extensive range of therapeutic molecules has been successfully loaded and distributed in a controlled or rapid burst release manner suitable for the researchers. The most extensively researched MSNs are genuinely 2D channeled MSNs, such as MCM-41 and SBA-15 (Figs. 13.4 and 13.5), which usually have required pore architectures of unidirectional channels.

The pore geometry, scale, architecture, and their general size/morphology are the main factors influencing drug filling, retention, and release habits. In the literature, it has been stated

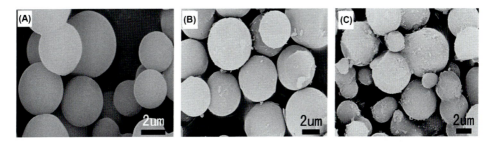

Figure 13.4 SEM micrographs of S16−80 (A), S16−120 (B), and S16−150 (C) (Hu, Wang, Zhi, Jiang, & Wang, 2011).

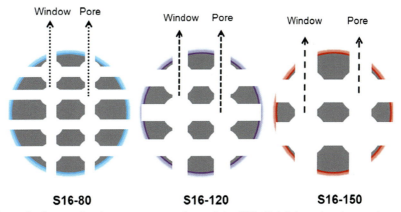

Figure 13.5 Schematic diagram for the pore constructions of the SBA-16 fabricated under varying temperature conditions (Hu et al., 2011).

that, due to an improvement in pore size, SBA-15, or MCM-41, the drug's release rate can be increased. However, the increase in particle size of MCM-41 considerably slowed the release rate due to the apparent extension of the drug pathway (Hu et al., 2011).

The SEM imaging of the MSNs shows a uniform size distribution in the 3–4 micrometer range with a spherical shape attributed to the template used, which is a surfactant which becomes spherical upon contact with water.

A correlation was found between the dissolution rate of indomethacin, IMC, and the pore architecture and size of the SBA-16 (Hu et al., 2011). The window and pore architecture of various SBA structures are given in Fig. 13.5. Higher pore number and the larger pore volume results in faster release of drugs from the system.

The rate of release of drugs has been shown to increase with respect to the pore size of the MSNs. IMC's increased delivery was derived from the noncrystalline structure of IMC and the high surface area required for rapid dissolution after being confined to the mesoporous structure and due to the 3D open pore network of SBA-16 (Figs. 13.4 and 13.5). Due to the relatively simple synthetic process, the specified morphology, the ability to easily change the pore size by merely varying the temperature during synthesis, and, in particular, the 3D interconnected pore networks which provide favorable mass transfer kinetics, the SBA-16 microspheres are excellent candidates for hydrophobic drug delivery as well as for biomolecules and biocatalysts immobilization.

Surface silanol groups of silica interact with drugs by either hydrogen bonding or Coulombic interactions with surface functional groups of MSNs. As mentioned above, the drugs' adsorption would influence the experimental parameters, the solution's pH, the solvent nature, the substance's concentration, and the MSN carriers' pore size. SBA-15 rods have a higher absorption rate and can immobilize more enzymes than SBA-15 with rope-like structures. These mesoporous materials' morphology plays a vital role in the drug's ingestion and release and is consequently beneficial over traditional drug delivery systems (Ganesh, Ubaidulla, Hemalatha, Peng, & Jang, 2015).

Mani et al. (2014). synthesized a new form of MSN using nonionic surfactants. The dual template pathway for MSN synthesis was accompanied by the use of the constant concentration of Triton X-100 as the main template and the use of Tween-40 as a secondary template or as a pore-modifying agent and the use of sol–gel and solvothermal synthesis. The sol–gel synthetic silica route had a high surface area and pore volume and was chosen for the drug release analysis. The XRD results revealed small improvements in the silica nanoparticles' composition after loading the drug molecules.

TGA confirmed 23%, 30%, and 34% of the doxorubicin (DX) drug uptake. Also, drug-loaded MSNs display an initial burst release pattern followed by an extended-release pattern up to 160 h, 180 h, and 210 h. It can be inferred that the combination of the two surfactants Triton X-100 and Tween-40 could lead to a successful dual template strategy for pore structure modification in silica (Fig. 13.6); additionally, the blend of the surfactants was

Figure 13.6 SEM images of Triton X-mediated synthesis of MSNs (Mani et al., 2014).

also seen as a promising drug reservoir candidate needing an extended-release sequence. Triton X forms a spherical micelle upon exposure to water and is able to act as a template for the silica monomers to deposit over them.

13.4 Drug loading for MSNs

13.4.1 Drug encapsulation mechanisms

Various methods are used to load APIs (Active Pharmaceutical Ingredients) (Fig. 13.7) into MSNPs (Trzeciak, Kaźmierski, Wielgus, & Potrzebowski, 2020). The present research focuses on wet procedures such as the organic solvent process (OSM), or the incipient wetness impregnation (IWI) is the most widely used approach. The first approach suggests that the silica matrix and API are applied to the solution and softly mixed for several hours. The suspension is then removed by filtration or centrifugation. The drug-loaded Silica can be dried for 24 h. This approach's key disadvantages are low drug filling, degradation of large amounts of medications through filtration, and time-consuming. It is, therefore, difficult to identify the percentage of medication loading that can be easily done. IWI (Incipient wetness impregnation) is used to prepare catalysts and is intended as a superior tool for filling APIs relative to OSM. The distinction is that the IWI uses a concentrated API solution. The solution is applied dropwise to the mesoporous Silica, producing a sticky powder. The volume of the solution is typically equal to or less than the volume of the silica pore. Once the mixture has been cleaned out, the solvent is removed by drying, not by filtration

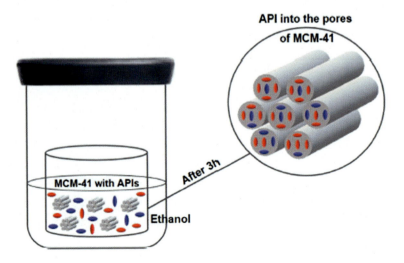

Figure 13.7 Schematic of APIs loading into the pores of MSNs by diffusion supported loading method (Trzeciak et al., 2020).

or centrifugation, as done in the OSM. In IWI, the volume of API loaded can be easily managed. Still, the possibility of residual API crystallization on the silica surface can be observed after the solvent has evaporated and the pores have been sealed off. The hexane solvent approach is a better OSM approach, but it is not recommended for biomedical applications because it is toxic. The APIs are able to load into the pore geometry as shown in the figure below in the presence of ethanol medium.

Diffusion assisted loading (DiSupLo) is another exciting way to load APIs into MSNs. In this procedure, the starting material is a homogenized mixture of two or more elements stored in an ethanol-containing tank. Initially, there would be no apparent interaction between the solid product and the solvent. The only interaction would be with the diffused ethanol vapor and the solid reagents. The next step follows with the vapors entering the mixture into the solid-state in the entire volume, condensing locally and dissolving the API. The API, which is dissolved in a minimum amount of solvent, would be transported to the pores of the MSNs.

Huanga et al. (2020) synthesized and integrated a multifunctional active molecule, 10-phenylphenothiazine (Ph-PTZ), into MSNs, containing luminescent MSNs. To synthesize copolymers over the surface of the luminescent MSNs through light irradiation using polyethylene glycol methyl acrylate (PEGMA) and itaconic acid (IA) as monomers a metal-free photocatalytic atom transfers radical polymerization (ATRP) was used (Fig. 13.8). Incorporating the copolymer increases the composites' aqueous stability and endows its ability to load a standard cis-diamine platinum dichloride (CDDP) anticancer molecule. The molecules are loaded on to the MSN structure as per the sequence described below in Fig. 13.8.

Immobilization of NH_2 to synthesize PTH@MSNs-NH2 on the surface of MSNs was performed by APTES, which by amidation can react further with alpha-bromoisobutyryl bromide. The resulting PTH@MSNs-NH_2-Br complex creates radicals and initiates ATRP to position the copolymers above the MSNs.

The ultrapure water was applied with CTAB and NaOH, and the resulting simple surfactant solution was stirred at 80°C for 30 min. The precursor TEOS was applied to the mixture dropwise. The mixture was isolated after 4 h of reaction by centrifugation at 8000 rpm for 3 min. The white precipitate obtained is impure MSNs and was rinsed three times with DI water and ethanol and dried for 24 h at 40°C in a vacuum drying oven. The dried impure MSNs were scattered in a methanolic solution of HCl and stirred at 65°C for 24 h to isolate CTAB. To acquire pristine pure MSNs, the resulting slurry was isolated and

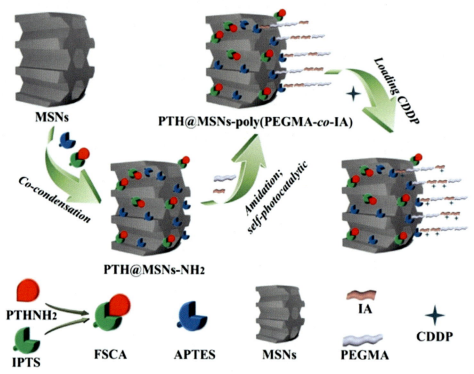

Figure 13.8 The synthesis method of PTH@MSNs-poly(PEGMA-co-IA) and their use for CDDP loading (Huanga et al., 2020).

washed twice with deionized water and methanol several times and again dried in a vacuum oven at 35°C.

Coordination bonds with the anticancer agent CDDP may be forged by the high number of carboxyl groups added on the surface of PTH@MSNs-poly(PEGMA-co-IA). By dissolving PEGMA, IA, PTH@MSNs-NH2-Br powder in anhydrous toluene, and irradiating LEDs for 12 h in a nitrogen atmosphere, the drug loading threshold of PTH@MSNs-poly (PEGMA-co-IA), formed by amination and addition of self-photocatalytic ATRP in a solution, was prepared. After centrifugation, the extraction of the pure MSNs was achieved through precipitation.

At a pH of 7.4, PTH@MSNs-poly(PEGMA-co-IA) and CDDP were stirred together in PBS. The solution was held under dark conditions and vigorously stirred for 48 h at 37°C, followed by centrifugation to isolate CDDP and free CDDP complexes from PTH@MSNs-poly(PEGMA-co-IA)-CDDP. Using UV–Vis spectroscopy, CDDP concentration in the supernatant was analyzed. The concentration of CDDP loaded on PTH@MSNs-poly

(PEGMA-co-IA) was deduced from the UV absorbance difference when the compound was loaded and before it was loaded.

The synthesis and characterization of new thermoresponsive polymer-coated Fe_3O_4 embedded hollow mesoporous silica ($HmSiO_2$)-based multifunctional superparamagnetic nanocarriers carrying doxorubicin to cancer cells was postulated by Asghar, Qasim, Dharmapuri, and Das (2020). P(NIPAM-MAm) a thermoresponsive polymer coated with a hollow mesoporous silica nanocomposite $HmSiO_2$-F-P(NIPAM-MAm) embedded Fe_3O_4 magnetoresponsive nanoparticles (NP) was synthesized by an in situ method (Fig. 13.9) in which NIPAM and MAm monomers, in the presence of magnetite nanoparticles (Fe_3O_4), an oxidizer and a cross-linker, are placed on the surface of hollow mesoporous silica NPs ($HmSiO_2$) TEM observations showed the almost spherical morphology of the $HmSiO_2$-F-P(NIPAM-MAm) structure with its radius in the 50–150 nm range. HRTEM has verified the thermoresponsive layer's coating and the embedding of iron nanoparticles on the surface of $HmSiO_2$ nanoparticles. The P(NIPAM-MAm) shells and Fe_3O_4 NPs on the hollow MSNs were verified by XRD and FTIR analysis. The $HmSiO_2$-F-P(NIPAM-MAm) nanoparticles have been proven by VSM to be superparamagnetic. A phase change at ~38°C was shown by a DSC study of $HmSiO_2$-F-P(NIPAM-MAm). The $HmSiO_2$-F-P(NIPAM-MAm) nanocarrier was tested through an in vitro assay for its ability to be used as a drug carrier using doxorubicin as a model drug. The system's encapsulation performance and encapsulation capability were observed to be 95% and 6.8%, respectively. $HmSiO_2$-F-P(NIPAM-MAm)-temperature-dependent and pH-dependent doxorubicin release profile was seen for Dox. At temperatures above the lower critical solution temperature (LCST), a quicker drug release was reported. The $HmSiO_2$-F-P(NIPAM-MAm) nanocarrier has been reported in in vitro cytotoxicity assays to be extremely biocompatible. The $HmSiO_2$-F-P(NIPAM-MAm)-Dox nanocomposite nanocarrier has also demonstrated intense anticancer activity against cancerous Hela cells. The $HmSiO_2$-F-P(NIPAM-MAm)-Dox nanocomposite nanocarrier in vitro cell uptake analysis shows high internalization levels in the Hela cells.

A combination of the sol–gel synthesis process for mesoporous silica and the coprecipitation method for iron oxide nanoparticles has prepared $HmSiO_2$ Fe_3O_4 particles. In situ polymerization of NIPAM and MAm monomers over the surface of MPS-modified $HmSiO_2$ NPs in the presence of iron oxide nanoparticles, MBA (crosslinker), SDS, and APS (oxidizer) are synthesized by thermoresponsive polymer (P(NIPAM-MAm)) coated and iron nanoparticles-

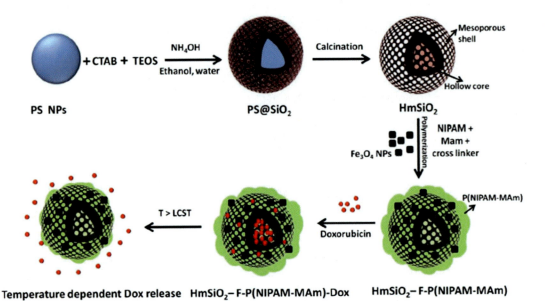

Figure 13.9 Schematic diagram of the synthesis of HmSiO$_2$-F-P(NIPAM-MAm) nanocarriers and the temperature-dependent release of doxorubicin from HmSiO$_2$-F-P(NIPAM-MAm)-Dox nanocarriers (Asghar et al., 2020).

embedded HmSiO$_2$-based magnetic (HmSiO$_2$-FP(NIPAM-MAm) nanocarrier).

Ultrasound of the Dox and HmSiO$_2$-F-P(NIPAM-MAm) nanocarrier was undertaken in a 4 mL distilled water beaker. The doxorubicin suspension and nanocarrier HmSiO$_2$-FP(NIPAM-MAm) was stirred for 24 h at 40°C. The shrunken polymer escaped the open pores above the LCST under dark conditions. The suspension, resulting in a bloated shell of the polymer, was cooled down. The doxorubicin-loaded composites were separated by centrifugation (15 min) and repeatedly washed with clean water extract the Doxorubicin's free molecules. For further study, the resulting HmSiO$_2$-F-P(NIPAM-MAm)-Doxorubicin nanocarrier and the supernatant with Doxorubicin's free molecules were deposited in a dark environment. The synthesis protocol and loading of targeting moieties and API is described in the Fig. 13.7.

The electron microscopy study has shown that the HmSiO$_2$-F-P (NIPAM-MAm) nanocomposite has hollow mesoporous Si in its center and thermoresponsive polymer embedding iron oxide nanoparticles in its shell. The HmSiO$_2$-F-P (NIPAM-MAm) composite was immune to pH and temperature, nontoxic, showed superparamagnetic properties, and was water-dispersible. The HmSiO$_2$-FP(NIPAM-MAm) composite has demonstrated good drug-loading capability with a high encapsulation efficiency of 95%. Effects of in vitro doxorubicin release experiments from nanocomposite revealed both a temperature-dependent and pH-dependent release activity critical for regulated drug delivery.

13.5 Gene delivery

Slowing, Vivero-Escoto, Wu, and Lin (2008) outlined recent advances in surface-functionalized MSNs for use as efficient drug delivery systems. The synthesis of these functionalized MSNs is developed along with the techniques for regulating the structural and chemical functions of the MSNs to be used in biomedical applications. Any drug/gene delivery vessel would have to conform with a series of necessary properties to achieve release at the required concentration, precisely at the targeted target region, and often within a predetermined period. MSNs may be used as stimulus-responsive gene carriers in the sense of a concept called gatekeeping fabricated to achieve these goals. These systems have been strategically built using various novel compounds, such as functionalized nanoparticles, organic molecules, or even supramolecular assemblies, which serve as gatekeepers to monitor and enhance drug encapsulation and subsequent release. The disulfide linker is kept as a trapping moiety for the guest molecule with the gatekeeper entities keeping the overall architecture of the assembly intact (Fig. 13.10).

Drug delivery mechanisms with "zero premature exposure" are useful when the drug is administered for highly dangerous targets, such as anticancer molecules (drugs). Obtaining the ability to monitor medication release location and timing accurately will lead to many breakthroughs in site-specific drug or gene delivery applications. Stimulus-responsive release mechanisms focused on MSNs have an immense ability to accomplish such a high goal (Table 13.2).

Mamaeva, Sahlgren, and Lindén (2013) summarized the use of MSNs in medicine in a comprehensive analysis, noting that MSNs are desirable materials for drug delivery and cancer treatment in particular. Alternative approaches such as photodynamic therapy and multidrug therapy can be used until hydrophobic materials are loaded and bioavailable. The delivery system can be designed in a way that they follow a particular pattern for their formulation, drug loading, release, and uptake by the cells of interest (target cells), as shown in Fig. 13.11.

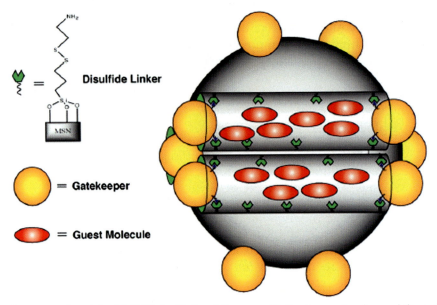

Figure 13.10 Representation of the MSN filled with the visitors, usually the drug molecules, and the surface was coated with the gatekeeper (Slowing et al., 2008).

Table 13.2 MSN-based nanodevices have been developed using photochemical, pH-responsive, and redox-active gatekeepers (Slowing et al., 2008).

Gatekeeper	Stimulus	Trigger	Guest molecule
CdS-NP	Redox	Reducing agent (DTT, ME)	ATP, vancomycin
Fe_3O_4-NP	Redox	Reducing agent (DTT, DHLA)	Fluorescein
Au-NP	Redox	Reducing agent (DTT)	β-oestradiol, DNA
Coumarin	Photo	UV light (λ = 310 nm)	Cholestane, Phenanthrene
Diethylenetriamine PAMAM	Ionic Redox	pH, anions Reducing agent (DTT, TCEP)	"Blue squaraine" ATP, DNA
[2]-Pseudorotaxane [DNPD-CBPQT]$^{4+}$	Redox	NaCNBH$_3$	Ir(ppy)$_3$
[2]-Rotaxane R^{4+}	Redox	Fe(ClO$_4$)$_3$, Ascorbic acid	Ir(ppy)$_3$, Rhodamine B
[2]-Pseudorotaxane PEI-CD	Ionic	pH	Calcein

Physical adsorption is the usual process for small molecules of drugs loaded into MSNs. Silanol groups ($\equiv$Si-OH), which are found on the silica surface and usually act as adsorption sites, are accessible at concentrations of 2–4 groups/nm.

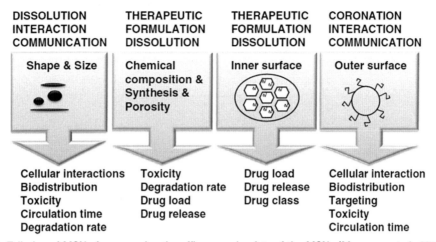

Figure 13.11 Tailoring of MSNs for assessing the efficacy and safety of the MSNs (Mamaeva et al., 2013).

The pH point for Silica to be zero is roughly 2–3, which means that the silica surface stays negatively charged under biologically important conditions without specific ion adsorption. This leaves an avenue for the adsorption of positively charged adsorbates by electrostatic adsorption to load hydrophilic molecules into MSNs. Adsorption can be further improved by functionalizing weak acids and bases such as carboxylic acids or amines on the silica surface, which can easily change the surface load under sufficient pH and provide additional adsorbent–adsorbent reactions. Hydrophobic medications are typically adsorbed by organic solvents, followed by vacuum drying to eliminate the remaining solvents' traces. Effectively, pore size plays an enormous role in adsorption if a drug's size is equivalent to or equal to the MSN pore diameter range. The physicochemical structure of the intended drug, in particular its polarity, its circulation duration, the carrier's deterioration/degradation rate, the relatively poor physical interactions between the pore surface, and the drug may not be successful enough to guarantee that the drug is not released prematurely until the target is approached. Therefore the emphasis was on designing a drug release technique activated by intracellular processes such as lysosomal pH sensitivity or external stimulation sensitivity, such as a magnetic field or light irradiation.

Only systemic delivery, using holds, can be realized by DNA, mRNA, and siRNA-dependent delivery systems, as the free molecules/drugs suffer due to low bioavailability and worse in vivo cellular absorption. Because of their potential negative charge, the molecules are not taken up readily and internalized by cells as they suffer more due to low stability and a very short half-life.

The resulting intracellular degradation significantly reduces its nuclear access even after cell internalization. In the current trend, gene distribution depends a lot on viral vectors, which are supposed to be used with strict care because of their potential side effects and mutations. Electroporation, sonoporation, and magnetoreception are unique nonviral transmission possibilities to be looked at. All three have their restrictions, which require comprehensive instrumentation for which a carrier is required. As of now, as possible suitors for gene transport, liposomes and polymers are under analysis as DNA carrier complexes or porous polymer nanoparticles. MSNs, however, demonstrate some remarkable properties that may be very suitable for gene transmission, including a minimal pore range that can effectively shield the cargo before release; if a plasmid is present inside the mesoporous structure, the surface's simple adjustment possibilities are still available to improve adsorption and release characteristics, effectively enhancing the efficiency of the surface. Several studies on plasmid adsorption inside and on the surface of MSNs have thus been documented and evaluated. However, it is not easy to load nucleic acids into the mesopores of MSNs. Several studies postulate that DNA has been adsorbed to the outer surface of cationic surface-functionalized MSNs for gene transmission.

The PEI layer's very high positive charge density will be useful for adsorption. The nucleic acid adsorption potential will be much smaller for particles covered with a layer of organophosphate. However, effective in-vitro transfection was observed, with DNA not being completely shielded against nuclease degradation in the body. In contrast, in vitro and in vivo transfection by nonporous silica carriers has been successfully demonstrated. Two methodologies can be used to increase the adsorption of genes into the mesopores by improving drug loading and using MSNs with optimal, typically larger mesopores.

The overall recorded load capacity is 121.6 mg DNA/g of MSN, which is higher than that reported in cases where only the DNA's outer surface was adsorbed. This is a rather large figure, given the anionic origin of both DNA and silica. Chaotropic conditions, however, mask the electrostatic push away from each other and allow high concentrations of loading. DNA desorption produced by the MSNs did not occur at temperatures below 20°C. However, a larger volume of DNA desorbed as the temperature increased above 20°C and near-complete desorption occurred within an hour, similar to physiological temperatures. Still, more functionalization of the MSNs by cationic polymers could achieve a slower release rate of DNA. The loading potential limit was decreased to around 22 mg DNA/g MSNs by using longer calf thymus DNA (20,000 base-pairs), which is attributed to the stiffness of the DNA under experimental conditions.

SiRNA was packaged into the mesopores of MSNs under conditions similar to salmon DNA adsorption, but siRNA was not adsorbed to the MSNs. However, repeating the process with 66.7% ethanol under dehydrating conditions led to a siRNA load of 13.5 mg siRNA/g MSN at an equilibrium siRNA concentration of 80 μg/mL. SiRNA or EGFP siRNA was used for experimentation, based on adsorption conditions' proportionality to the cargo's hydrophilicity. SiRNA concentrations of 170 μg/mL, up to 27.5 mg siRNA/g MSNs, were able to be adsorbed on the mesopores' insides at equilibrium. 25 kDa PEI was adsorbed from pure ethanol to the MSNs to cap the mesopores. In vitro experiments projected that the siRNA trapped in the interiors of the PEI-capped MSNs are effectively secured against enzymatic degradation, resulting in long-term systemic circulation stability. SiRNA could be readily released into the cytoplasm and, after internalization by A549 cells, could escape from the interior of the MSNs. Fluorescently labeled siRNA and MSNs confirmed this and were meant to effectively knock down target genes. This siRNA delivery system displayed less cytotoxicity when used at adequate concentrations to exhibit essential gene-silencing levels.

Using amine-functionalized MSNs with large cage-like pores, with a diameter of 20 nm and a size of 70–300 nm, demonstrated the advantages of using MSNs with larger pores. The MSNs could consume 150 μg/g MSN firefly luciferase plasmid DNA (5256 base-pairs) from the suspension buffer. Interaction with the negatively charged DNA was made possible by the surface amino groups. The plasmid adsorbed was shielded from enzymatic degradation, but the plasmid's exact position remained unknown on the particles' surface and within the mesopores. Reports have emerged of high loading levels of luciferase plasmid DNA achieved by PBS adsorption using monodisperse MSNs of 250 nm in diameter with a mean mesopore diameter of 23 nm. These MSNs have enabled in vitro, proficient gene delivery of plasmids encoding two proteins, luciferase and Green Fluorescent Protein.

13.6 Protein absorption and separation

Liu and Xu (2019) detailed the different factors influencing the protein loading on MSNs and techniques for targeted protein delivery and their controlled release using MSNs in a study

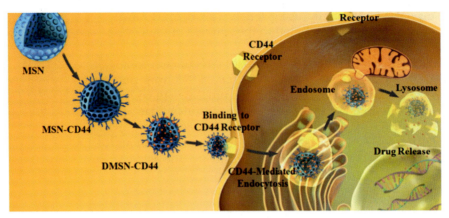

Figure 13.12 Schematic of an MSN-loaded protein (CD44 McAb) with a codelivery device targeted by a chemotherapy drug to address multidrug resistance in breast cancer cells of MCF-7/MDR1 (Wang, Liu, Wang, Shi, & Zhou, 2015).

on protein delivery and the difficulties faced in the clinical translation of protein loaded MSNs (Fig. 13.12). The breast cancer cell which is exorbitantly multidrug resistant can be targeted and destroyed by an MSN-based delivery system which can bind to the CD44 receptors in its surface. The drug will be released once the lysosomal pH drops to acidic levels releasing the cargo.

MSNs have gained significant interest in delivery vectors for drugs, genetic materials, and protein biomacromolecules after Mobil synthesized the first MCM-41 form in 1992 due to ease of synthesis and fast surface functionalization with different molecules, excellent biocompatibility, tunable physicochemical properties, and good thermal stability. Also, only by optimizing their scale, form, and alteration of their surfaces can MSNs achieve controlled release and targeted delivery of payloads. The synthesis of MSN and drug delivery applications have been summarized by several scholars, although relatively few have been interested in protein delivery. An overview of the applications of MSNs for protein delivery for therapeutic applications, a discussion of the factors affecting protein loading, and general strategies for targeted delivery and controlled release of proteins with MSNs are discussed.

Enzymes in the human body are essential in metabolic and pathological processes. Due to its high sensitivity and selectivity, an enzyme in the target site makes itself an endogenous trigger due to irregular expression or upregulated expression. Matrix metalloproteinases-2 (MMP-2) is one of the most widely found overexpressed proteins closely related to tumor invasion and metastasis in specific tumor types. This phenotype was used by Junjie Liu et al. (2015) to develop a codelivery mechanism of MMP-2 responsive Bovine Serum Albumin and Doxorubicin (Fig. 13.13). Phenylboronic acid (PBA) has a high sialic acid (SA) affinity, a well-known overexpressed protein proxy for metastatic tumors of HepG2 cells. They decorated PBA with human serum albumin (PBA-HSA). They then used an MMP-2 substrate peptide (PVGLIG) sequence and polyarginine sequence (RRRRRRRR) intermediate linker to bind it to the MSN surface, forming SNs-HSA-PBA@DOX. The HAS model protein stopped DOX from premature release and adhered to the PBA ligand on the surface of the MSNs for successful targeting until MMP-2 cut off the linker. Once the linker was cleaved, the opened pore entrance would open to release the drugs.

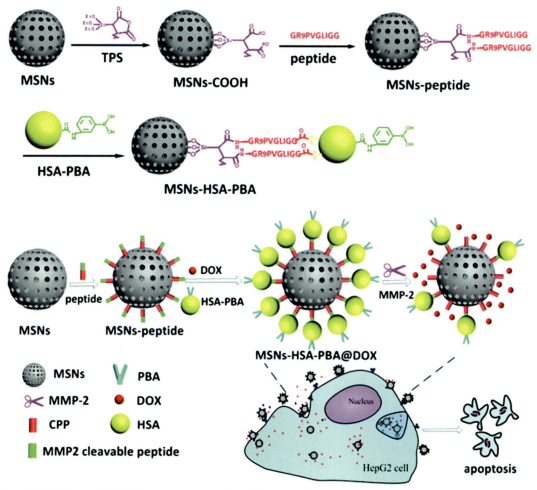

Figure 13.13 Schematic MMP-2 enzyme-responsive MSNs for codelivery of HAS and Doxorubicin (Liu et al., 2015).

MSNs-HSA-PBA@DOX demonstrated high effectiveness in vitro and in vivo against tumors by combining aggressive targeting and enzyme regulation drug release strategies. The surface functionalization of MSNs using a carboxylic group for creating a peptide bond with macromolecules of interest and subsequent drug loading in its cavities for inducing apoptosis in HepG2 cell is described in the following Fig. 13.13.

There are some significant constraints for MSNs-based protein delivery systems that should be investigated to develop several practical applications (Table 13.3).

Table 13.3 Modified MSNs for protein delivery (Liu & Xu, 2019).

The type of MSN	Modification	Proteins	Cell or disease models
MSN	Propylthiol	Cytochrome C	HeLa cells
MSN	Amino (—NH2), carboxyl (—COOH)	Glucose oxidase (GOX) and glucose isomerase (GI)	No
FDU-12	Aminopropyltriethoxysilane (APTES), 3-mercaptopropyltrimethoxysilane (MPTMS), vinyltrimethoxysilane (VTMS), and phenyltrimethoxysilane (PTMS)	Bovine serum albumin	No
BSA-15	Aminosilanes	Lysozyme (LYS) and myoglobin (MYO)	No
BSA-15	Aminosilanes	Bovine serum albumin (BSA)	No
MSN	Boronic acid	Insulin and cyclic adenosine monophosphate (cAMP)	Rat pancreatic RIN-5F cells
MSN	Glycidoxypropyltrimethoxysilane (GPTMS), chitosan	Bone morphogenetic protein-2 (BMP-2)	Bone mesenchymal stem cells (bMSCs), Bone regeneration
MSN	3-Aminopropyltrimethoxysilane (APTMS)	Superoxide dismutase (SOD) and glutathione peroxidase (GPx)	Inflammation and oxidative stress
MSN	3-Aminopropyltrimethoxysilane (APTMS)	Superoxide dismutase (SOD)	HeLa cells
BSA-15	3-Aminopropyltriethoxysilane (APTMS)	Bone morphogenetic protein 2 (BMP-2)	Bone marrow stromal cells (BMSCs)
MSN	2-[methoxy(polyethylenoxy)-propyl] trimethoxysilane (PEG-silane)	Luciferase	Hela cells
Hollow mesoporous silica capsules (HMSCs)	Carboxyl, amino, 5-aminofuorescein (AFL)	BSA, Goat IgG	HeLa cells

(Continued)

Table 13.3 (Continued)

The type of MSN	Modification	Proteins	Cell or disease models
MSN	2-(Methoxy [polyethyleneoxy]propyl) trimethoxysilane	BSA, macrophage colony-stimulating factor, and receptor	Zebrafish embryos/larvae
MSN	Soluble CD4 ("sCD4"), amide-immobilized sCD4, 18-peptide CD4 fragment	HIV-1 gp120 Glycoprotein	No
SBA-15	Unmodified	Porcine pancreas lipase (PPL)	Catalyst
MSN	Unmodified	β-galactosidase	N2a cell
MSN	Gold nanoparticle	BSA and enhanced green fluorescent protein (eGFP)	Onion epidermis cells (plant cell)
MSN	PEGylated	BSA	No
SBA-15	3-aminopropyltriethoxysilane (APTES), 3-mercaptopropyltrimethoxysilane (MPTMS), phenyltrimethoxysilane (PTMS), vinyltriethoxysilane (VTES), and 4-(triethoxysilyl)butyronitrile (TSBN)	Penicillin G acylase (PGA)	No
MSN	3-Aminopropyltriethoxysilane (APTES)	Carbonic anhydrase (CA)	Human cervical cancer (HeLa) cells
MSN	Poly(ethyleneimine)-b-poly(N-isopropylacrylamide) (PEI/NIPAM)	Trypsin inhibitor protein (type II − S), catalase	No

In vitro protein release profile analysis alone will not be adequate; much further attention is required to monitor the transmitted proteins' in vivo navigation, such as their release into target tissues, using combined multifunctional or multimodal imaging methods. There are records of the start of research into the manufacture of multifunctional MSNs, and there are also very few creative routes for the functionalization of MSN surfaces. The structure of the protein is the basis of its functionality. Owing to the sheer vulnerability of the proteins and their entanglement in different pathological and physiological environments, the characterization of their structures and the assessment of their biofunction after release into the target cells or organs should be carried out with caution. Toxicity is the most troublesome question for translating any new substance from preclinical studies to intended clinical applications. Most recent studies have shown that MSNs have relatively little cytotoxicity and exhibit strong in vivo biocompatibility, but further work needs to be done to explain the thorough biosafety of MSNs, such as the end of the duration of MSNs following their in vivo use.

Yang, Lin, Mou, and Tung (2019), who documented a mesoporous ultrafiltration membrane inspired by nature with excellent permeability and high selectivity for protein selective separation and isolation of nanoparticles, detail the potential of the MSNs, in particular protein recognition and separation. For ultrafiltration applications, a diatom-mimicking membrane with hierarchical pores in a single-layer mesoporous silica thin film (MSTF) with perpendicular pores assisted by a macroporous anodic aluminum oxide (AAO) membrane (MSTF-AAO) was developed (Fig. 13.14). The AAO membrane was positioned over the standardized vertical straight-through mesoporous channels with a pore diameter of 5.9 ± 0.4 nm, exhibited superhydrophilicity, and constant nonbreaking to centimeter range area of 17.3 cm^2, showing both ultrahigh water permeability of $1027 \pm 20 \, L \, m^{-2} \, h^{-1} \, bar^{-1}$ and an exceptional rejection rate of >99% for molecules greater than 7 nm. The permeability is considerably higher than that of the nanochanneled membranes previously mentioned. Diatomic frustules have been exhaustively used to exclude micrometer-sized particles in water filtration. The silicate and alumina composition of the diatom frustules has multilayers of an epitheca and a hypotheca. With variable pore sizes, these layers' overlap each other; all pores have uniform periodic macropores. A dual-layer MSTF-AAO membrane was designed using highly ordered mesoporous silica thin-film (MSTF) communication, standing on anodic aluminum oxide (AAO) membrane with periodic straight-through macropores, by mimicking the hierarchically periodic porous form of a standard diatom frustule to attain selective ultrafiltration. A flat and smooth-surfaced substrate is essential for producing the MSTF. The macroporous AAO was then spin-coated with cross-linked polystyrene (PS) to form a smooth surface working with PS/AAO.

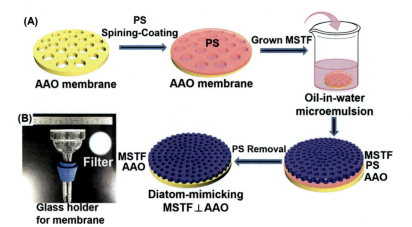

Figure 13.14 Synthesis and fabrication of mesoporous MSTF⊥AAO filter membrane (Yang et al., 2019).

By immersion of the PS/AAO membrane into an emulsion of CTAB, ethanol, and decane, the growth of MSTF over the PS/AAO surface was carried out. A complex heating process eliminated the polystyrene coating by annealing the MSTF/PS/AAO, accompanied by UV ozone washing. With outstanding thermostability, the diatom-mimicking membrane was then obtained.

With a touch angle of 8.8°, consistently nonbreaking to the centimeter range area of 17.3 cm^2, ultrahigh-flux water permeability is up to 1027 ± 20 L m^{-2} h^{-1} bar^{-1}, the MSTF$\perp$AAO membrane demonstrated outstanding hydrophilicity. Permeability is comparatively higher than that reported for other ultrafiltration membranes. In protein isolation and quantum dot ultrafiltration, the MSTF$\perp$AAO membrane exhibited a superior molecular-sieving capacity (Fig. 13.15). A rejection rate of up to 99% was recorded for molecular-sieving of protein mixtures. In addition to the excellent permeability and thermal stability, the excellent size-exclusivity makes this diatom-mimicking MSTF$\perp$AAO membrane an outstanding nanofiltration candidate.

13.7 Nucleic acid detection and purification

Bitar, Ahmad, Fessi, and Elaissari (2012) have developed silica-based nanomaterials in protein separation/adsorption, nucleic acid identification and purification, drug delivery and gene delivery, bioimaging, and pharmaceutical applications. DNA contains valuable knowledge about the compounds used for developmental, diagnosis, and gene therapy research. A significant step in DNA manipulation is the isolation of DNA and its subsequent purification.

DNA purification has thus become a major pillar of modern-day molecular biology, gene therapy, and genetics. The new techniques developed have enhanced the ability, expertise, and facility for the separation techniques of DNA molecules, followed by their purification. Using silica-based nanoparticles for DNA detection, separation, and purification, the various methodologies and techniques for preparing nanoparticles with multifunctionalized surfaces have advanced biomedical science. In general, weak electrostatic repulsion forces, dehydration, and hydrogen bonding govern the adsorption of DNA on the surface of SiNPs. Raman and FTIR spectroscopy can recognize the interaction of DNA with silica through hydrogen bonds. Researchers have developed tunable silica surfaces for more precise and effective interactions by perceiving these interactions' essence. Kneuer et al. (2000) synthesized amine-functionalized SiNPs, N-(2-aminoethyl)-3-aminopropyltrimethoxysilane(AEAPS), or N-(6-aminohexyl)-3-aminopropyltrimethoxysilane (AHAPS)

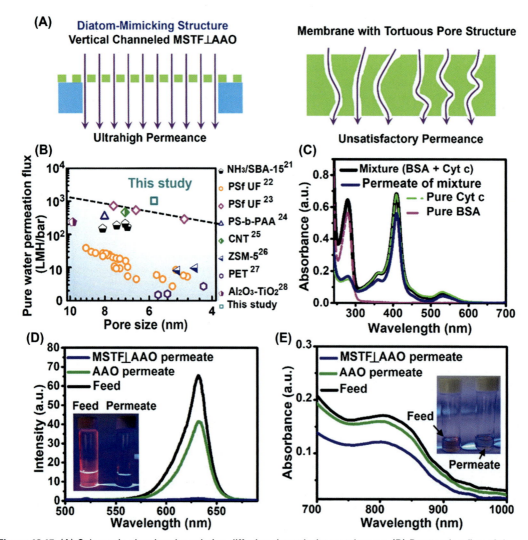

Figure 13.15 (A) Schematic showing the solution diffusion through the membranes. (B) Permeation flux of the MSTF⊥AAO membrane compared with the permeation flux of other membranes. (C) UV − Vis absorption spectra before and after Cyt c and BSA was filtered through the MSTF⊥AAO nano filter device. (D) Fluorescence spectra before and after the filtration of CdSe@ZnS quantum dots. (E) UV − Vis absorption spectra before and after the filtration of PbS quantum dots (Yang et al., 2019).

modified SiNPs. The mean size of the SiNPs was 10–100 nm, and the surface at 7.4 pH was +7 mV to +31 mV. To form a complex shielded from DNase I degradation plasmid DNAs were also used to observe their association with SiNPs. The presence of SiNPs (ten parts) almost preserves the plasmid DNA entirely, unlike free plasmid DNA, which is completely degraded by DNase I, with only a small amount of supercoiled DNA ending up as nicked circular DNA. Although 30 components of SiNPs completely covered the DNA, its separation at this pace was difficult.

In order to achieve variable fluorescence, SiNPs are used to produce biosensors by coupling with oligonucleotides through hybridization with the target complementary DNA or RNA probes. The immobilization of oligonucleotides over SiNPs using disulfide conjugation was reported by Hilliard, Zhao, and Tan (2002). 3-mercaptopropyltrimethoxysilane (MPTS) was silanized with 60 nm silica particles, and oligonucleotides were immobilized on the incubated SiNPs. To test the hybridization quality, fluorescence was observed at 520 nm. SiNPs have also helped construct and improve fast, inexpensive, and robust DNA isolation, purification, and analysis techniques. The kinetics and structural changes of plasmid DNA binding to silica in monovalent and divalent salts were studied by Nguyen et al. To be considered, two kinds of electrostatic interactions were found, one between plasmid DNA and the silica surface and the other being the one inside the plasmid DNA subunits that regulate molecule conformation.

As carriers for use as drug delivery devices, mesoporous SiNPs are also used. The drug's intake and release process rely on its contact with the SiNPs' pores; gold NPs can be used to regulate this interaction as gatekeepers. The synthesis of MSNs with removable CdS nanoparticle caps as drug delivery systems was reported by Lai et al. (2003). After the drug loading is completed within the pores, the covalent interaction between the pore surface and CdS nanoparticles ensures the drugs remain within (Fig. 13.16).

Disulfide bond reducing molecules, such as dithiothreitol (DTT) or mercaptoethanol (ME), may be used to cleave the covalent bond and open the pores in order to release the loaded material. Thanks to their significant stimulus-controlled release, zero external modifications needed for drug preparation, and their ability to enable the movement of a wide range of drugs, these systems can be highly helpful for drug distribution.

13.8 Remarkable trends in mesoporous silica nanoparticles toward cancer therapeutic applications

The ingestion of many cancer drugs that hamper successful cancer treatment is impaired by low bioavailability, volatility, and hydrophobicity. Many lifesaving medications' therapeutic potential

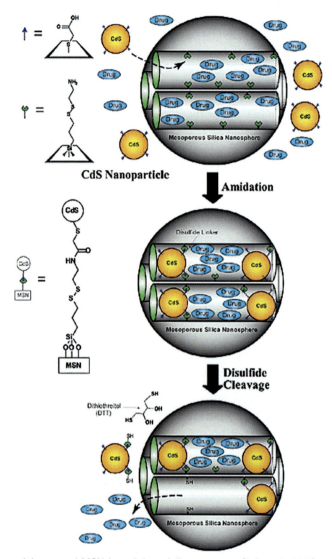

Figure 13.16 CdS nanoparticle-capped MSN-based drug delivery system (Lai et al., 2003).

has been reduced because of the apparent side effects of normal cells. Therefore, therapeutic applicability needs to recognize carriers that can carry a high payload of drugs, those that shield the drug from premature degradation, promote improved cellular penetration, and those that target individual cells. Due to their ability to hold a high payload of hydrophobic drugs with low solubility in water, MSNs have always been ideal for small molecule delivery. Since over 40% of molecules discovered have low water solubility by combinatorial screening schemes, it is no surprise to see MSNs attracting strong attention as drug delivery systems. The bulk of studies to date have concentrated on the delivery of anticancer drugs such as doxorubicin (DOX). Lee et al. (2010) administered doxorubicin to tumor sites and detected fluorescence and apoptosis caused by DOX, ex vivo, 48 h after injection, in tumor cells. The free drug was not used as a monitor, and the therapeutic advantage of the distribution mediated by particles remained uncertain. Hillegass et al. (2011) applied silica microparticles as doxorubicin carriers to treat malignant mesothelioma (MM). The microparticles filled with DOX and placebo were inserted subcutaneously, s.c. (160 mg/kg) into the tumor directly, i.e., intraperitoneally, i.p. (104 mg/mg) in mice in peritoneal MMs and contrasted with s.c. or i.v. free DOX administration. Once a week, the particles were inserted: s.c. for 3 weeks and 3 days a week; i.p. for a total of 1 week. The findings revealed a high cellular absorption of Doxorubicin, increased overall therapeutic effectiveness, and improved MM due to transmission mediated by particles. In particle-mediated delivery of PEG–PEI-coated 50 nm MSNs after intravenous administration (50 mg/kg, once a week for 3 weeks), Meng et al. (2011) demonstrated improved efficacy of doxorubicin. In contrast with that of free drug administration, tumor regression was found to be considerably greater. Doxorubicin-induced particles displayed decreased toxicity in the systemic, hepatic, and renal regions. It was found that aggregation at tumor sites was due to the usual EPR effect. To improve the overall therapeutic effectiveness, successful targeting may be accomplished by adding tumor-targeting ligands to the particles. CMPT has demonstrated impressive anticancer capacity in clinical trials but has poor solubility and toxicity in large doses. Owing to these drawbacks, to optimize the effects, different variants have been identified, and two CMPT analogs, topotecan, and irinotecan are now commonly used in cancer therapy. CMPT-loaded MSNs was inserted into the tail vein of mice carrying MCF-7 breast tumors in experiments by the Tamanoi group (Lu, Liong, Li, Zink, & Tamanoi, 2010). The scale of the tumor was larger than that of the free substance. The same CMPT-loaded MSNs were injected in the nude, and successful outcomes were also obtained for SCID mice with pancreatic cancer. Owing to the involvement of surfactants and the systematic toxicity docetaxel (Dtxl) itself, or the therapeutic formulation, Taxotere has had adverse side effects. Li et al. (2010) showed increased anticancer efficacy and low systematic toxicity of Dxtl encapsulated in 125 nm diameter PEGylated silica nanorattles using the liver cancer subcutaneous mice model. For every fourth day (days 1, 5, 9, 13), the IV dosage was 20 mg/kg before the animals were slaughtered, and the tumor load was measured on day 17. These findings show that MSNs promise drug delivery candidates to boost bioavailability, increase medications' effectiveness, and minimize adverse side effects.

The use of MSNs for cancer treatment by Moreira, Dias, and Correia (2016) has many benefits over traditional clinical approaches. The hydrophobicity, accelerated degradation, low bioavailability, and nonspecificity of highly cytotoxic drugs are influenced by their administration, and responsible for the adverse side effects. These challenges illustrate the clinic's use of anticancer medications, and there are limitations on the dosage amount and amount of therapeutic cycles that should be prescribed to each patient. Therefore, dosage reduction results in a lower clinical effect, poor bioavailability, and cancer cells' resistance. Consequently, it is vital to establish carriers that can shield the drugs from degradation and distribute them in a spatial and temporally regulated manner that enhances their therapeutic outcome without the cancer cells eliciting resistance. Of late, there has been a rise in MSN use for anticancer applications. These nanoparticles have been used in diagnostics (fluorescence imaging or magnetic resonance imaging), in medicine (drug distribution or photothermal therapy), or as theranostics agents (nanocarriers capable of mixing diagnostic and therapeutic functions). These revolutionary ideas led to the optimization of the size, architecture, and surface properties of MSNs, including the use of stealth agents and targeting ligands along with MSNs to enhance biocompatibility, biodistribution, and bioavailability in the vicinity of cancer cells. To map the nanoparticles' fate in the human body, quantum dots, iron oxide nanoparticles, or fluorescent dyes have recently been introduced into MSNs. The cargo of anticancer entities, macromolecules, are usually loaded in the cavity of the drug delivery system (Fig. 13.17). The functionalization agent is tagged on the surface of the delivery system for targeting specific cell types and the compounds which render stealth properties by imparting a negative charge as well as by their hydrophilic nature are coated on the topmost layer. Adjuvants which can function as a tracking moiety by emitting some sort of signal from the system like magnetic susceptibility, florescence, or metal-enhanced florescence can be added to the system to function as theranostics, which renders therapy and diagnoses the particular target cell type.

By adsorption, different biomolecules may be loaded onto the mesopores of the MSNs. For the development of stimulus-responsive MSN systems, other capping agents may be used. Also, with a polymer, such as PEG and PHEMA, the surface of

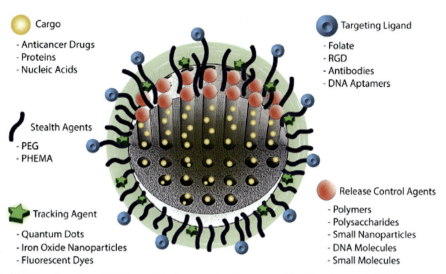

Figure 13.17 Multifunctionality of MSNs and their cargo-loading capability (Moreira et al., 2016).

the MSNs may be functionalized/changed to endow stealth properties, preventing early removal from the body. Through targeting ligands such as antibodies, DNA, peptides, aptamers, etc. the MSN method can be further changed to improve the sensitivity of the tumor tissue, and the use of tracking agents such as contrast molecules and dyes is also commonly used to allow particle tracking and tumor imaging.

One of the significant obstacles to designing nanocarriers for cancer therapy remains the ability to monitor a delivery system's release profile and distribute particular drugs to a specific cell or tissue. Stimuli-responsive carriers are currently being developed to meet these aims. These systems operate on triggers such as temperature, metabolites, pH, electromagnetic field, near-infrared radiation, redox potential, ultrasound, and even ATP to react to local or external stimuli (Fig. 13.18). This technique decreases opioid damage to healthy tissues throughout the human body's systemic circulation, reducing the side effects. Therefore, a considerable effort has been made to build MSN-based systems capable of distributing bioactive molecules in a regulated and sustained manner. The cancer cells exhibit very fast metabolism which can work on the treatments strategy. High amounts of lysosomal activity will render the extracellular matrix around the cancer cell in an acidic environment. Using an acid-sensitive polymer, the therapy can be provided in a controlled release where the payload is released only in an acidic environment.

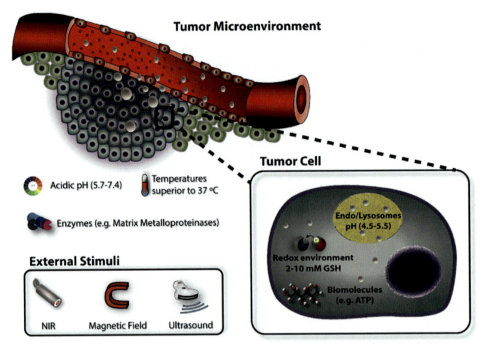

Figure 13.18 Stimuli-responsive cargo release in the tumor tissue (Moreira et al., 2016).

13.9 Biosafety of mesoporous silica nanoparticles

In vivo procedures for nanomaterials' ingestion, delivery, metabolism, excretion (ADME), and toxicity should discuss nanoparticle-based therapy's protection component. Like every pharmaceutical, according to Paracelsus, "the dosage makes the toxin," and doses must be defined as having no adverse effects and inducing life-threatening toxicity for each nanoparticle following single administration (acute toxicity) and repeated administration (chronic toxicity). Also, the separate delivery paths for ADME and toxicity assays must be analyzed. Furthermore, the functionalities and adjuvants of the nanoparticles may also affect total toxicity and ADME. Like the distribution of the organ, targetability may differ, thereby eventually impacting the safety profile. Because of the sheer variety of features in this category of products, checking the compatibility and biosafety of MSNs is intensely challenging. In designing secure and efficient distribution systems, studies on modulations of different parameters such as form, size, and chemical composition are necessary. Yu et al. (2013) studied the acute toxicity of MSNs in competent immune mice. They observed that in vivo toxicity is primarily influenced by the porosity and surface characteristics of MSNs after intravenous injection.

The overall tolerated dosage (MTD) increased in the following order: mesoporous silica particles with an aspect ratio of 1, 2, 8 with an aspect ratio of 30–65 mg/kg, amine-functionalized particles with an aspect ratio of 1, 2, 8 with an aspect ratio of 100–150 mg/kg, unmodified or amine-functionalized nonporous particles with an aspect ratio of 450 mg/kg. It was noted that the adverse reactions occurred due to mechanical vasculature obstruction accompanied by organ failure. The particles' hydrodynamic size could be related to the particles' influence on the vasculature and the resulting tolerance threshold. It was deduced that the greater the particles' hydrodynamic scale, the lower the MTD. The routine toxicity tests for MSNs, however, are somewhat uncommon. Liu, Lin, and Mou (2004) showed an example of laboratory design where the investigators used 110 nm FITC-doped MSNs for single and repetitive dose toxicity, which were investigated through detailed hematology analyses and histopathology, and fluorescent microscopy, transmission electron microscopy (TEM), and inductively coupled plasma optical emission spectrometry (ICP-OES) distribution were tested. The study described a single dose of acute toxicity found to be 1280 mg/kg and a 20 mg/kg dose of no reported adverse reaction level (NOAEL) for continuous i.v. 14-day administration. This research produced crucial statistics on the biodistribution and compartmentalization of MSNs in organs such as spleen and liver macrophages. It defined the 4-week period as adequate for 50% body clearance of nanoparticles, as determined by reducing silica content in different tissues. In preclinical studies of NPDD systems, this work is part of necessary pharmacokinetics studies. It improves scientific reliability for the following experiments using various modifications/decorations of the nanoparticles provided.

Two essential techniques are envisaged in the creation of nanopharmacology: developing a unique class of drugs and diagnostic agents focused on pharmacogenomics and drug exploration and advancing old pharmaceuticals and diagnostics by envisioning flexible delivery options for theranostics. The second approach indicates the competitive examination of multiple carriers. The plan to show the pros of nanoparticulate devices can be contrasted with commercially marketed drugs. The research by Li et al. (2010) reexamines where PEGylated hollow MSNs filled with FITC-labeled docetaxel are tested for systemic toxicity and therapeutic effectiveness combined with

Taxotere, a formulation containing docetaxel available on the market since 1996. The specific precaution was taken to circumvent the hematological side effects and liver toxicity caused by Taxotere using the equal dosage of docetaxel in MSNs comparable to the single dose (60–100 mg/m^2) given to humans during chemotherapy. The administration was 20 mg/kg, three times in 9 days, indicating lower systemic toxicity with MSNs.

Furthermore, after a single dose of 40 mg/kg, empty MSNs and docetaxel-loaded MSNs did not alter the organ morphology. A substantial decrease in xenotransplanted hepatocarcinoma was seen in both Taxotere and docetaxel-loaded MSNs, with the overall tumor weight inhibition rate measured in the MSN community being 15% higher. Although these studies are essential for the future of theranostic nanoparticulate distribution, they tend to be very small in number.

13.10 Summary

After hitting the target cells and releasing the drugs in a very controlled way, MSNs are already proven candidates capable of holding the drugs tightly in their interior during circulation. Such response-controlled release mechanisms can minimize the adverse side effects of cancer diagnostics and treatments' effectiveness. However, these drug delivery mechanisms are often complicated because of these systems' versatility and the potential to resist premature drug release, and many are only proof of concept. To allow technology transfer from the laboratory to the clinic, it is also essential to develop novel methods for the synthesis and functionalization of MSNs that conform to the reasonable manufacturing practical requirements (Fig. 13.19).

The nanoparticles need to have a proper sequence to be planned before synthesis. The formulation involved identifying a synthesis strategy to keep the particles in nanosize and identifying its size and functionalization options and then moving on to surface ligands which perform activities like tracking and tagging to cell receptors, followed by the mechanisms in which the delivery system is to be taken up by the target cells and how they behave alongside the cell's usual cycle. This is followed by strategies to avoid the immune system and the blood-brain barrier, depending on the area of action of the drug delivery system, and calculating its mode of emission from the body or degeneration by stomach acids or bodily enzymes. Finally, the pharmacokinetic outcomes are calculated for adsorption,

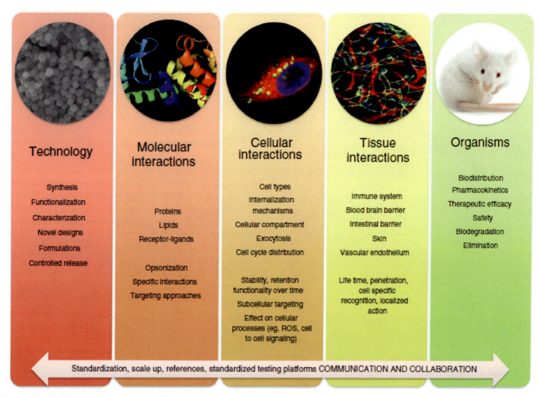

Figure 13.19 Nanotechnology and biological complexity to be considered for particle design (Mamaeva et al., 2013).

diffusion, metabolism, and elimination of the drug as well the delivery system.

The majority of mesoporous silica nanoparticle-based structures currently being built are only tested by incorporating 2D cell cultures that do not align with the tumors' complexity and other conditions to be considered in the human body. The particles' actual antitumor ability can also be tested practically in in vitro models such as 3D cell cultures, which have a dynamic structure and cellular activity close to that found on real tissues. Furthermore, it is still essential to develop the validation of MSN systems, primarily in the fields of biodistribution, biodegradation due to different body causes, excretion from the body, and clearance percentages in conjunction with the colloidal stability of MSNs in physiological media, as well as the impact of the route of administration on the efficacy of nanoparticles. The details gained from these detailed and systematic studies will allow the production of more well-designed therapies focused on MSNs

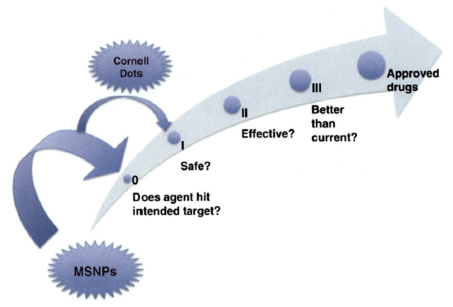

Figure 13.20 Schematic presentation of the current status of MSNPs as the current status of MSNPs' pharmacology development relative to significant stages of clinical trials (Mamaeva et al., 2013).

capable of hitting the market, offering a new generation of therapeutic agents without adverse side effects that are more appropriate for cancer use (Fig. 13.20).

At the end of the day the market gets the drug only after the evaluation of the efficacy of the delivery system. It starts with the bioavailability percentage of the drug, how aggressive it is toward normal cells and the diseased cell of interest, how effective it is for the therapy, and whether it is better in terms of efficiency and available at an affordable price for the consumer. An MSN-based delivery system will achieve all these factors due to its intricate design to deliver more than 90% of its payload as well as excellent biocompatibility and reduced toxicity of the entire delivery system.

References

Almeida, J. P., Chen, A. L., Foster, A., & Drezek, R. (2011). In vivo biodistribution of nanoparticles. *Nanomedicine (London, England)*, 6(5), 815–835.

Argyo, C., Weiss, V., Bräuchle, C., & Bein, T. (2014). Multifunctional mesoporous silica nanoparticles as a universal platform for drug delivery. *Chemistry of Materials: A Publication of the American Chemical Society*, 26(1), 435–451.

Asgari, M., Soleymani, M., Miri, T., & Barati, A. (2019). A robust method for fabrication of monodisperse magnetic mesoporous silica nanoparticles with core-shell structure as anticancer drug carriers. *Journal of Molecular Liquids*, *292*, 111367.

Asghar, K., Qasim, M., Dharmapuri, G., & Das, D. (2020). Thermoresponsive polymer gated and superparamagnetic nanoparticle embedded hollow mesoporous silica nanoparticles as smart multifunctional nanocarrier for targeted and controlled delivery of doxorubicin. *Nanotechnology, 31*, 455604, 21pp.

Baek, S., Singh, R. K., Khanal, D., Patel, K. D., Lee, E.-J., Leong, K. W., ... Kim, H.-W. (2015). Smart multifunctional drug delivery towards anticancer therapy harmonized in mesoporous nanoparticles. *Nanoscale, 7*, 14191–14216.

Baeza, A., Vallet-Regí, M., & Mesoporous, S. (2015). Silica nanocarriers for antitumoral therapy. *Current Topics in Medicinal Chemistry, 15*, 2306.

Bitar, A., Ahmad, N. M., Fessi, H., & Elaissari, A. (2012). Silica-based nanoparticles for biomedical applications. *Drug Discovery Today, 17*, 1147–1154.

Butler, K. S., Durfee, P. N., Theron, C., Ashley, C. E., Carnes, E. C., & Brinker, C. J. (2016). Protocells: modular mesoporous silica nanoparticle-supported lipid bilayers for drug delivery. *Small (Weinheim an der Bergstrasse, Germany), 12*(16), 2173–2185, Apr 27.

Castillo, R. R., Baeza, A., & Vallet-Regi, M. (2017). Recent applications of the combination of mesoporous silica nanoparticles with nucleic acids: development of bioresponsive devices, carriers and sensors. *Biomaterials Science, 5*, 353–377.

Castillo, R. R., Colilla, M., & Vallet-Regi, M. (2017). Advances in mesoporous silica-based nanocarriers for codelivery and combination therapy against cancer. *Expert Opinion on Drug Delivery, 14*(2), 229–243, Feb.

Castillo, R. R., Lozano, D., Gonzalez, B., Manzano, M., Izquierdo-Barba, I., & Vallet-Regi, M. (2019). Advances in mesoporous silica nanoparticles for targeted stimuli-responsive drug delivery. *Journal of Expert Opinion on Drug Delivery, 16*(4), Issue.

Cheng, Y., Zhang, Y., Deng, W., & Hu, J. (2020). Antibacterial and anticancer activities of asymmetric lollipop-like mesoporous silica nanoparticles loaded with curcumin and gentamicin sulfate. *Colloids and Surfaces B: Biointerfaces, 186*, 110744.

Chen, F., Hong, H., Shi, S., Goel, S., Valdovinos, H. F., Hernandez, R., ... Cai, W. (2014). Engineering of hollow mesoporous silica nanoparticles for remarkably enhanced tumor active targeting efficacy. *Scientific Reports, 30*(4), 5080.

Chen, F., Ma, M., Wang, J., Wang, F., Chern, S.-X., Zhao, E. R., ... Jokerst, J. V. (2017). Exosome-like silica nanoparticles: a novel ultrasound contrast agent for stem cell imaging. *Nanoscale., 9*(1), 402–411.

Chen, Y., & Shi, J. (2016). Chemistry of mesoporous organosilica in nanotechnology: Molecularly organic–inorganic hybridization into frameworks. *Advanced Materials, 28*(17), 3235–3272, May.

Cotí, K. K., Belowich, M. E., Liong, M., Ambrogio, M. W., Lau, Y. A., Khatib, H. A., ... Stoddart, J. F. (2009). Mechanised nanoparticles for drug delivery. *Nanoscale, 1*, 16–39.

Couleaud, P., Morosini, V., Frochot, C., Richeter, S., Raehm, L., & Duran, J.-O. (2010). Silica-based nanoparticles for photodynamic therapy applications. *Nanoscale, 2*, 1083–1095.

Croissant, J. G., Cattoen, X., Man, M. W. C., Durand, J.-O., & Khashab, N. M. (2015). Syntheses and applications of periodic mesoporous organosilica nanoparticles. *Nanoscale, 7*, 20318–20334.

Ding, X., Yu, W., Wan, Y., Yang, M., Hua, C., Peng, N., & Liu, Y. (2020). A pH/ROS-responsive, tumor-targeted drug delivery system based on carboxymethyl chitin gated hollow mesoporous silica nanoparticles for anti-tumor chemotherapy. *Carbohydrate Polymers, 245*, 116493.

Du, X., Li, X., Xiong, L., Zhang, X., Kleitz, F., & Qia, S. Z. (2016). Mesoporous silica nanoparticles with organo-bridged silsesquioxane framework as innovative platforms for bioimaging and therapeutic agent delivery. *Biomaterials, 91*, 90127.

Panzarini, E., Inguscio, V., Tenuzzo, B. A., Carata, E., & Dini, L. (2013). Nanomaterials and autophagy: New insights in cancer treatment. *Cancers, 5*, 296–319. Available from https://doi.org/10.3390/cancers5010296.

Feng, Y., Panwar, N., Tng, D. J. H., Tjin, S. C., Wang, K., & Yong, K.-T. (2016). The application of mesoporous silica nanoparticle family in cancer theranostics. *Coordination Chemistry Reviews, 319*, 86109.

Ganesh, M., Ubaidulla, U., Hemalatha, P., Peng, M. M., & Jang, H. T. (2015). Development of duloxetine hydrochloride loaded mesoporous silica nanoparticles: characterizations and in vitro evaluation. *AAPS PharmSciTech, 16*, 944–951.

Giret, S., Man, M. W. C., & Carcel, C. (2015). Mesoporous-silica-functionalized nanoparticles for drug delivery. *Chemistry A European Journal, 21*(40), 13850–13865, Issue.

Gisbert-Garzaran, M., Manzano, M., & Vallet-Regi, M. (2017). pH-responsive mesoporous silica and carbon nanoparticles for drug delivery. *Bioengineering (Basel), 4*(1), 3, Jan 18.

Guo, D., Yang, H., Zhang, Y., & Chen, L. (2019). Constructing mesoporous silica-grown reduced graphene oxide nanoparticles for photothermal-chemotherapy. *Microporous and Mesoporous Materials, 288*, 109608.

Hao, N., Lia, L., & Tang, F. (2016). Shape matters when engineering mesoporous silica-based nanomedicines. *Biomaterials Science, 4*, 575–591.

Ha, S.-W., Viggeswarapu, M., Habib, M. M., & Beck Jr, G. R. (2018). Bioactive effects of silica nanoparticles on bone cells are size, surface, and composition dependent. *Acta Biomaterialia, 82*, 184–196.

He, Q., & Shi, J. (2011). Mesoporous silica nanoparticle based nano drug delivery systems: synthesis, controlled drug release and delivery, pharmacokinetics and biocompatibility. *Journal of Materials Chemistry, 21*, 5845–5855.

He, Q., & Shi, J. (2014). MSN anticancer nanomedicines: Chemotherapy enhancement, overcoming of drug resistance, and metastasis inhibition. *Advanced Materials, 26*(3), 391–411, Jan 22.

Hillegass, J. M., et al. (2011). Increased efficacy of DOXorubicin delivered in multifunctional microparticles for mesothelioma therapy. *International Journal of Cancer. Journal International du Cancer, 129*(1), 233–244.

Hilliard, L. R., Zhao, X., & Tan, W. (2002). Immobilization of oligonucleotides onto silica nanoparticles for DNA hybridization studies. *Analytica Chimica Acta, 470*, 51–56.

Huanga, L., Yua, S., Long, W., Huang, H., Wen, Y., Deng, F., ... Wei, Y. (2020). The utilization of multifunctional organic dye with aggregation-induced emission feature to fabricate luminescent mesoporous silica nanoparticles based polymeric composites for controlled drug delivery. *Microporous and Mesoporous Materials, 308*, 110520.

Hu, Y., Wang, J., Zhi, Z., Jiang, T., & Wang, S. (2011). Facile synthesis of 3D cubic mesoporous silica microspheres with a controllable pore size and their application for improved delivery of a water-insoluble drug. *Journal of Colloid and Interface Science, 363*, 410–417.

Jain, B., Reeja, K. V., Mondal, P., & Sinha, A. K. (2018). Luminescent mesoporous silica nanoparticles for biomedical applications:Synthesis and characterization. *Journal of Luminescence, 200*, 200–205.

Khosravian, P., Ardestani, M. S., Khoobi, M., Ostad, S. N., Dorkoosh, F. A., Javar, H. A., & Amanlou, M. (2016). Mesoporous silica nanoparticles functionalized with folic acid/methionine for active targeted delivery of docetaxel. *OncoTargets and Therapy, 9*, 7315–7330.

Kneuer, C., Sameti, M., Haltner, E. G., Schiestel, T., Schirra, H., Schmidt, H., & Lehr, C. M. (2000). Silica nanoparticles modified with aminosilanes as carriers for plasmid DNA. *International Journal of Pharmaceutics, 196*, 257–261.

Knezevic, N. Z., & Durand, J.-O. (2015). Large pore mesoporous silica nanomaterials for application in delivery of biomolecules. *Nanoscale, 7*, 2199–2209.

Knezevic, N. Z., Ruiz-Hernandez, E., Henninkc, W. E., & Vallet-Regíde, M. (2013). Magnetic mesoporous silica-based core/shell nanoparticles for biomedical applications. *RSC Advances, 3*, 9584–9593.

Kolbe G. Das komplexchemische verhalten der kieselsäure (Ph.D.Thesis). 1956.

Kresge, C. T., Leonowicz, M. E., Roth, W. J., Vartuli, J. C., & Beck, J. S. (1992). Ordered mesoporous molecular sieves synthesized by a liquid-crystal template mechanism. *Nature Materials, 359*, 710–712.

Lai, C.-Y., Trewyn, B. G., Jeftinija, D. M., Jeftinija, K., Xu, S., Jeftinija, S., & Lin, V. S.-Y. (2003). A mesoporous silica nanosphere-based carrier system with chemically removable CdS nanoparticle caps for stimuli-responsive controlled release of neurotransmitters and drug molecules. *Journal of the American Chemical Society, 125*(15), 4451–4459.

Lee, J. E., Lee, N., Kim, T., Kim, J., & Hyeon, T. (2011). Multifunctional mesoporous silica nanocomposite nanoparticles for theranostic applications. *Accounts of Chemical Research, 44*(10), 893–902.

Lee, J. E., et al. (2010). Uniform mesoporous dye-doped silica nanoparticles decorated with multiple magnetite nanocrystals for simultaneous enhanced magnetic resonance imaging, fluorescence imaging, and drug delivery. *Journal of the American Chemical Society, 132*(2), 552–557.

Lin, W., Huang, Y.-wern, Zhou, X.-D., & Ma, Y. (2006). In vitro toxicity of silica nanoparticles in human lung cancer cells. *Toxicology and Applied Pharmacology, 217*, 252–259.

Lin, Y.-S., Hurley, K. R., & Haynes, C. L. (2012). Critical considerations in the biomedical use of mesoporous silica nanoparticles. *Journal of Physical Chemistry Letters, 3*(3), 364–374.

Liu, Y.-H., Lin, H.-P., & Mou, C.-Y. (2004). Direct method for surface silyl functionalization of mesoporous silica. *Langmuir: The ACS Journal of Surfaces and Colloids, 20*, 3231–3239.

Liu, J., Li, S., Fang, Y., & Zhu, Z. (2019). Boosting antibacterial activity with mesoporous silica nanoparticles supported silver nanoclusters. *Journal of Colloid and Interface Science, 555*, 470–479.

Liu, H.-J., & Xu, P. (2019). Smart Mesoporous Silica Nanoparticles for Protein Delivery. *Nanomaterials, 9*, 511.

Liu, J., Zhang, B., Luo, Z., Ding, X., Li, J., Dai, L., ... Cai, K. (2015). Enzyme responsive mesoporous silica nanoparticles for targeted tumor therapy in vitro and in vivo. *Nanoscale, 7*, 3614–3626.

Li, B., Harlepp, S., Gensbittel, V., Wells, C. J. R., Bringel, O., Goetz, J. G., ... Mertz, D. (2020). Near infra-red light responsive carbon nanotubes@mesoporous silica for photothermia and drug delivery to cancer cells. *Materials Today Chemistry, 17*, 100308.

Li, L., Tang, F., Liu, H., Liu, T., Hao, N., Chen, D., ... He, J. (2010). In vivo delivery of silica nanorattle encapsulated docetaxel for liver cancer therapy with low toxicity and high efficacy. *ACS Nano, 4*(11), 6874–6882.

Li, S., Zhang, Y., He, X.-W., Li, W.-Y., & Zhang, Y.-K. (2020). Multifunctional mesoporous silica nanoplatform based on silicon nanoparticles for targeted two-photon-excited fluorescence imaging-guided chemo/photodynamic synergetic therapy in vitro. *Talanta, 209*, 120552.

Li, L., et al. (2010). In vivo delivery of silica nanorattle encapsulated docetaxel for liver cancer therapy with low toxicity and high efficacy. *ACS Nano, 4*(11), 6874–6882.

Llopis-Lorente, A., Lozano-Torres, B., Bernardos, A., Martinez-Manez, R., & Sancenon, F. (2017). Mesoporous silica materials for controlled delivery based on enzymes. *Journal of Materials Chemistry B, 5*, 3069–3083.

Lu, J., Liong, M., Li, Z., Zink, J. I., & Tamanoi, F. (2010). Biocompatibility, biodistribution, and drug-delivery efficiency of mesoporous silica nanoparticles for cancer therapy in animals. *Small (Weinheim an der Bergstrasse, Germany), 6*(16), 1794–1805.

Mamaeva, V., Sahlgren, C., & Lindén, M. (2013). Mesoporous silica nanoparticles in medicine—Recent advances. *Advanced Drug Delivery Reviews, 65*(5), 689–702, May.

Mamaeva, V., Sahlgren, C., & Lindén, M. (2013). Mesoporous silica nanoparticles in medicine—Recent advances. *Advanced Drug Delivery Reviews, 65*, 689–702.

Mani, G., Pushpараj, H., Peng, M. M., Muthiahpillai, P., Udhumansha, U., & Jang, H. T. (2014). Synthesis and characterization of pharmaceutical surfactant templated mesoporous silica: Its application to controlled delivery of duloxetine. *Materials Research Bulletin, 51*, 228–235.

Meng, H., et al. (2011). Use of size and a copolymer design feature to improve the biodistribution and the enhanced permeability and retention effect of DOXorubicin-loaded mesoporous silica nanoparticles in a murine xenograft tumor model. *ACS Nano, 5*(5), 4131–4144.

Mody, K. T., Popat, A., Mahony, D., Cavallaro, A. S., Yu, C., & Mitter, N. (2013). Mesoporous silica nanoparticles as antigen carriers and adjuvants for vaccine delivery. *Nanoscale, 5*(12), 5167–5179, Jun 21.

Moreira, A. F., Dias, D. R., & Correia, I. J. (2016). Stimuli-responsive mesoporous silica nanoparticles for cancer therapy: A review. *Microporous and Mesoporous Materials, 236*, 141–157.

Morey, G. W., Fournier, R. O., & Rowe, J. J. (1964). The solubility of amorphous silica at $25°C$. *Journal of Geophysical Research, 69*, 1995–2002.

Napierska, D., Thomassen, L. C. J., Rabolli, V., Lison, D., Gonzalez, L., Kirsch-Volders, M., ... Hoet, P. H. (2009). Size-dependent cytotoxicity of monodisperse silica nanoparticles in human endothelial cells. *Small (Weinheim an der Bergstrasse, Germany), 5*(7), 846–853.

Neda, Alasvand, Aleksandra, M. Urbanska, et al. (2017). Chapter 13 - Therapeutic Nanoparticles for Targeted Delivery of Anticancer Drugs. In Grumezescu Alexandru Mihai (Ed.), *Multifunctional Systems for Combined Delivery, Biosensing and Diagnostics* (pp. 245–259). Elsevier.

Paris, J. L., Mannaris, C., & Cabanas, M. V. (2017). Ultrasound-mediated cavitation-enhanced extravasation of mesoporous silica nanoparticles for controlled-release drug delivery. *Chemical Engineering Journal, 340*.

Park, S. S., Jung, M. H., Lee, Y.-S., Bae, J.-H., Kim, S.-H., & Ha, C.-S. (2019). Functionalised mesoporous silica nanoparticles with excellent cytotoxicity against various cancer cells for pH-responsive and controlled drug delivery. *Materials & Design, 184*, 108187.

Pattnaik, S., & Pathak, K. (2017). Mesoporous silica molecular sieve based nanocarriers: transpiring drug dissolution research. *Current Pharmaceutical Design, 23*(3), 467–480.

Riva, B., Bellini, M., Corvi, E., Verderio, P., Rozek, E., Colzani, B., ... Prosperi, D. (2018). Impact of the strategy adopted for drug loading in nonporous silica nanoparticles on the drug release and cytotoxic activity. *Journal of Colloid and Interface Sciecne, 519*, 18–26.

Rosenholm, J. M., Zhang, J., Linden, M., & Sahlgren, C. (2016). Mesoporous silica nanoparticles in tissue engineering–A perspective. *Nanomedicine (Lond), 11*(4), 391–402, Feb.

Ruhle, B., Saint-Cricq, P., & Zink, J. I. (2016). Externally controlled nanomachines on mesoporous silica nanoparticles for biomedical applications. *Chemphyschem: A European Journal of Chemical Physics and Physical Chemistry, 17*(12), 1769–1779, Jun 17.

Samykutty, A., Grizzle, W. E., Fouts, B. L., McNally, M. W., Chuong, P., Thomas, A., ... McNall, L. R. (2018). Optoacoustic imaging identifies ovarian cancer using a microenvironment targeted theranostic wormhole mesoporous silica nanoparticle. *Biomaterials, 182*, 114–126.

Santra, S., Dutta, D., & Moudgil, B. M. (2005). Functional dye-doped silica nanoparticles for bioimaging, diagnostics and therapeutics. *Trans Ichem E, Part C, Food and Bioproducts Processing, 83*(C2), 136–140.

Schmidt-Winkel, P., Lukens, W. W., Zhao, D., Yang, P., Chmelka, B. F., & Stucky, G. D. (1999). Mesocellular siliceous foams with uniformly sized cells and windows. *Journal of the American Chemical Society, 121*(1), 254–255.

Singh, R. K., Patel, K. D., Leong, K. W., & Kim, H.-W. (2017). Progress in nanotheranostics based on mesoporous silica nanomaterial platforms. *ACS Applied Materials & Interfaces, 9*(12), 10309–10337.

Slowing, I. I., Trewyn, B. G., Giri, S., & Lin, V. S.-Y. (2007). Mesoporous silica nanoparticles for drug delivery and biosensing applications. *Advanced Functional Materials, 17*, 1225–1236.

Slowing, I. I., Vivero-Escoto, J. L., Wu, C.-W., & Lin, V. S.-Y. (2008). Mesoporous silica nanoparticles as controlled release drug delivery and gene transfection carriers. *Advanced Drug Delivery Reviews, 60*, 1278–1288.

Slowing, I. I., Vivero-Escoto, J. L., Wu, C.-W., & Lin, V. S.-Y. (2008). Mesoporous silica nanoparticles as controlled release drug delivery and gene transfection carriers. *Advanced Drug Delivery Reviews, 60*, 1278–1288.

Smith, J. E., Wang, L., & Tan, W. (2006). Bioconjugated silica-coated nanoparticles for bioseparation and bioanalysis. *Trends in Analytical Chemistry, 25*(9), 848–855.

Songa, N., & Yang, Y.-W. (2015). Molecular and supramolecular switches on mesoporous silica nanoparticles. *Chemical Society Reviews, 44*, 3474–3504.

Stöber, W., Fink, A., & Bohn, E. (1968). Controlled growth of monodisperse silica spheres in the micron size range. *Journal of Colloid and Interface Science, 26*(1), 62–69.

Sun, B., Zhou, G., & Zhang, H. (2016). Synthesis, functionalization, and applications of morphology-controllable silica-based nanostructures: a review. *Progress in Solid State Chemistry, 44*(1), 119, Issue.

Tang, F., Li, L., & Chen, D. (2012). Mesoporous silica nanoparticles: Synthesis, biocompatibility and drug delivery. *Advanced Materials, 24*, 1504–1534.

Tarn, D., Ashley, C. E., Xue, M., Carnes, E. C., Zink, J. I., & Jeffrey Brinker, C. (2013). Mesoporous silica nanoparticle nanocarriers: biofunctionality and biocompatibilityAcc. *Chemical Research, 46*(3), 792–801.

Trewyn, B. G., Slowing, I. I., Giri, S., Chen, H.-T., & Lin, V. S.-Y. (2007). Synthesis and functionalization of a mesoporous silica nanoparticle based on the sol–gel process and applications in controlled release. *Accounts of Chemical Research, 40*, 846–853.

Trzeciak, K., Kaźmierski, S., Wielgus, E., & Potrzebowski, M. J. (2020). DiSupLo - New extremely easy and efficient method for loading of active pharmaceutical ingredients into the pores of MCM-41 mesoporous silica particles. *Microporous and Mesoporous Materials, 308*, 110506.

Vivero-Escoto, J. L., Huxford-Phillips, R. C., & Lin, W. (2012). Silica-based nanoprobes for biomedical imaging and theranostic applications. *Chemical Society Reviews, 41*, 2673–2685.

Vivero-Escoto, J. L., Slowing, I. I., Trewyn, B. G., & Lin, V. S. -Y. (2010). Mesoporous silica nanoparticles for intracellular controlled drug delivery. *Small (Weinheim an der Bergstrasse, Germany), 6*, 1952–1967.

Vogelsberger, W., Seidel, A., & Rudakoff, G. (1992). Solubility of silica gel in water. *Journal of the Chemical Society, Faraday Transactions, 88*, 473–476.

Vogt, C., Toprak, M. S., Muhammed, M., Laurent, S., Bridot, J.-L., & Muller, R. N. (2010). High quality and tuneable silica shell–magnetic core nanoparticles. *Journal of Nanoparticle Research, 12*, 1137–1147.

Wang, X., Liu, Y., Wang, S., Shi, D., Zhou, X., et al. (2015). CD44-engineered mesoporous silica nanoparticles for overcoming multidrug resistance in breast cancer. *Applied Surface Science, 332*, 308–317.

Wang, Y., Zhao, Q., Han, N., Bai, L., Li, J., Liu, J., ... Wang, S. (2015). Mesoporous silica nanoparticles in drug delivery and biomedical applications. *Nanomedicine: Nanotechnology, Biology and Medicine, 11*(2), 313327.

Wan, Y., & Zhao. (2007). On the controllable soft-templating approach to mesoporous silicates. *Chemical Reviews, 107*(7), 2821–2860.

Yang, Y.-W. (2011). Towards biocompatible nanovalves based on mesoporous silica nanoparticles. *Medicinal Chemistry Communications., 2*, 1033–1049.

Yang, P., Gai, S., & Lin, J. (2012). Functionalized mesoporous silica materials for controlled drug delivery. *Chemical Society Reviews, 41*, 3679–3698.

Yang, J., Lin, G.-S., Mou, C.-Y., & Tung, K.-L. (2019). Diatom-mimicking ultrahigh-flux mesoporous silica thin membrane with straight-through channels for selective protein and nanoparticle separations. *Chemistry of Materials: a Publication of the American Chemical Society, 31*, 1745–1751.

Yang, L., Yin, T., Liu, Y., Sun, J., Zhou, Y., & Liu, J. (2016). Gold nanoparticle-capped mesoporous silica-based H_2O_2-responsive controlled release system for Alzheimer's disease treatment. *Acta Biomaterialia, 46*, 177–190.

Yang, K.-N., Zhang, C.-Q., Wang, W., Wang, P. C., Zhou, J.-P., & Liang, X.-J. (2014). pH-responsive mesoporous silica nanoparticles employed in controlled drug delivery systems for cancer treatment. *Cancer Biology & Medicine, 11*(1), 34–43, Mar.

Yu, M., Jambhrunkar, S., Thorn, P., Chen, J., Gu, W., & Yu, C. (2013). Hyaluronic acid modified mesoporous silica nanoparticles for targeted drug delivery to CD44-overexpressing cancer cells. *Nanoscale, 5*, 178–183.

Yu, T., Tong, L., Ao, Y., Zhang, G., Liu, Y., & Zhang, H. (2019). Novel design of NIR-triggered plasmonic nanodots capped mesoporous silica nanoparticles loaded with natural capsaicin to inhibition of metastasis of human papillary thyroid carcinoma B-CPAP cells in thyroid cancer chemo-photothermal therapy. *Journal of Photochemistry and Photobiology. B, Biology, 197*, 111534, Aug.

Zhang, K., Chen, H., Guo, X., Zhang, D., Zheng, Y., Zheng, H., & Shi, J. (2015). Double-scattering/reflection in a single nanoparticle for intensified ultrasound imaging. *Scientific Reports, 5*, 8766–8776.

Zhou, S., Ding, C., Wang, C., & Fu, J. (2020). UV-light cross-linked and pH de-cross-linked coumarin-decorated cationic copolymer grafted mesoporous silica nanoparticles for drug and gene co-delivery in vitro. *Materials Science & Engineering C-Materials for Biological Applications, 108*, 110469.

Zhou, W., Gao, P., & Shao, L. (2005). Daniela Caruntu, Minghui Yu, Jianfeng Chen, Charles J. O'Connor, Drug-loaded, magnetic, hollow silica nanocomposites for nanomedicine. *Nanomedicine: Nanotechnology, Biology, and Medicine, 1*, 233–237.

Zhu, J., Niu, Y., Li, Y., Gong, Y., Shi, H., Huo, Q., ... Xu, Q. (2017). Stimuli-responsive delivery vehicles based on mesoporous silica nanoparticles: recent advances and challenges. *Journal of Materials Chemistry B, 5*, 1339–1352.

Zhu, C.-L., Wang, X.-W., Lin, Z.-Z., Xie, Z.-H., & Wang, X.-R. (2014). Cell microenvironment stimuli-responsive controlled-release delivery systems based on mesoporous silica nanoparticles. *Journal of Food and Drug Analysis, 22*(1), 18–28, Mar.

Application in hyperthermia treatment

Sabrina A. Camacho[1,2], J.J. Hernández-Sarria[1],
Josino Villela S. Neto[3], M. Montañez-Molina[4],
F. Muñoz-Muñoz[5], H. Tiznado[6], J. López-Medina[7],
O.N. Oliveira Jr[1] and J.R. Mejía-Salazar[3]

[1]Instituto de Física de São Carlos, Universidade de São Paulo, CP 369, São Carlos, Brasil [2]São Paulo State University (UNESP), School of Sciences, Humanities and Languages, Assis, Brazil [3]Instituto Nacional de Telecomunicações (Inatel), Santa Rita do Sapucaí, Brasil [4]Centro de Investigación Científica y Educación Superior de Ensenada-CICESE, Ensenada, México [5]Facultad de Ingeniería, Arquitectura y Diseño, Universidad Autonoma de Baja California (UABC), Ensenada BC, Mexico [6]Centro de Nanociencias y Nanotecnología, Universidad Nacional Autónoma de México, Ensenada BC, México [7]CONACYT — Centro de Nanociencias y Nanotecnología, UNAM, Ensenada, México

14.1 Introduction

Resonantly coupled electromagnetic fields and nanostructures have been used to generate heat for thermal ablation of dangerous cells in the human body (Hirsch et al., 2003; Koohi et al., 2018). In this approach, electromagnetic energy is absorbed by plasmonic or magnetic nanoparticles, and then locally dissipated to increase the temperature around the nanoparticle. In contrast to conventional therapeutics, for example for cancer metastasis, nanoparticle-based hyperthermia enables high specificity, minimal invasiveness, and precise selectivity through the attachment of nanoheaters to targeted malignant cells by using selective biomolecular linkers (Abadeer & Murphy, 2016). Efficient thermal delivery with plasmonic nanoparticles is reached by proper tuning of localized surface plasmon polaritons (LSPPs), that is resonantly coupled light-charge oscillations, with near-infrared (NIR) light (Cortie, Cortie, & Timchenko, 2018; Govorov & Richardson, 2007). Plasmonic nanoparticles should be tailored to absorb light in the wavelength range from 670 nm to 890 nm, where biological tissues exhibit high transparency. This tuning of LSPPs can be achieved with hybrid plasmonic nanoshells and nanorods (Ma, Bendix, & Oddershede, 2012), as it will be discussed later on. Furthermore, alternating magnetic fields can be applied to change the magnetization sense/direction in magnetic nanoparticles continuously, which induces absorption of energy by hysteresis loss, and Brownian and Néel thermal relaxation processes (Hergt et al., 1998; Li et al., 2010). The frequency and amplitude of the alternating magnetic fields are selected according to the material and geometrical properties of the nanoparticles in order to induce heating without affecting the human body (Chandrasekharan et al., 2020; Jose et al., 2020; Kucharczyk et al., 2020).

In this chapter, we introduce LSPP- and magnetic-based heating mechanisms, outlining the physics behind the increase in thermal energy in Section 14.2. Then, we describe experimental techniques for the synthesis and characterization of plasmonic and magnetic nanoheaters in Section 14.3. Recent applications of magnetic and plasmonic nanoparticles in hyperthermia treatment are highlighted in Section 14.4.

14.2 Nanoparticle heating: fundamentals

There are two main approaches through which Si (silicon) hybrid nanoparticles are used for hyperthermia applications. The first one exploits the ability of hybrid plasmonic nanoparticles to enhance and localize optical fields through the excitation of LSPPs, that is optical fields resonantly coupled to collective harmonic oscillation of conduction electrons (at the metal surface), to turn light into heat by Ohmic losses. An important advantage of this approach is the notable structural tunability of LSPP frequencies through the design of noble metal-coated SiO_2 nanoparticle geometries. The second approach uses magnetic properties at the nanoscale, for example with single domain nanostructures, to convert magnetic energy into thermal energy. This latter mechanism is mainly based on core-ferromagnetic nanoparticles covered by SiO_2 nanoshells. The SiO_2 outer layer improves chemical stability, enabling functionalization for specific targeting and avoiding toxicity to the human body. High-frequency magnetic fields are then used to heat the nanoparticles, where two main independent mechanisms contribute to heating: Brownian and Néel relaxation. Hysteresis loss may induce very small or negligible contribution, mainly for small nanoparticles, due to a dominating superparamagnetic behavior of magnetic nanoparticles. In this section, we discuss the fundamentals of heating mechanisms via LSPP and magnetic nanoparticles.

14.2.1 LSPP-based heating mechanism

Hybrid plasmonic nanoparticles, that is dielectric core–plasmonic shell nanoparticles (or plasmonic core–dielectric shell), possess far greater structural tunability than their solid nanoparticle counterparts, along with much larger field enhancements. These unique features open the possibility to design plasmonic scatterers and absorbers for visible and near-infrared wavelengths. In particular, plasmonic absorbers to convert light into heat are used in local hyperthermia treatment of tumor cells. The design of such nanoparticles requires the solution of Maxwell's electromagnetic wave equations to tune the corresponding resonances for combinations of materials and geometries. This set of equations can only be solved under the proper definition of electromagnetic boundary conditions, thus limiting the analytical solutions to spherical, cylindrical, and ellipsoidal geometries, whereas the treatment of more complicated nanoparticle geometries is done with numerical methods such as discrete-dipole approximation (DDA), finite-difference time-domain (FDTD) technique, and finite element method (FEM). The heating mechanism can be determined analytically only for spherical nanoparticles much smaller than the incident electromagnetic wavelength (Govorov & Richardson, 2007), for which the heat transfer equation

$$\rho(\mathbf{r})c(\mathbf{r})\frac{\partial T(\mathbf{r},t)}{\partial t} = \nabla k(\mathbf{r})\nabla T(\mathbf{r},t) + Q(\mathbf{r},t), \quad (14.1)$$

can be considered in the steady-state regime. r and t are used to denote the coordinate and time. T(r,t) is the local temperature, and $\rho(\mathbf{r})$, $c(\mathbf{r})$, and $k(\mathbf{r})$ are the mass density, specific heat, and thermal conductivity, respectively. $Q(\mathbf{r},t)$ is the heat source from the optical excitation, which for an isolated plasmonic nanoparticle can be described by

$$Q = \int_{\lambda_1}^{\lambda_2} C_{abs}(\lambda)I(\lambda)d\lambda, \quad (14.2)$$

where $C_{abs}(\lambda)$ and $I(\lambda)$ are the absorption cross section of the nanoparticle and the spectral irradiance of the light source, respectively. λ_1 and λ_2 are the limiting wavelengths for the operation spectra of the light source. Eq. (14.2) indicates that efficient light to heat conversion can be reached by suitable design of plasmonic nanoparticles with a strong absorption peak (due to LSPP resonance) at the laser wavelength. The optimum range in clinical applications is the so-called tissue window, or near-infrared window, from 670 nm to 890 nm (Cortie et al., 2018; Govorov & Richardson, 2007).

Fig. 14.1(A) shows the absorption spectra for spherical hybrid SiO_2-Au shell nanoparticles in an aqueous environment. Results are presented for four external Au layer thicknesses (t) and a fixed inner SiO_2 core radius, $r = 70$ nm. The strong electromagnetic field confinement and enhancement at the nanoparticle surface due to the excitation of a LSPP is illustrated in Fig. 14.1(B) for a metallic shell thickness $t = 15$ nm under illumination at $\lambda = 880$ nm. The nanoparticle temperature increases with time, as shown in Fig. 14.1(C), and the geometric temperature distribution is given in Fig. 14.1(d). Calculations were performed for spherical hybrid nanoparticles immersed in

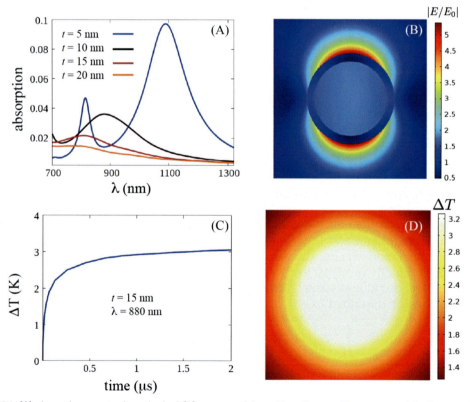

Figure 14.1 (A) absorption spectra for spherical SiO_2 nanoparticles with radius $r = 70$ nm, covered by Au nanoshells with thicknesses $t = 5$ nm, $t = 10$ nm, $t = 15$ nm, and $t = 20$ nm. (B) Electric near-field profile associated to the LSPP at 880 nm for the nanoparticle with $t = 15$ nm. (C) local temperature increase, as function of time, for the nanoparticle in (B). (D) geometrical temperature distribution around the hybrid SiO_2-Au nanoshell in (B).

water under ideal conditions. Eq. (14.1) was also solved numerically with COMSOL Multiphysics software. LSPP resonances can also be tuned by changing the SiO_2 radius, or by varying the SiO_2 radius and shell thickness simultaneously to reach the desired resonance wavelength.

The suitability of hybrid spherical nanoparticles for hyperthermia can be inferred from Fig. 14.1, but the maximum temperature increases of only c.3.5 K points to a low efficiency in the treatment or limits the application to a few analytes. Anisotropic nanoparticles are thus necessary (Ma et al., 2012), such as the hybrid nanorods whose absorption spectra are shown in Fig. 14.2(A). These consist of Au nanorods with a tip-

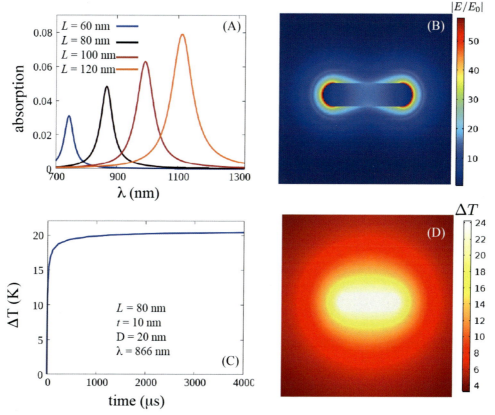

Figure 14.2 (A) absorption spectra for cylindrical Au nanoparticles with tip-to-tip length L, covered by a SiO_2 nanoshell of thickness $t = 10$ nm. (B) Electric near-field profile associated to LSPP at 866 nm for the nanoparticle with $L = 80$ m. (C) Local temperature increases as a function of time for the nanoparticle in (B). (D) Geometrical temperature distribution around the hybrid Au-SiO_2 nanorod in (B). All the calculations were made for the electric field polarized along the rod axis.

to-tip length L covered by a SiO$_2$ shell of thickness $t = 10$ nm. The electric field in the calculations was polarized along the nanorod axis. The straight part of the nanorods was taken with a circular cross section with diameter $D = 20$ nm. The nanorod tips were semispheres with $D = 20$ nm. The electric near-field profile in Fig. 14.2(B) corresponds to the LSPP mode at $\lambda = 866$ nm for $L = 80$ nm. By comparing Fig. 14.1(B) and 14.2(B), one notes a one-order of magnitude enhancement in the near-field at the nanorod tips compared to the spherical nanoparticle. These higher field enhancements are reflected in sevenfold temperature increases for the nanorods, as seen from Fig. 14.1(C)–(D) and 14.2(C)–(D).

Despite the clear advantages of asymmetric hybrid nanoparticles for light-to-heat conversion, the need for a specific polarization of light in relation to the orientation of the nanoparticles can represent a great challenge for practical applications. In particular, LSPP resonances due to longitudinal electric field configurations (parallel to the rod axis) show higher absorption efficiencies than the ones for transversal configuration (perpendicular to the rod axis). If the polarization is not along the nanorods, for instance, the temperature increase will be much lower. Indeed, Fig. 14.3(A) shows the absorption spectra and the local temperature increase around the nanoparticle as a function of the angle between the electric field of the incident light and the rod's axis. For $\theta = 0°$, the 3D color map in the upper panel of Fig. 14.3(B) shows a maximum increase in temperature

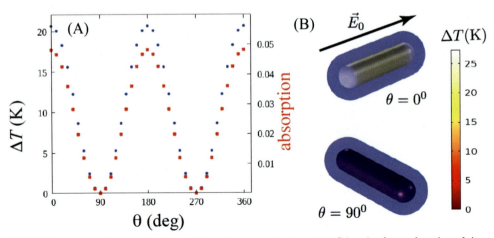

Figure 14.3 (A) The absorption spectra (red dots) and temperature increase (blue dots) as a function of the relative angle (θ) between the electric field of light and the rod's axis. (B) 3D color map for the temperature increases around the nanorod for $\theta = 0°$ and $\theta = 90°$.

at the tips of the nanorod, corresponding to the positions of maximum electromagnetic field enhancement. In contrast, the lower panel of Fig. 14.3(B) exhibits a near-zero ΔT for $\theta = 90°$. All these simulations were made using a laser with a paraxial Gaussian beam profile with an intensity $I = 10^8$ W/m^2 and a spot-size $\omega_0 = 4$ μm.

14.2.2 Magnetic-based heating mechanism

Differently from plasmonic nanoparticles, where the resonant coupling of light to collective oscillation of charge carriers is used to convert light into heat, with magnetic nanoparticles one exploits losses due to reversal of magnetization for absorption of electromagnetic energy (Blanco-Gutiérrez, Climent-Pascual, Sáez-Puche, & Torralvo-Fernández, 2016; Deatsch & Evans, 2014; Manohar, Geleta, Krishnamoorthi, & Lee, 2020). In magnetic hyperthermia, the heat source Q(**r**, t) in Eq. (14.1) is the power dissipated after each magnetic cycle (Cortie et al., 2018)

$$P = \mu_0 f \oint M(t) dH, \qquad (14.3)$$

where H and $M(t)$ are the external magnetic field intensity and dynamic magnetization, respectively. f is the frequency of the alternating applied magnetic field and μ_0 is the vacuum permeability. At the nanoscale, hysteresis losses due to the shifting of magnetic domain walls only occur for nanoparticles larger than 100 nm (Li et al., 2010). For smaller nanoparticles, structural changes from multidomain to single-domain (superparamagnetic) and the coercivity and remanence vanish, then the heating is dominated by the Brownian and Néel relaxation processes. The Brownian relaxation mechanism releases thermal energy through the shear stress in the surrounding fluid, whereas the Néel relaxation dissipates energy by the rotation of magnetic moments within the nanoparticle (Hergt et al., 1998).

14.3 Synthesis of hyperthermia agents

Plasmonic and magnetic silicon hybrids are suitable hyperthermia nanoagents owing to their ability to generate heat under an external stimulus (electromagnetic field, ultrasound, radiofrequency, or alternating magnetic field) (Cheung & Neyzari, 1984; Skinner, Iizuka, Kolios, & Sherar, 1998). Furthermore, they exhibit relevant features such as biocompatibility (Alhmoud, Cifuentes-

Rius, Delalat, Lancaster, & Voelcker, 2017; Majeed et al., 2014; Starsich et al., 2016), limited aggregation (Camacho et al., 2020; Majeed et al., 2014), dielectric properties (Hirsch et al., 2006; Jankiewicz, Jamiola, Choma, & Jaroniec, 2012), and can be used as drug carriers (Alhmoud et al., 2015; Shen, Gao, Yu, Ma, & Ji, 2016) and in surface functionalization (Khosroshahi & Ghazanfari, 2012). In this section we discuss the synthesis and characterization of plasmonic and magnetic hyperthermia nanoagents made with silicon hybrid nanoparticles.

14.3.1 Plasmonic hyperthermia nanoagents

Plasmonic hyperthermia nanoagents made with silicon–gold hybrids include nanopillars (Alhmoud et al., 2017; Convertino et al., 2018), nanoshells (Camacho et al., 2020; Kong et al., 2020; Koohi et al., 2018), and nanorods (Dickerson et al., 2008; MacKey, Ali, Austin, Near, & El-Sayed, 2014; Wang et al., 2018b). The synthetic route to prepare nanoshells is the well-known Stöber method, which allows for adjusting the core-to-shell ratio and tuning LSPP (Hirsch et al., 2003; Hirsch et al., 2006; Jankiewicz et al., 2012). Koohi et al. (2018) fabricated silica nanospheres (SiNPs) with this method using 30% ammonia to alkalinize the medium followed by addition of tetraethyl orthosilicate (TEOS) (Hirsch et al., 2006). These SiNPs had an average diameter of 137 ± 26 nm, but different diameters could be synthesized by changing the amounts of TEOS (Abdollahi, Naderi, & Amoabediny, 2012; Hirsch et al., 2003; Pham, Jackson, Halas, & Lee, 2002). Small gold nanoparticles (AuNPs, 1–3 nm) were grown according to Duff and Baiker's method (Duff, Baiker, & Edwards, 1993) by quickly mixing tetrakis (hydroxymethyl) phosphonium chloride (THPC) and sodium hydroxide (NaOH) with gold (III) chloride trihydrate ($HAuCl_4$). Au nanoshells were formed by shaking SiNPs with the Au colloidal suspension after functionalizing SiNPs with 3-aminopropyl trimethoxysilane (APTMS). A more artful strategy was developed by Kong et al. (2020) to produce nanoshells with selenium (Se). $Cu_{2-x}Se$ nanocrystals were prepared via the thermal injection method and coated with silica (SiO_2) through reverse microemulsion, resulting in Se@SiO_2 nanospheres c.60 nm in diameter (Liu et al., 2013, 2016a). Then, Au seeds were grown to create the nanoshell and obtain the nanocomposite Se@SiO_2@Au (~100 nm). A schematic cartoon of the preparation and TEM images are shown in Fig. 14.4(A). The SiO_2 layer protects Se from degrading and behaves as a linker, while the Au shell works as a photothermal agent. Upon combining Se with the Au shell, different

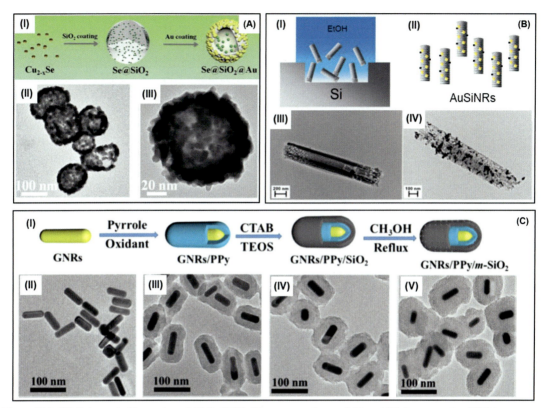

Figure 14.4 (A) Schematic illustration of (I) Se@SiO$_2$@Au preparation and (II and III) its respective TEM images (Kong et al., 2020). (B) Designing of the (I) free-standing SiNPs, (II) decorated with AuNPs and TEM images of (III) a SiNP and (IV) an AuSiNP decorated with AuNPs (Alhmoud et al., 2017). (C) (I) Preparation steps of GNRs/PPy/m-SiO$_2$ and (II-V) their respective TEM images (Wang et al., 2018b). (14.4A) Reproduced with the permission of The Royal Society of Chemistry. (14.4B,C) Reproduced with the permission of American Chemistry Society.

amounts of reactive oxygen species (ROS) can be produced to kill cancer cells, while protecting the normal cells from the harmful ROS (Kong et al., 2020). Silicon nanopillars decorated with gold nanoparticles (AuSiNPs), with excellent hyperthermia biocompatibility, were fabricated according to a methodology by Alhmoud et al. (2017) A polystyrene monolayer was built on a flat silicon wafer and exposed to an HF/H$_2$O$_2$ solution. The reagents were removed, creating silicon nanopillars (SiNPs) with 200 nm of diameter and 1200 nm long, which were detached by applying current and ultrasonication. Gold deposition was done by immersing the SiNPs in a HAuCl$_4$ hot solution, reducing the Au^{3+} into Au^0 on the SiNPs surface. Fig. 14.4(B) displays the illustrations and TEM images of the free-standing

SiNPs and SiNPs decorated with AuNPs of c.2−50 nm (Alhmoud et al., 2017). Gold nanorods (GNRs) are also promising hyperthermia agents due to their tunable LSPP and high conversion efficiency in the near infrared (NIR) (Huang, Wu, Ke, Li, & Du, 2017; Wang et al., 2011; Zhang, L. et al., 2016). LSPP can be tuned by adjusting the size (length × width) of GNRs, as demonstrated by changing the volumes of Au seeds, where lesser amounts of Au seeds resulted in larger GNRs (Ming et al., 2009; Zhang, Li, Xiao, Li, & Sun, 2013). Wang et al. (2018b) prepared GNRs (25 nm × 75 nm of diameter × length) containing polypyrrole (∼20 nm of thickness) and a mesoporous SiO_2 shell (m-SiO_2, c.15 nm). The synthetic route involves GNRs preparation using a binary surfactant mixture, where thin and thicker GNRs could be prepared by controlling the volumes of Au seeds (Ye, Zheng, Chen, Gao, & Murray, 2013). The polymeric layer is formed over the GNRs through chemical polymerization of pyrrole using ammonium persulfate as oxidant. Then, the m-SiO_2 shell is made via CTAB sol−gel method by a process of hydration and condensation of TEOS, leading to GNRs/PPy/m-SiO_2. The preparation steps and the TEM images are exhibited in Fig. 14.4(C). With a combination of GNRs and a conducting polymer, both photothermal conversion efficiency and stability are maintained (Wang et al., 2018a), while m-SiO_2 works as a protective barrier and drug carrier (Liu et al., 2016b; Yang et al., 2016).

14.3.2 Magnetic hyperthermia nanoagents

Magnetic nanomaterials from silicon hybrids used as hyperthermia nanoagents are formed by a magnetic core wrapped with a m-SiO_2 shell, named magnetic mesoporous silica system (MMSS) (Majeed et al., 2014; Shen et al., 2016; Tao & Zhu, 2014; Zhang, Y. et al., 2016). These magnetic nanoparticles are normally magnetite (Fe_3O_4) and related spinels with cobalt, nickel, zinc, magnesium, or other substitutions (Kamzin, Das, Wakiya, & Valiullin, 2018; Kumar & Mohammad, 2011). The coating SiO_2 shell can provide increased colloidal stability by preventing anisotropic magnetic dipolar attraction between the magnetic nanoparticles, chemical stability by avoiding any further oxidation, biocompatibility, and reduced toxicity (Coşkun & Korkmaz, 2014; Digigow et al., 2014; Hervault & Thanh, 2014; Toropova et al., 2017). In addition, the SiO_2 shell provides a chemically inert surface, protecting nanoparticles from leaching in an acid medium. It has a hydrophilic surface that enables functionalization and bioconjugation with cell targeting agents through covalent bonds with the silane groups (Hui et al., 2011;

Khosroshahi & Ghazanfari, 2012; Stjerndahl et al., 2008). Zhang et al. (2015) synthesized Fe_3O_4 crystalline nanoparticles (c.14 nm) inside a m-SiO_2 nanosphere, resulting in a magnetic nanomaterial (Fe_3O_4@m-SiO_2) with 20–50 nm of shell thickness and 100–200 nm of diameter. The monodisperse Fe_3O_4 nanoparticles were prepared by thermal decomposition using oleic acid as stabilizer (Park et al., 2004). Next, a modified sol–gel procedure was applied (Che et al., 2015) to assemble the Fe_3O_4 nanoparticles and produce m-SiO_2. CTAB and TEOS were mixed with the magnetic fluid in a medium with pH adjusted to 12 (by addition of NaOH), forming m-SiO_2 via hydrolysis and condensation of the alkoxysilane TEOS with a basic catalyst. The m-SiO_2 shell improved the hyperthermia efficiency and allowed the magnetic nanomaterial to function as a drug carrier (Zhang et al., 2015). A similar nanocomposite was synthesized by Shen et al. (2016) via encapsulation of the magnetic Fe_3O_4 nanoparticles with poly(N-isopropylacrylamide) (PNIPAM) and coating with m-SiO_2. The super paramagnetic Fe_3O_4 nanoparticles were synthesized by mixing $FeCl_3$ with ethyleneglycol under ultrasonication for 30 min. Then, polyethylene glycol (PEG) and sodium acetate (CH_3COONa) were added, mixed for 30 min, and placed in an autoclave at 180°C for 19 h to obtain the Fe_3O_4 nanoparticles (Xiong et al., 2015). For encapsulation, PNIPAM and 5-Fu were dispersed with Fe_3O_4 nanoparticles in a solution of ethanol and distilled water. The m-SiO_2 was created by adding CTAB and ammonia under vigorous stirring, followed by dropwise volumes of TEOS. Fe_3O_4/PNIPAM/5-Fu@mSiO_2 exhibited excellent capability of generating heat under an alternate magnetic field. The thermoresponse of PNIPAM facilitates the 5-Fu drug release through the m-SiO_2 shell (Shen et al., 2016). SEM images of Fe_3O_4 nanoparticles and Fe_3O_4/PNIPAM/5-Fu@mSiO_2 nanocomposites are displayed in Fig. 14.5(A). Starsich et al. (2016) also used iron oxide matrices to create highly magnetic nanocrystals with enhanced hyperthermia performance based on the incorporation of Zn^{2+}. The $Zn_{0.4}Fe_{2.6}O_4$ spinel structure and SiO_2 coating were prepared by scalable flame technology (FPS) (Teleki et al., 2009). Precursor solutions made by dissolving zinc and iron(III) nitrates in pure tetrahydrofuran (THF) were fed through a capillary and dispersed by O_2 into a premixed supporting CH_4/O_2 flame. The SiO_2 nanolayer was then created via swirl injection of N_2 carrying hexamethyldisiloxane (HMDSO). N_2 gas was fed through a bubbler filled with HMDSO and placed in a water bath (20°C). Thereafter the HMDSO-laden N_2 stream was mixed with additional N_2 and was swirl-injected through a ring on the top of a

Figure 14.5 (A) SEM images of (I) Fe_3O_4 nanoparticles and (II) Fe_3O_4/PNIPAM/5-Fu@mSiO$_2$ nanocomposites (Shen et al., 2016). (B) TEM images of (I) $Zn_{0.4}Fe_{2.6}O_4$ bare and (II) coated with SiO_2 nanolayer (Starsich et al., 2016). (14.5A) Reproduced with the permission of The Royal Society of Chemistry. (14.5B) Reproduced with the permission of WILEY-VCH Verlag GmbH & Co. KGaA, Weinheim.

quartz tube. The SiO_2 nanolayer allows functionalization with polymers, further improving the biocompatibility (Starsich et al., 2016). Fig. 14.5(B) brings TEM images of $Zn_{0.4}Fe_{2.6}O_4$ bare and coated with SiO_2 nanolayer, with final size between 5–35 nm.

Cobalt ferrites ($CoFe_2O_4$) are also interesting for hyperthermia applications due to their chemical and mechanical stability. The size and stabilization (prevention of aggregation) of these nanoparticles can be controlled during their synthesis using a thermal decomposition method with organometallic precursors (CoFe-oleate) in a solvent (1-octadecene) with high boiling point, using oleic acid as a surfactant agent. These cobalt ferrites can be coated with silicon oxides through a reverse microemulsion method, which consists of the internalization of the nanoparticles in the hydrophilic phase of micelles (Igepal CO-520) prepared in cyclohexane, and then incorporated into the polar phase of an aqueous domain (NH_4OH). This promotes a micellar nanoreactor, in which the hydrolysis and condensation of TEOS (Si $(OC_2H_5)_4$) occurs in a confined manner, leading to amorphous silicon oxide layers on the surface of cobalt nanoferrites.

Following the method described above, we developed $CoFe_2O_4$ and $CoFe_2O_4@SiO_2$ magnetic nanoparticles shown in Fig. 14.6, where high-resolution transmission electron microscopy (HRTEM) images are presented. $CoFe_2O_4$ nanoparticles in Fig. 14.6(A) and (B), from HRTEM images at 50 nm and 5 nm scales, exhibit an average size between 15 nm and 20 nm. They are monocrystalline nanoparticles with well-defined crystalline planes. Fig. 14.6(C) and (D) show the morphology of magnetic nanoparticles coated and embedded into the silica beads via the reverse microemulsion method ($CoFe_2O_4@SiO_2$) described above. The total average size of $CoFe_2O_4@SiO_2$ nanoparticles

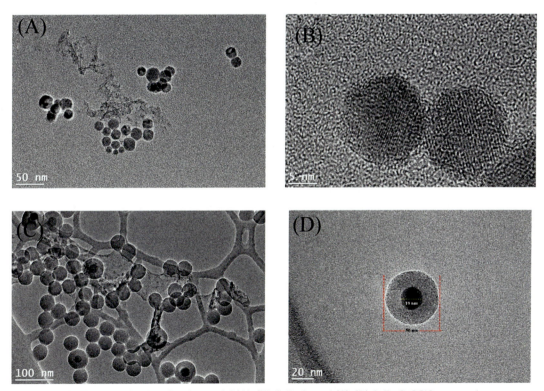

Figure 14.6 HRTEM image for functional material (A), (B) $CoFe_2O_4$ and (C), (D) $CoFe_2O_4@SiO_2$ magnetic nanoparticles.

increased up to 50 nm, that is 2.5-fold the value of the magnetic cores. This change of size slightly affected the magnetic properties, as will be shown below. All these characterizations were made for samples in powder using a JEOL JEM-2100F scanning transmission electron microscope (STEM) with Schottky-field emission electron gun microscope at 200 keV.

We also used energy dispersive X-ray spectroscopy (EDS) combined with scanning electron microscopy (STEM–EDS) to study the segregation of elements in the nanoparticle. Fig. 14.7(A) shows the STEM image of a magnetic nanoparticle in a powder sample. Fig. 14.7(B)–(E) show the false-color X-ray emission of Co ions (in red color), Fe ions (in blue color), Si (in green color), and O (in gray color), respectively, from which the distribution of elements along the nanoparticle can be inferred. The EDS spectra are shown in Fig. 14.7(F).

The magnetic behavior of these nanoparticles is shown in Fig. 14.8, where the M versus H hysteresis loops at room temperature are presented for $CoFe_2O_4$ (red color) and $CoFe_2O_4@SiO_2$

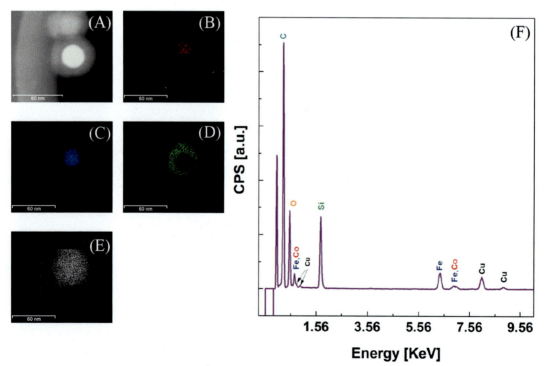

Figure 14.7 (A) STEM image of $CoFe_2O_4@SiO_2$ (white). Images (B) to (E) corresponding to elemental mapping images of (B) Co (red), (C) Fe (blue), (D) Si (green), and (E) O (gray scale) on the nanoparticle. (F) EDS spectra show like Co, Fe, Si, O, C, and elements in the magnetic nanoparticles.

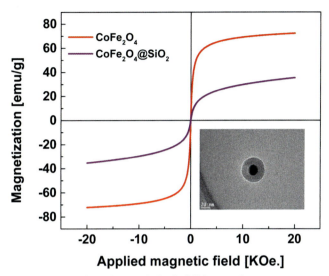

Figure 14.8 M vs. H hysteresis loop for $CoFe_2O_4$ and $CoFe_2O_4@SiO_2$ samples at room temperature.

(violet) magnetic nanoparticles. The saturation magnetization, M_s, decreases when the magnetic nanoparticles are encapsulated in the SiO_2 beads. The samples exhibited a small hysteresis loop with coercive fields (H_c) of 13 and 16 Oe for $CoFe_2O_4$ and $CoFe_2O_4@SiO_2$, respectively. The remanence and coercivity are close to zero, as expected. Therefore, these samples behave like soft magnetic materials, that is they can be easily demagnetized owing to their low coercive field and monotonic increase of magnetization up to saturation. Moreover, these magnetic characteristics indicate a superparamagnetic behavior which can be exploited for hyperthermia. The maximum M_s was 72 and 35 emu/g for $CoFe_2O_4$ and $CoFe_2O_4@SiO_2$ samples, respectively. A single magnetic domain was observed from the ratio $M_r/M_s = 0.9 \times 10^{-3}$ for these samples. This kind of magnetic nanoparticle was used for magnetic hyperthermia applications (Manohar et al., 2020), with a heating efficiency of 185.32 W/g, making them promising for in vivo application in thermoablation of cancer cells (Blanco-Gutiérrez et al., 2016).

14.4 Hyperthermia applications

Hyperthermia treatment consists in the injection or implantation of nanoagents into a tumor area and increasing the temperature to 42–45°C, several degrees above the physiological temperature (37°C) (Andrade, Fabris, Domingues, & Pereira, 2015; Cavaliere et al., 1967; Hynynen & Lulu, 1990). It is believed that tumor tissues are more hypoxic, more acidic, and nutrient-deficient than healthy tissues (Wust et al., 2002). These traits may render some cancer cells more sensitive to heat (Hegyi, Szigeti, & Szász, 2013). Hence, plasmonic and magnetic hyperthermia nanoagents based on silicon hybrid nanoparticles are suited to thermal destruction of cancer cells and microorganisms due to their induced heating ability and ease of surface functionalization (Dreaden, Alkilany, Huang, Murphy, & El-Sayed, 2012; Jain, Hirst, & O'Sullivan, 2012; Kim, Jung, Lee, & Kim, 2017). In plasmonic hyperthermia therapy the nanoagents are subjected to light irradiation at a specific wavelength, in resonance with the plasmon excitation. The latter induces LSPP, a coherent oscillation of free electrons in the conduction band (Mie, 1908). The consequence is a strong electromagnetic field and enhanced light absorption (Akhter, Ahmad, Ahmad, Storm, & Kok, 2012; Aroca, 2007; Beik et al., 2019), which is converted into heat by the nanoagents (Cabral & Baptista, 2013; Mendes, Pedrosa, Lima, Fernandes, & Baptista, 2017; Qin et al., 2016). In magnetic hyperthermia therapy the nanoagents are submitted to an alternating magnetic field increasing the temperature between 42–45°C or over 46°C. The mechanism of rising temperature involves losses such as eddy current loss, hysteresis loss, and relaxation losses (Bañobre-López, Teijeiro, & Rivas, 2013; Lima et al., 2013; Rosensweig, 2002; Thakur et al., 2006).

14.4.1 Plasmonic hyperthermia applications

Plasmonic nanoagents based on silicon hybrids have been designed for hyperthermia to treat cancer and bacterial infections (Alhmoud et al., 2017; Camacho et al., 2020; Convertino et al., 2018; Kong et al., 2020). In a pioneering work in 2003 (Hirsch et al., 2003), SiNPs with 110 nm of diameter, 10 nm of Au shell thickness and functionalized with PEG were applied against SK-BR-3 human breast carcinoma cells. The irradiated area (7 min with 35 W/cm^2 at 820 nm) of cancer cells incubated with the nanoshells for 1 h was damaged, as seen with fluorescence images. Cell membrane integrity was lost and cell death was observed, which did not occur in the controls (only light irradiation or nanoshells). A nanoshell system made of AuNPs cores (16 nm of diameter) and SiO_2 shells (5.5 nm thick)—referred to as gold shell-isolated nanoparticles (AuSHINs)—was employed to demonstrate photoinduced heating in cells derived from oropharyngeal (HEp-2), glandular (MCF-7), and ductal (BT-474) breast carcinomas (Camacho et al., 2020). MTT assays showed a significant population reduction down to 50% with photoactivation (525 nm, 32.8 mW/cm^2, 1h) of 2.2×10^{12} AuSHINs/mL incubated for 6 h. In contrast, in control experiments with AuNPs alone the reduction of cell viability was smaller owing to nanoparticle aggregation in the ionic medium. Moreover, subsidiary experiments pointed to generation of ROS during local heating, confirming that oxidative reactions play a role in the cellular phototoxicity (Camacho et al., 2020). The cell viability of the carcinoma cells incubated for 6 h with the concentrations of AuSHINs (nonirradiated and irradiated) is presented in Fig. 14.9(A). Silicon nanowires (SiNWs) coated with a thin Au film were also applied to induce localized photothermal death of cancer cells. The capability of cylindrical Au/SiNWs (120–180 nm of size) to kill human colon adenocarcinoma cells (Caco-2) was reported by Convertino et al. (2018). The cancer cells were seeded on Au/SiNWs substrates and cultured for up to 3 days followed by laser light irradiation (785 nm, 5–24 mW, 20 min). Fluorescence staining images showed dead cells in the irradiated region using the higher Raman laser powder, which did not happen in the control experiment (Caco-2 cell monolayer cultured only on planar Au film and irradiated with 24 mW for 20 min) (Fig. 14.9(B)). Furthermore, cell death was monitored by Raman spectroscopy, collecting spectra from 0 to 20 min of irradiation, and checking intensity changes of some spectral components that are enhanced or reduced during the death process. A similar approach based on silicon nanopillars decorated with Au (AuSiNPs) was demonstrated against Gram-positive *Staphylococcus aureus* (*S. aureus*) and Gram-negative *Escherichia coli* (*E. coli*) bacteria. Alhmoud et al. (2017) showed a reduction in bacterial viability of up to 99% after laser irradiation (808 nm, 5 mW, 1.25 W/cm^2, 10 min). They also demonstrated an increase in antibacterial performance after modifying the AuSiNPs with *S. aureus* targeting antibodies, causing up to a 10-fold increase in bactericidal efficiency compared to *E. coli*. Fig. 14.9(C) exhibits O.D.$_{600}$ measurements along 16 h of *S. aureus* growth after incubation with AuSiNPs and AuSiNPs-Ab followed by irradiation and incubation with the controls (Alhmoud et al., 2017).

14.4.2 Magnetic hyperthermia applications

Magnetic nanomaterials for hyperthermia treatment have been used mainly for the reduction and/or elimination of human malignant tumors (Das et al., 2015; Misra et al., 2016; Theis-Bröhl et al., 2018). Of particular interest are magnetic nanoagents based on silicon hybrids owing to their low toxicity, strong response to magnetic fields, and superparamagnetic relaxation (Giordano, Gutierrez, & Rinaldi, 2010). A magnetic

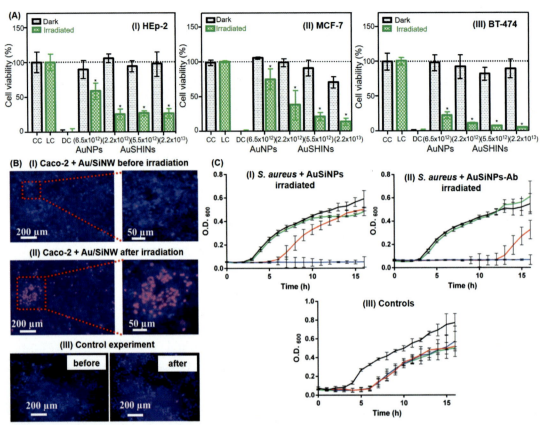

Figure 14.9 (A) Toxic (dark) and phototoxic (irradiated) effects of 6.5×10^{12} AuNPs/mL and 2.2×10^{12}, 5.5×10^{12}, and 2.2×10^{13} AuSHINs/mL on (I) HEp-2, (II) MCF-7 and (III) BT-474 cells after 6 h of incubation. CC correspond to the cellular controls; DC is the death control and LC is the light control (Camacho et al., 2020). (B) Fluorescence staining of CaCo-2 cells on Au/SiNWs substrates (I) before and (II) after NIR irradiation, respectively. Red fluorescence reveals the presence of dead cells. (III) Fluorescence staining of CaCo-2 cells plated on the reference Au film substrate, before and after irradiation (control experiment) with no visible area of dead cells (Convertino et al., 2018). (C) O.D.$_{600}$ measurements over 16 h of *S. aureus* growth after irradiation. (I) and (II) depict 10^8 CFU/mL *S. aureus* grown with 0 (black), 0.1 (green), 0.5 (red), and 1.0 mg/mL (blue) of AuSiNPs and AuSiNPs-Ab, respectively. (III) depicts 10^8 CFU/mL *S. aureus* grown with Ab only (0.03 mg/mL) (black), AuSiNPs (0.5 mg/mL) + free antibody (0.03 mg/mL) (red), AuSiNPs-Ab conjugated with non-*S. aureus* specific antibody (0.5 mg/mL) (blue), and AuSiNPs (0.5 mg/mL) (green) (Alhmoud et al., 2017). (14.9A) Reproduced with the permission of Elsevier. (14.9B) Reproduced with the permission of IOP Publishing. (14.9C) Reproduced with the permission of American Chemistry Society.

mesoporous silica system (MMSS) made of 5 nm maghemite nanoparticles (γ-Fe$_2$O$_3$) embedded in a m-SiO$_2$ matrix (mesoporous with 4.6 nm of diameters) was applied to induce heating cell death in human carcinoma cells (lung (A549), cervical (HeLa), osteosarcoma (Saos-2), and hepatocellular (HepG2)) after exposing them to an alternating magnetic field (AMF) (Martín-Saavedra et al., 2010). A cell viability reduction was reported as a function of the intensity of the heat treatment achieved by adjusting the amount of MMSS and the time of exposure to AMF. Heat treatments of mild to very high intensities could be obtained, as can be seen in the cell viability results for M1 cultures treated with different amounts of MMSS for 24 h and exposure times to the AMF (100 kHz/200 Oe) (Fig. 14.10(A)) (Martín-Saavedra et al., 2010). Another superparamagnetic nanoparticle containing SiO_2 shell was used against human cervical cancer cells (HeLa) (Majeed et al., 2014). The core–shell magnetite (Fe_3O_4-SiO_2) was evaluated in terms of its biocompatibility and hyperthermic killing ability. The effect of surface coating on specific absorption rate was also investigated, where samples prepared at 80°C (Fe_3O_4-SiO_2 80) exhibited greater heating efficiency than samples prepared at 100°C (Fe_3O_4-SiO_2 100). Under an AMF (induction heating), the HeLa cells treated with SiO_2 80-coated Fe_3O_4 reduced the cell viability in 42%, which is more efficient than the uncoated Fe_3O_4 and Fe_3O_4-SiO_2 100 (decrease of 15% and 22%, respectively) for the same number of nanoparticles (2 mg/mL), incubation time (2 h), and applied current (400 A, f = 250 kHz) (Fig. 14.10(B)) (Majeed et al., 2014). Silicon naphthalocyanine loaded nanoclusters of iron oxide doped with Zn and Mn encapsulated with PEG-PCL [methoxy poly(ethylene glycol)-b-poly(-caprolactone)] (SiNc-loaded ZnMn-IONCs) were applied in in vivo hyperthermia treatment of prostate cancer cells (DU145) in nude mice (Albarqi et al., 2020). Fluorescence imaging of the whole mice at various times demonstrated that SiNc-loaded ZnMn-IONCs started to accumulate in cancer tumors within 1 h postinjection and reached a maximum accumulation at 7 h (Fig. 14.10(C) I). Then, the therapeutic efficacy of the systemically delivered magnetic hyperthermia was assessed by an intravenous injection of the SiNc-loaded ZnMn-IONCs (10 g of Fe/kg) applying an AMF (420 kHz, 26.9 kA/m) for 30 min once per week for 4 weeks. The results revealed that four cycles of magnetic hyperthermia significantly reduced the growth of prostate cancer tumors. The tumor volume was approximately two times smaller than the tumor volume exposed to only iron oxide nanoclusters (IONC). Moreover, the nanoparticles and AMF alone did not show considerable effects on tumor growth (Fig. 14.10(C) II) (Albarqi et al., 2020).

14.5 Final remarks

Silicon-based hybrid nanoparticles may act as hyperthermia nanoagents to induce malignant cell damage through an increased

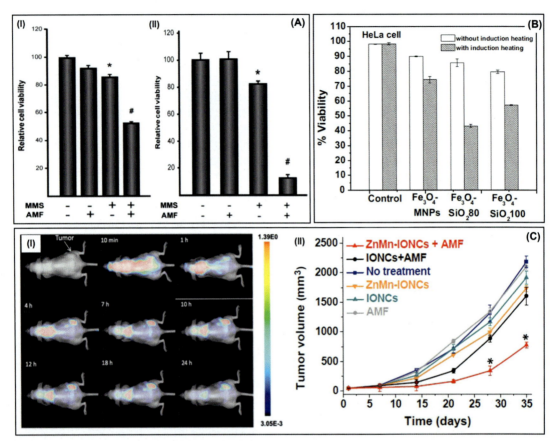

Figure 14.10 (A) Induced hyperthermia response of M1 cells (I) incubated with 48 mg of γ-Fe$_2$O$_3$@m-SiO$_2$ for 24 h and subjected to an AMF at 100 kHz/200 Oe for 60 min and (II) incubated with 80 mg of γ-Fe$_2$O$_3$@m-SiO$_2$ for 24 h and subjected to an AMF at 100 kHz/200 Oe for 45 min (Martín-Saavedra et al., 2010). (B) Percentage viability of HeLa cells treated with 2 mg/mL of Fe$_3$O$_4$, Fe$_3$O$_4$@SiO$_2$ 80 and Fe$_3$O$_4$@SiO$_2$ 100 for 2 h followed with and without induction heating (applied current = 400 A, f = 250 kHz, t = 20 min) (Majeed et al., 2014). (C) (a) Fluorescence images of a mouse with DU145 xenograft at different times following intravenous administration of SiNc-loaded ZnMn-IONCs and (b) tumor growth profiles after four cycles of no treatment; ZnMn-IONCs + AMF, mice injected with ZnMn-IONCs and subjected to AMF; IONCs + AMF, mice injected with IONCs and subjected to AMF; ZnMn-IONCs, mice injected with ZnMn-IONCs; IONCs, mice injected with IONCs; and AMF, mice injected with 5% dextrose and exposed to AMF. AMF = 420 kHz, 26.9 kA/m and exposure time = 30 min (Albarqi et al., 2020). (14.10A) Reproduced with the permission of Elsevier. (14.10B) Reproduced with the permission of Elsevier. (14.10C) Reproduced from the Open Access Journal MDPI.

local temperature. In this chapter, our first goal was to describe the fundamentals of heating mechanisms via LSPP and magnetic nanoparticles. Then, we reviewed the synthesis and characterization of plasmonic and magnetic nanoheaters made with silicon hybrids, and showcased recent applications in hyperthermia treatment.

Plasmonic and magnetic silicon hybrids are suitable nanoheaters for in vivo and in vitro applications due to their properties, which include the ability to generate heat, biocompatibility, limited aggregation, dielectric properties, and protective barriers. It is also significant that mesoporous SiO_2 shells around plasmonic and magnetic nanoparticles may function as drug carriers. This allows for combining hyperthermia and chemotherapy in a synergistic treatment. The plasmonic and magnetic silicon hybrids must be specific for targeting molecules, tissues, and cells, which is a challenge that may be overcome via functionalization and bioconjugation.

Another challenge for hyperthermia nanoagents is in their incorporation into cancer cells through intravenous injection. Even after functionalization with targeting molecules, this incorporation is limited owing to toxicity and dosage. In this scenario, minimally invasive surgical laser tools could be useful for removing residual or hard-to-reach microtumors, with no need of injection into the human body. These tools may contain plasmonic and magnetic silicon hybrid nanoparticles integrated into optical fiber lasers. In contrast to other therapeutic modalities such as chemotherapy, hyperthermia treatment assisted by plasmonic and magnetic silicon hybrids is a minimally invasive approach capable of reducing or eliminating malignant cells with a low systemic toxicity to avoid side effects.

14.6 Acknowledgments

The authors thank Eloisa Aparicio, Eduardo Murillo, David Dominguez, Israel Gradilla, Francisco Ruíz, and Jaime Mendoza, for their valuable technical support.

14.6.1 Funding

This work was partially supported by the Basic Science projects 2017−2018 A1-S-21323 and A1-S-21084, UABC-PTC-595 PRODEP-SEP, FORDECYT − CONACYT 272894 and DGAPA-UNAM, through research projects: PAPIIT IN103220, IG200320, IN110018, IN113219. The authors also acknowledge the financial support from the Brazilian agencies CAPES, FAPESP (2018/22214−6, 2018/14692−5, 2017/25587−5), and CNPq (429496/2018−4, 305958/2018−6). Partial financial support was also received from RNP, with resources from MCTIC, Grant No. 01245.010604/2020-14, under the 6G Mobile Communications Systems project of the Radiocommunication Reference Center (Centro de Referência em Radiocomunicações - CRR) of the National Institute of Telecommunications (Instituto Nacional de Telecomunicações - Inatel), Brazil.

References

Abadeer, N. S., & Murphy, C. J. (2016). Recent progress in cancer thermal therapy using gold nanoparticles. *Journal of Physical Chemistry C, 120*, 4691−4716.

Abdollahi, S. N., Naderi, M., & Amoabediny, G. (2012). Synthesis and physicochemical characterization of tunable silica-gold nanoshells via seed growth method. *Colloids Surfaces A Physicochemical and Engineering Aspects*. Available from https://doi.org/10.1016/j.colsurfa.2012.08.043.

Akhter, S., Ahmad, M. Z., Ahmad, F. J., Storm, G., & Kok, R. J. (2012). Gold nanoparticles in theranostic oncology: Current state-of-the-art. *Expert Opinion on Drug Delivery*. Available from https://doi.org/10.1517/17425247.2012.716824.

Albarqi, H. A., et al. (2020). Systemically delivered magnetic hyperthermia for prostate cancer treatment. *Pharmaceutics*. Available from https://doi.org/10.3390/pharmaceutics12111020.

Alhmoud, H., et al. (2015). Porous silicon nanodiscs for targeted drug delivery. *Advanced Functional Materials*. Available from https://doi.org/10.1002/adfm.201403414.

Alhmoud, H., Cifuentes-Rius, A., Delalat, B., Lancaster, D. G., & Voelcker, N. H. (2017). Gold-decorated porous silicon nanopillars for targeted hyperthermal treatment of bacterial infections. *ACS Applied Materials & Interfaces*. Available from https://doi.org/10.1021/acsami.7b13278.

Andrade, Â. L., Fabris, J. D., Domingues, R. Z., & Pereira, M. C. (2015). Current status of magnetite-based core@shell structures for diagnosis and therapy in oncology. *Current Pharmaceutical Design*. Available from https://doi.org/10.2174/1381612821666150917093543.

Aroca, R. (2007). Surface-enhanced vibrational spectroscopy. *Surface-Enhanced Vibrational Spectroscopy*. Available from https://doi.org/10.1002/9780470035641.

Bañobre-López, M., Teijeiro, A., & Rivas, J. (2013). Magnetic nanoparticle-based hyperthermia for cancer treatment. *Reports of Practical Oncology and Radiotherapy*. Available from https://doi.org/10.1016/j.rpor.2013.09.011.

Beik, J., et al. (2019). Gold nanoparticles in combinatorial cancer therapy strategies. *Coordination Chemistry Reviews*. Available from https://doi.org/10.1016/j.ccr.2019.02.025.

Blanco-Gutiérrez, V., Climent-Pascual, E., Sáez-Puche, R., & Torralvo-Fernández, M. J. (2016). Temperature dependence of superparamagnetism in $CoFe_2O_4$ nanoparticles and $CoFe_2O_4/SiO_2$ nanocomposites. *Physical Chemistry Chemical Physics: PCCP*, *18*(13), 9186–9193. Available from https://doi.org/10.1039/c6cp00702c.

Cabral, R. M., & Baptista, P. V. (2013). The chemistry and biology of gold nanoparticle-mediated phototermal therapy promises and challenges. *Nano Life*, *03*, 1330001.

Camacho, S. A., et al. (2020). Molecular-level effects on cell membrane models to explain the phototoxicity of gold shell-isolated nanoparticles to cancer cells. *Colloids Surfaces B: Biointerfaces*. Available from https://doi.org/10.1016/j.colsurfb.2020.111189.

Cavaliere, R., et al. (1967). Selective heat sensitivity of cancer cells. Biochemical and clinical studies. *Cancer*, doi:10.1002/1097–0142(196709)20:9 < 1351::AID-CNCR2820200902 > 3.0.CO;2-#.

Chandrasekharan, P., et al. (2020). Using magnetic particle imaging systems to localize and guide magnetic hyperthermia treatment: Tracers, hardware, and future medical applications. *Theranostics*, *10*(7), 2965–2981. Available from https://doi.org/10.7150/thno.40858.

Che, E., et al. (2015). Paclitaxel/gelatin coated magnetic mesoporous silica nanoparticles: Preparation and antitumor efficacy in vivo. *Microporous and Mesoporous Materials*. Available from https://doi.org/10.1016/j.micromeso.2014.11.013.

Cheung, A. Y., & Neyzari, A. (1984). Deep local hyperthermia for cancer therapy: External electromagnetic and ultrasound techniques. *Cancer Research*.

Convertino, A., et al. (2018). Array of disordered silicon nanowires coated by a gold film for combined NIR photothermal treatment of cancer cells and Raman monitoring of the process evolution. *Nanotechnology*. Available from https://doi.org/10.1088/1361-6528/aad6cd.

Cortie, M. B., Cortie, D. L., & Timchenko, V. (2018). Heat Transfer from Nanoparticles for Targeted Destruction of Infectious Organisms. *International Journal of Hyperthermia: the Official Journal of European Society for Hyperthermic Oncology, North American Hyperthermia Group, 34*, 157–167.

Coşkun, M., & Korkmaz, M. (2014). The effect of SiO_2 shell thickness on the magnetic properties of $ZnFe_2O_4$ nanoparticles. *Journal of Nanoparticle Research*. Available from https://doi.org/10.1007/s11051-014-2316-3.

Das, H., et al. (2015). Investigations of superparamagnetism in magnesium ferrite nano-sphere synthesized by ultrasonic spray pyrolysis technique for hyperthermia application. *Journal of Magnetism and Magnetic Materials*. Available from https://doi.org/10.1016/j.jmmm.2015.05.029.

Deatsch, A. E., & Evans, B. A. (2014). Heating efficiency in magnetic nanoparticle hyperthermia. *Journal of Magnetism and Magnetic Materials*. Available from https://doi.org/10.1016/j.jmmm.2013.11.006.

Dickerson, E. B., et al. (2008). Gold nanorod assisted near-infrared plasmonic photothermal therapy (PPTT) of squamous cell carcinoma in mice. *Cancer Letters*. Available from https://doi.org/10.1016/j.canlet.2008.04.026.

Digigow, R. G., et al. (2014). Preparation and characterization of functional silica hybrid magnetic nanoparticles. *Journal of Magnetism and Magnetic Materials*. Available from https://doi.org/10.1016/j.jmmm.2014.03.026.

Dreaden, E. C., Alkilany, A. M., Huang, X., Murphy, C. J., & El-Sayed, M. A. (2012). The golden age: Gold nanoparticles for biomedicine. *Chemical Society Reviews*. Available from https://doi.org/10.1039/c1cs15237h.

Duff, D. G., Baiker, A., & Edwards, P. P. (1993). A new hydrosol of gold clusters. 1. Formation and particle size variation. *Langmuir: the ACS Journal of Surfaces and Colloids*. Available from https://doi.org/10.1021/la00033a010.

Giordano, M. A., Gutierrez, G., & Rinaldi, C. (2010). Fundamental solutions to the bioheat equation and their application to magnetic fluid hyperthermia. *International Journal of Hyperthermia*. Available from https://doi.org/10.3109/02656731003749643.

Govorov, A. O., & Richardson, H. H. (2007). Generating heat with metal nanoparticles. *Nano Today, 2*, 30–38.

Hegyi, G., Szigeti, G. P., & Szász, A. (2013). Hyperthermia versus oncothermia: Cellular effects in complementary cancer therapy. *Evidence-based Complementary and Alternative Medicine*. Available from https://doi.org/10.1155/2013/672873.

Hergt, R., Andra, W., d'Ambly, C. G., Hilger, I., Kaiser, W. A., Richter, U., & Schmidt, H. G. (1998). Physical limits of hyperthermia using magnetite fine particles. *IEEE Transactions on Magnetics, 34*, 3745–3754.

Hervault, A., & Thanh, N. T. K. (2014). Magnetic nanoparticle-based therapeutic agents for thermo-chemotherapy treatment of cancer. *Nanoscale*. Available from https://doi.org/10.1039/c4nr03482a.

Hirsch, L. R., et al. (2003). Nanoshell-mediated near-infrared thermal therapy of tumors under magnetic resonance guidance. *Proceedings of the National Academy of Sciences, 100*, 13549–13554.

Hirsch, L. R., et al. (2006). Metal nanoshells. *Annals of Biomedical Engineering*. Available from https://doi.org/10.1007/s10439-005-9001-8.

Huang, X., Wu, S., Ke, X., Li, X., & Du, X. (2017). Phosphonated pillar[5]arene-valved mesoporous silica drug delivery systems. *ACS Applied Materials & Interfaces*. Available from https://doi.org/10.1021/acsami.7b04015.

Hui, C., et al. (2011). Core–shell Fe_3O_4@SiO_2 nanoparticles synthesized with well-dispersed hydrophilic Fe_3O_4 seeds. *Nanoscale*. Available from https://doi.org/10.1039/c0nr00497a.

Hynynen, K., & Lulu, B. A. (1990). Hyperthermia in cancer treatment. *Investigative Radiology*. Available from https://doi.org/10.1097/00004424-199007000-00014.

Jain, S., Hirst, D. G., & O'Sullivan, J. M. (2012). Gold nanoparticles as novel agents for cancer therapy. *The British Journal of Radiology, 85*, 101–113.

Jankiewicz, B. J., Jamiola, D., Choma, J., & Jaroniec, M. (2012). Silica-metal core-shell nanostructures. *Advances in Colloid and Interface Science*. Available from https://doi.org/10.1016/j.cis.2011.11.002.

Jose, J., et al. (2020). Magnetic nanoparticles for hyperthermia in cancer treatment: An emerging tool. *Environmental Science and Pollution Research, 27*(16), 19214–19225. Available from https://doi.org/10.1007/s11356-019-07231-2.

Kamzin, A. S., Das, H., Wakiya, N., & Valiullin, A. A. (2018). Magnetic core/shell nanocomposites $MgFe_2O_4$/SiO_2 for biomedical application: synthesis and properties. *Physics of the Solid State*. Available from https://doi.org/10.1134/S1063783418090147.

Khosroshahi, M. E., & Ghazanfari, L. (2012). Synthesis and functionalization of SiO_2 coated Fe_3O_4 nanoparticles with amine groups based on self-assembly. *Materials Science and Engineering C*. Available from https://doi.org/10.1016/j.msec.2011.09.003.

Kim, J., Jung, B., Lee, S. & Kim, K. Hyperthermic effects of FeCoNi coated glass fibers in alternating magnetic field. in: *2017 IEEE International Magnetics Conference, INTERMAG 2017* (2017). Available from: https://doi.org/10.1109/INTMAG.2017.8007821.

Kong, W., et al. (2020). Se@SiO_2@Au-PEG/DOX NCs as a multifunctional theranostic agent efficiently protect normal cells from oxidative damage during photothermal therapy. *Dalton Transactions Home*. Available from https://doi.org/10.1039/c9dt04867g.

Koohi, S. R., et al. (2018). Plasmonic photothermal therapy of colon cancer cells utilising gold nanoshells: An in vitro study. in: *IET Nanobiotechnology*. Available from https://doi.org/10.1049/iet-nbt.2017.0144.

Kucharczyk, K., Kaczmarek, K., Jozefczak, A., Slachcinski, M., Mackiewicz, A., & Dams-Kozlowska, H. (2020). "Hyperthermia treatment of cancer cells by the application of targeted silk/iron oxide composite spheres. *Materials Science and Engineering C, 111654*. Available from https://doi.org/10.1016/j.msec.2020.111654, no. May.

Kumar, C. S. S. R., & Mohammad, F. (2011). Magnetic nanomaterials for hyperthermia-based therapy and controlled drug delivery. *Advanced Drug Delivery Reviews*. Available from https://doi.org/10.1016/j.addr.2011.03.008.

Li, Z. X., Kawashita, M., Araki, N., Mitsumori, M., Hiraoka, M., & Doi, M. (2010). Magnetite nanoparticles with high heating efficiencies for application in the hyperthermia of cancer. *Materials Science and Engineering C: Materials for Biology Applications, 30*, 990–996.

Lima, E., et al. (2013). Heat generation in agglomerated ferrite nanoparticles in an alternating magnetic field. *Journal of Physics D: Applied Physics*. Available from https://doi.org/10.1088/0022-3727/46/4/045002.

Liu, X., et al. (2013). Size-controlled synthesis of $Cu_{2-x}E$ (E = S, Se) nanocrystals with strong tunable near-infrared localized surface plasmon resonance and high conductivity in thin films. *Advanced Functional Materials*. Available from https://doi.org/10.1002/adfm.201202061.

Liu, X., et al. (2016a). A novel and facile synthesis of porous SiO$_2$-coated ultrasmall Se particles as a drug delivery nanoplatform for efficient synergistic treatment of cancer cells. *Nanoscale*. Available from https://doi.org/10.1039/c6nr02298g.

Liu, H., et al. (2016b). Hyperbranched polyglycerol-doped mesoporous silica nanoparticles for one- and two-photon activated photodynamic therapy. *Advanced Functional Materials*. Available from https://doi.org/10.1002/adfm.201504939.

Ma, H., Bendix, P. M., & Oddershede, L. B. (2012). Large-scale orientation dependent heating from single irradiated gold nanorods. *Nano Letters, 12*, 3954–3960.

MacKey, M. A., Ali, M. R. K., Austin, L. A., Near, R. D., & El-Sayed, M. A. (2014). The most effective gold nanorod size for plasmonic photothermal therapy: Theory and in vitro experiments. *The Journal of Physical Chemistry. B*. Available from https://doi.org/10.1021/jp409298f.

Majeed, J., et al. (2014). Enhanced specific absorption rate in silanol functionalized Fe$_3$O$_4$ core-shell nanoparticles: Study of Fe leaching in Fe$_3$O$_4$ and hyperthermia in L929 and HeLa cells. *Colloids Surfaces B Biointerfaces*. Available from https://doi.org/10.1016/j.colsurfb.2014.07.019.

Manohar, A., Geleta, D. D., Krishnamoorthi, C., & Lee, J. (2020). Synthesis, characterization and magnetic hyperthermia properties of nearly monodisperse CoFe$_2$O$_4$ nanoparticles. *Ceramics International, 46*(18), 28035–28041. Available from https://doi.org/10.1016/j.ceramint.2020.07.298.

Martín-Saavedra, F. M., et al. (2010). Magnetic mesoporous silica spheres for hyperthermia therapy. *Acta Biomaterialia*. Available from https://doi.org/10.1016/j.actbio.2010.06.030.

Mendes, R., Pedrosa, P., Lima, J. C., Fernandes, A. R., & Baptista, P. V. (2017). Photothermal enhancement of chemotherapy in breast cancer by visible irradiation of Gold Nanoparticles. *Scientific Reports, 7*.

Mie, G. (1908). Beiträge zur Optik trüber Medien, speziell kolloidaler Metallösungen. *Annals of Physics, 330*, 377–445.

Ming, T., et al. (2009). Growth of tetrahexahedral gold nanocrystals with high-index facets. *Journal of the American Chemical Society*. Available from https://doi.org/10.1021/ja907549n.

Misra, S. K., et al. (2016). Study of paramagnetic defect centers in as-grown and annealed TiO$_2$ anatase and rutile nanoparticles by a variableerature X-band and high-frequency (236 GHz) EPR. *Journal of Magnetism and Magnetic Materials*. Available from https://doi.org/10.1016/j.jmmm.2015.10.072.

Park, J., et al. (2004). Ultra-large-scale syntheses of monodisperse nanocrystals. *Nature Materials*. Available from https://doi.org/10.1038/nmat1251.

Pham, T., Jackson, J. B., Halas, N. J., & Lee, T. R. (2002). Preparation and characterization of gold nanoshells coated with self-assembled monolayers. *Langmuir: the ACS Journal of Surfaces and Colloids*. Available from https://doi.org/10.1021/la015561y.

Qin, Z., et al. (2016). Quantitative comparison of photothermal heat generation between gold nanospheres and nanorods. *Scientific Reports, 6*.

Rosensweig, R. E. (2002). Heating magnetic fluid with alternating magnetic field. *Journal of Magnetism and Magnetic Materials*. Available from https://doi.org/10.1016/S0304-8853(02)00706-0.

Shen, B. B., Gao, X. C., Yu, S. Y., Ma, Y., & Ji, C. H. (2016). Fabrication and potential application of a di-functional magnetic system: Magnetic hyperthermia therapy and drug delivery. *CrystEngComm*. Available from https://doi.org/10.1039/c5ce02267c.

Skinner, M. G., Iizuka, M. N., Kolios, M. C., & Sherar, M. D. (1998). A theoretical comparison of energy sources – Microwave, ultrasound and laser – For interstitial thermal therapy. *Physics in Medicine and Biology*. Available from https://doi.org/10.1088/0031-9155/43/12/011.

Starsich, F. H. L., et al. (2016). Silica-coated nonstoichiometric nano Zn-ferrites for magnetic resonance imaging and hyperthermia treatment. *Advanced Healthcare Materials*. Available from https://doi.org/10.1002/adhm.201600725.

Stjerndahl, M., et al. (2008). Superparamagnetic Fe_3O_4/SiO_2 nanocomposites: Enabling the tuning of both the iron oxide load and the size of the nanoparticles. *Langmuir: the ACS Journal of Surfaces and Colloids*. Available from https://doi.org/10.1021/la7035604.

Tao, C., & Zhu, Y. (2014). Magnetic mesoporous silica nanoparticles for potential delivery of chemotherapeutic drugs and hyperthermia. *Dalton Transactions Home*. Available from https://doi.org/10.1039/c4dt01984a.

Teleki, A., et al. (2009). Hermetically coated superparamagnetic Fe_2O_3 particles with SiO_2 nanofilms. *Chemistry of Materials: A Publication of the American Chemical Society*. Available from https://doi.org/10.1021/cm803153m.

Thakur, M., et al. (2006). Interparticle interaction and size effect in polymer coated magnetite nanoparticles. *Journal of Physics. Condensed Matter: An Institute of Physics Journal*. Available from https://doi.org/10.1088/0953-8984/18/39/035.

Theis-Bröhl, K., et al. (2018). Self-assembled layering of magnetic nanoparticles in a ferrofluid on silicon surfaces. *ACS Applied Materials & Interfaces*. Available from https://doi.org/10.1021/acsami.7b14849.

Toropova, Y. G., et al. (2017). In vitro toxicity of FemOn, FemOn-SiO_2 composite, and SiO_2-FemOn core-shell magnetic nanoparticles. *International Journal of Nanomedicine*. Available from https://doi.org/10.2147/IJN.S122580.

Wang, J., et al. (2018a). Controllable synthesis of gold nanorod/conducting polymer core/shell hybrids toward in vitro and in vivo near-infrared photothermal therapy. *ACS Applied Materials & Interfaces*. Available from https://doi.org/10.1021/acsami.7b16784.

Wang, J., et al. (2018b). Gold nanorods/polypyrrole/m-SiO_2 core/shell hybrids as drug nanocarriers for efficient chemo-photothermal therapy. *Langmuir: the ACS Journal of Surfaces and Colloids*. Available from https://doi.org/10.1021/acs.langmuir.8b02667.

Wang, T. T., et al. (2011). Fluorescent hollow/rattle-type mesoporous Au@SiO_2 nanocapsules for drug delivery and fluorescence imaging of cancer cells. *Journal of Colloid and Interface Science*. Available from https://doi.org/10.1016/j.jcis.2011.02.023.

Wust, P., et al. (2002). Hyperthermia in combined treatment of cancer. *Lancet Oncology*. Available from https://doi.org/10.1016/S1470-2045(02)00818-5.

Xiong, Y., et al. (2015). Preparation of magnetic core-shell nanoflower Fe_3O_4@MnO_2 as reusable oxidase mimetics for colorimetric detection of phenol. *Analytic Methods Home*. Available from https://doi.org/10.1039/c4ay02687j.

Yang, Y., et al. (2016). Complex assembly of polymer conjugated mesoporous silica nanoparticles for intracellular pH-responsive drug delivery. *Langmuir: the ACS Journal of Surfaces and Colloids*. Available from https://doi.org/10.1021/acs.langmuir.6b01845.

Ye, X., Zheng, C., Chen, J., Gao, Y., & Murray, C. B. (2013). Using binary surfactant mixtures to simultaneously improve the dimensional tunability and monodispersity in the seeded growth of gold nanorods. *Nano Letters*. Available from https://doi.org/10.1021/nl304478h.

Zhang, B. Q., Li, S. B., Xiao, Q., Li, J., & Sun, J. J. (2013). Rapid synthesis and characterization of ultra-thin shell Au@SiO_2 nanorods with tunable SPR for shell-isolated nanoparticle-enhanced Raman spectroscopy (SHINERS). *Journal of Raman Spectroscopy*. Available from https://doi.org/10.1002/jrs.4336.

Zhang, L., et al. (2016). Tailored synthesis of octopus-type janus nanoparticles for synergistic actively-targeted and chemo-photothermal therapy. *Angewandte Chemie International Edition*. Available from https://doi.org/10.1002/anie.201510409.

Zhang, Y., et al. (2015). A novel magnetic mesoporous silicon composite combining the function of magnetic target drug delivery and magnetic-induction hyperthermia. *Materials and Technology*. Available from https://doi.org/10.1179/17535557B15Y.000000010.

Zhang, Y., et al. (2016). A novel magnetic mesoporous silicon composite combining the function of magnetic target drug delivery and magnetic-induction hyperthermia. *Materials and Technology*. Available from https://doi.org/10.1080/10667857.2015.1117217.

15

Silicon–metal hybrid nanoparticles as nanofluid scale inhibitors in oil/gas applications

Yasser Mahmoud A. Mohamed[1] and Mohamed F. Mady[2,3]
[1]*Photochemistry Department, National Research Center, Cairo, Egypt* [2]*Green Chemistry Department, National Research Center, Cairo, Egypt* [3]*Chemistry Department, Bioscience and Environmental Engineering, Faculty of Science and Technology, University of Stavanger, Stavanger, Norway*

15.1 Introduction

Silicon nanoparticles have provided potential solutions for various problems in the petroleum industry, such as enhanced oil recovery, well drilling, fracturing, completion, and flow assurance (Yang, Ji, Li, Qin, & Lu, 2015; Alsaba, Al Dushaishi, & Abbas, 2020; Boul & Ajayan, 2020). In recent years, the use and application of silicon nanoparticles have become an attractive research direction in the upstream and downhole oil and gas field due to their potential to allow nanoparticles to travel through a porous rock without significant risks of formation damage in the petroleum reservoir (Fakoya & Shah, 2017). Table 15.1 summarizes some of the potential applications of silicon nanoparticles in the oil and gas industry.

In this section, we will focus on the application of silicon nanoparticles in oilfield scale inhibition. Oilfield scale is the precipitation of sparingly inorganic salts from aqueous solutions in the petroleum reservoir (Mady & Kelland, 2017; Mady, Charoensumran, Ajiro, & Kelland, 2018; Mady, Bayat, & Kelland, 2020). Scale formation is one of the biggest water-based production problems in the oil and gas industry. It is very important to determine and control in advance the scale deposition. A widely used technique for controlling mineral precipitation is the use of scale inhibitors (SIs). There are various conventional scale inhibitors to mitigate scale formation in the field. These classes are water-soluble polymers

Table 15.1 A list of the applications of the silicon nanoparticles in the petroleum industry.

Area	Potential applications	References
Production processes and flow assurance	Emulsion treating	Alade, Al Shehri, and Mahmoud (2019), Hassan, Abdalla, and Mustafa (2019)
	Corrosion inhibition	Jeong, Youm, and Yun (2018), Yen et al. (2012)
	Wax inhibition	Lim, Al Salim, Ridzuan, Nguele, and Sasaki (2018)
	Hydrate inhibition	Zhang et al. (2019)
	Asphaltene inhibition	López et al. (2020)
	Scale inhibition	Mady and Kelland (2020)
	Well stimulation	Singh, Panthi, and Mohanty (2017), Singh, Tong, Panthi, and Mohanty (2018)
Enhanced oil recovery	Improving CO_2 foam stability	Dickson, Binks, and Johnston (2004), Worthen et al. (2013)
	Improving emulsions stability	Griffith, Ahmad, Daigle, and Huh (2016), Kim et al. (2016)
	Wettability alteration	Dehghan Monfared and Ghazanfari (2019), Hou et al. (2019)
Drilling and completions	Drilling fluids applications: preventing water intrusion, improving shale inhibition, improving wellbore stability	Cai, Chenevert, Sharma, and Friedheim (2012), Yang, Shang, Liu, Cai, and Jiang (2017)
	Cementing applications: improving mechanical properties, accelerating the setting time, improving the filtration properties, improving the compressive strength	Jalal, Mansouri, Sharifipour, and Pouladkhan (2012), Pang, Boul, and Cuello Jimenez (2014)
	Fracturing fluids applications, enhancing hydraulic fracturing stimulation, improving the rheological properties of fracturing fluid	Al-Muntasheri, Liang, and Hull (2017), Fakoya and Shah (2018)

and/or nonpolymers containing phosphonate, sulfonate, and carboxylate groups. However, most traditional scale inhibitors have several serious drawbacks, such as short squeeze lifetime, poor biodegradation, weak thermal stability, and intolerant to calcium ions Kelland (2014). In order to overcome these challenges, a new generation of scale inhibitors-based nanotechnology has been developed. Interestingly, most nanoparticles showed high oilfield scale inhibition efficiency and extended squeeze lifetime compared to the conventional scale inhibitors (Mady & Kelland, 2020).

Silicon dioxide nanoparticles (also known as silica nanoparticles, SiO_2 NPs) play an important role in the future of oilfield scale inhibitor squeeze treatment. Scale inhibitors-based silicon nanoparticles can be adsorbed on the reservoir rock by squeezing the inhibitor inside the reservoir rocks in the near-wellbore region. It was found that silica nanoparticles provided a slow and sustained scale inhibitor release when the well is switched back to production, leading to a prolonged squeeze life span (Zhang, Ruan, Kan, & Tomson, 2016; Kumar, Chishti, Rai, & Patwardhan, 2012).

A series of silicon nanoparticles were synthesized and investigated as oilfield scale inhibitors, especially for carbonate and Group II sulfate scales. For example, Safari et al. studied various silicon nanoparticles as scale inhibitors for gypsum scale using a static test method (Safari, Golsefatan, & Jamialahmadi, 2014). The scale inhibition performance tests showed that the ideal size and concentration of silicon nanoparticles minimized the scale formation. Furthermore, silicon nanoparticles-capped scale inhibitor diethylenetriamine pentamethylene phosphonic acid (DETPMP) was prepared and used to enhance the inhibition rate of gypsum oilfield scale (Golsefatan, Safari, & Jamialahmadi, 2016). The results showed that this class of silicon nanoparticles with a particle size of 20 − 30 nm provided a better inhibition performance than the conventional scale inhibitor DETPMP. In addition, silicon nanoparticles-capped metal oxides, for example gadolinium oxide (Gd_2O_3) in the presence of copoly(4-styrenesulfonic acid-co-maleic acid) sodium salt, were synthesized and tested as an antiscaling agent for zinc sulfide (ZnS) and lead sulfide (PbS) scales (Baraka-Lokmane et al., 2016). The high-pressure dynamic tube blocking test results showed that functionalized silicon nanoparticles improved adsorption properties compared with conventional scale inhibitors using bench coreflood experiments, affording an increase in the lifetime of scale inhibitor treatments.

Silicon−metal hybrid nanoparticles have been developed as nanofluid scale inhibitors in oil and gas applications. These nanofluids have drawn much recent attention because of their unique physicochemical properties as well as potential oilfield scale inhibitor squeeze treatment performance. Zhang et al. synthesized silicon−zinc hybrid nanoparticles in the presence of diethylene triamine pentamethylene phosphonate (DTPMP) via a silica-templated process, giving new scale inhibitor nanoparticles (Si-Zn-DTPMP SINPs) (Zhang et al., 2010). It was proposed that divalent metal zinc can improve scale inhibitor retention and effectiveness properties. In this study, Si-Zn-DTPMP SINPs were synthesized by reacting silicon nanoparticles with zinc chloride

(ZnCl$_2$) in deionized water, and the pH was adjusted to 4–4.5 with 1 N HCl. The reaction mixture was heated to 70°C and vigorously stirred for 15 mins, giving silicon–zinc hybrid nanoparticles. The obtained nanoparticles were then functionalized with DTPMP in deionized water at pH 9 to produce silicon nanofluid Si-Zn-DTPMP SINPs (Fig. 15.1). The particle size of the final nanofluid was around 100 nm in diameter with a zeta potential of −55 mV. In addition, the particle size of Si-Zn-DTPMP SINPs was controlled in the presence of sodium dodecylbenzene sulfonate (SDBS) surfactant.

The new Si-Zn-DTPMP SINPs were used to improve the propagation and transportation of scale inhibitor-based phosphonates onto the carbonate and sandstone porous media using a column breakthrough test. Moreover, the adsorption properties of this class of hybrid nanofluid onto the formation rock in sandstone and calcite reservoir rocks have been investigated. Fig. 15.2 shows the schematic representation of the mechanism of silicon–metal hybrid nanoparticles in an oilfield scale squeeze treatment to evaluate the transport and inhibitor return properties of the new nanofluid (Zhang, Shen, Kan, & Tomson, 2016). This protocol can be illustrated in two stages as follows: (1) injecting stage of nanofluid into downhole formation; and (2) returning scale inhibitor after the shut-in process. As shown in Fig. 15.2, Si-Zn-DTPMP SINPs can be injected into the downhole formation, where the nanofluid can transport through the near-well formation rock. These nanoparticles-based phosphonated

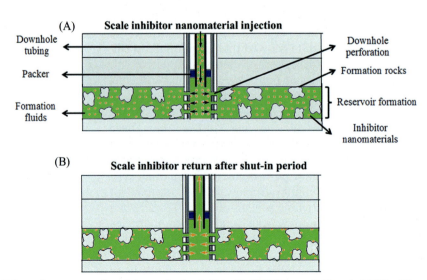

Figure 15.1 Schematic representation of the synthesis of silicon-Zn-DTPMP nanofluid scale inhibitor (Zhang et al., 2010).

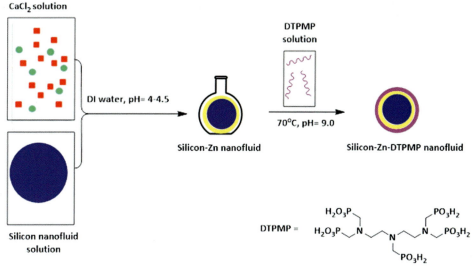

Figure 15.2 Schematic diagram of the plausible mechanism of nanofluid in an oilfield scale squeeze application (Zhang et al., 2016).

scale inhibitors will retain and adsorb onto the formation rock during a shut-in stage. When the oil well is switched back to production, Si-Zn-DTPMP SINPs will dissolve in the formation water to prevent scale deposition.

The results of transportation efficiency tests showed that Si-Zn-DTPMP SINPs in the presence of SDBS improved the diffusion rate of the phosphonated scale inhibitor in crushed columns of calcite and sandstone compared to conventional scale inhibitors. The long-term flowback performance test indicated that Si-Zn-DTPMP SINPs gave excellent prolonged squeeze life span treatment compared to DTPMP itself.

Another class of silicon—metal nanoparticles for improved squeeze lifetime treatment has been reported by Zhang et al. (Zhang, Fan, Lu, Kan, & Tomson, 2011; Zhang, Kan, & Tomson, 2016) Crystalline silicon—calcium phosphonate scale inhibitor nanoparticles (Si-Ca-DTPMP SINPs) were prepared from amorphous silica-templated calcium phosphonate precipitates via a diafiltration process. The obtained Si-Ca-DTPMP SINPs were surface-functionalized with an anionic surfactant, sodium dodecylbenzene sulfonate (SDBS), under ultrasonic irradiation. Carbonate and sandstone porous cores were tested with these nanofluids to evaluate their transportation and retention activities. It was found that Si-Ca-DTPMP SINPs showed improved travel and adsorption activities of the inhibitor in calcite and

Louise sandstone formation media, leading to prolonged squeeze time treatment. Long-term squeeze treatment experiments of the Si-Ca-DTPMP NPs afforded significant retention performance of returned phosphonate scale inhibitor concentrations over thousands of pore volumes (PVs) in comparison with conventional inhibitors.

More recently, Ali et al. synthesized a series of silicon–metal nanoparticles and investigated their adsorption morphology in comparison with commercial phosphonate scale inhibitors, such as DTPMP and Ca-DTPMP (Ali, Masuri, Kaniappan, Abu, & Zahari, 2020). Transmission electron microscopy (TEM) analyses were used to detect the particle size ranges of all synthesized nanoparticles scale inhibitors. It was found that Si-Zn-DTPMP SINPs gave the smallest particle diameter in the range of 31–37 nm. The adsorption morphology of these nanoparticle scale inhibitors onto the Kaolinite surface have been studied via static adsorption experiments. The static adsorption test results indicated that Si-Zn-DTPMP SINPs (31–37 nm) gave better adsorption performance onto the Kaolinite surface than other nanofluids.

15.2 Conclusions and future perspectives

The silicon-metal nanostructures are displaying ever-increasing research interests for their important properties and diverse applications, especially for oilfield scale inhibition. The controlled release of silicon-metal hybrid nanoparticles under mild conditions through porous rocks has been demonstrated. Currently, the large-scale production of various silicon nanoparticles, in addition to their surface-functionalization still remains challenge for practical commercialization. Hence, multi-functional silica-nanoparticles are highly demanded for creating new opportunities for novel applications. With the great application potential, silica-templated process is expected to receive a more intensified research interest as alternatives of silica-coated nanostructures.

References

Al-Muntasheri, G. A., Liang, F., & Hull, K. L. (2017). Nanoparticle-enhanced hydraulic-fracturing fluids: A review. *SPE-98275-PA*, *32*, 186–195.

Alade, O. S., Al Shehri, D. A., & Mahmoud, M. (2019). Investigation into the effect of silica nanoparticles on the rheological characteristics of water-in-heavy oil emulsions. *Petroleum Science*, *16*, 1374–1386.

Ali, F. N. M., Masuri, S. U., Kaniappan, L., Abu, R. H., & Zahari, N. I. (2020). Adsorption morphology of nano scale inhibitors for oilfield mineral scale mitigation. *Journal of Advance Research in Fluid Mechanics and Thermal Science, 68*, 22–33.

Alsaba, M. T., Al Dushaishi, M. F., & Abbas, A. K. (2020). A comprehensive review of nanoparticles applications in the oil and gas industry. *Journal of Petroleum Exploration and Production Technology, 10*, 1389–1399.

Baraka-Lokmane, S., Hurtevent, C., Rossiter, M., Bryce, F., Lepoivre, F., Marais, A., ... Graham, G. M. (2016). Design and performance of novel sulphide nanoparticle scale inhibitors for north sea HP/HT fields. *SPE International Oilfield Scale Conference and Exhibition* (p. 20) Aberdeen, Scotland, UK: Society of Petroleum Engineers.

Boul, P. J., & Ajayan, P. M. (2020). Nanotechnology research and development in upstream oil and gas. *Energy Technology, 8*, 1901216.

Cai, J., Chenevert, M. E., Sharma, M. M., & Friedheim, J. E. (2012). Decreasing water invasion into atoka shale using nonmodified silica nanoparticles. *SPE-146979-PA, 27*, 103–112.

Dehghan Monfared, A., & Ghazanfari, M. H. (2019). Wettability alteration of oil-wet carbonate porous media using silica nanoparticles: electrokinetic characterization. *Industrial & Engineering Chemistry Research, 58*, 18601–18612.

Dickson, J. L., Binks, B. P., & Johnston, K. P. (2004). Stabilization of carbon dioxide-in-water emulsions with silica nanoparticles. *Langmuir, 20*, 7976–7983.

Fakoya, M. F., & Shah, S. N. (2017). Emergence of nanotechnology in the oil and gas industry: Emphasis on the application of silica nanoparticles. *Petroleum, 3*, 391–405.

Fakoya, M. F., & Shah, S. N. (2018). Effect of silica nanoparticles on the rheological properties and filtration performance of surfactant-based and polymeric fracturing fluids and their blends. *SPE-146979-PA, 33*, 100–114.

Golsefatan, A. R., Safari, M., & Jamialahmadi, M. (2016). Using silica nanoparticles to improve DETPMP scale inhibitor performance as a novel calcium sulfate inhibitor. *Desalin Water Treatment, 57*, 20800–20808.

Griffith, N., Ahmad, Y., Daigle, H., & Huh, C. (2016). Nanoparticle-stabilized natural gas liquid-in-water emulsions for residual oil recovery. *SPE Improved Oil Recovery Conference* (p. 22) Tulsa, Oklahoma, USA: Society of Petroleum Engineers.

Hassan, S. A., Abdalla, B. K., & Mustafa, M. A. (2019). Addition of silica nanoparticles for the enhancement of crude oil demulsification process. *Petroleum Science and Technology, 37*, 1603–1611.

Hou, B., Jia, R., Fu, M., Wang, Y., Jiang, C., Yang, B., & Huang, Y. (2019). Wettability alteration of oil-wet carbonate surface induced by self-dispersing silica nanoparticles: Mechanism and monovalent metal ion's effect. *Journal of Molecular Liquids, 294*, 111601.

Jalal, M., Mansouri, E., Sharifipour, M., & Pouladkhan, A. R. (2012). Mechanical, rheological, durability and microstructural properties of high performance self-compacting concrete containing SiO_2 micro and nanoparticles. *Materials & Design, 34*, 389–400.

Jeong, Y. J., Youm, K. S., & Yun, T. S. (2018). Effect of nano-silica and curing conditions on the reaction rate of class G well cement exposed to geological CO_2-sequestration conditions. *Cement and Concrete Research, 109*, 208–216.

Kelland, M. A. (2014). *Production Chemicals for the Oil and Gas Industry* (2nd ed.). Boca Raton, FL: CRC Press (Taylor & Francis Group).

I. Kim, A.J. Worthen, M. Lotfollahi, K.P. Johnston, D.A. DiCarlo, C. Huh, 2016. Nanoparticle-stabilized emulsions for improved mobility control for adverse-mobility waterflooding, in: *SPE Improved Oil Recovery Conference, Society of Petroleum Engineers*, Tulsa, Oklahoma, USA, pp. 10.

D. Kumar, S.S. Chishti, A. Rai, S.D. Patwardhan, 2012, Scale inhibition using nano-silica particles, in: *SPE Middle East Health, Safety, Security, and Environment Conference and Exhibition, Society of Petroleum Engineers*, Abu Dhabi, UAE, p. 7.

Lim, Z. H., Al Salim, H. S., Ridzuan, N., Nguele, R., & Sasaki, K. (2018). Effect of surfactants and their blend with silica nanoparticles on wax deposition in a Malaysian crude oil. *Petroleum Science, 15*, 577–590.

López, D., Jaramillo, J. E., Lucas, E. F., Riazi, M., Lopera, S. H., Franco, C. A., & Cortés, F. B. (2020). Cardanol /SiO_2 nanocomposites for inhibition of formation damage by asphaltene precipitation/deposition in light crude oil reservoirs. Part II: Nanocomposite evaluation and coreflooding test. *Acs Omega, 5*, 27800–27810.

Mady, M. F., Bayat, P., & Kelland, M. A. (2020). Environmentally friendly phosphonated polyetheramine scale inhibitors—excellent calcium compatibility for oilfield applications. *Industrial & Engineering Chemistry Research, 59*, 9808–9818.

Mady, M. F., Charoensumran, P., Ajiro, H., & Kelland, M. A. (2018). Synthesis and characterization of modified aliphatic polycarbonates as environmentally friendly oilfield scale inhibitors. *Energy & Fuels, 32*, 6746–6755.

Mady, M. F., & Kelland, M. A. (2020). Review of nanotechnology impacts on oilfield scale management. *ACS Applied Nano Materials, 3*, 7343–7364.

Mady, M. F., & Kelland, M. A. (2017). Overview of the synthesis of salts of organophosphonic acids and their application to the management of oilfield scale. *Energy & Fuels, 31*, 4603–4615.

Pang, X., Boul, P. J., & Cuello Jimenez, W. (2014). Nanosilicas as accelerators in oilwell cementing at low temperatures. *SPE-146979-PA, 29*, 98–105.

Safari, M., Golsefatan, A., & Jamialahmadi, M. (2014). Inhibition of scale formation using silica nanoparticle. *Journal of Dispersion Science and Technology, 35*, 1502–1510.

Singh, R., Panthi, K., & Mohanty, K. K. (2017). Microencapsulation of acids by nanoparticles for acid treatment of shales. *Energy & Fuels, 31*, 11755–11764.

Singh, R., Tong, S., Panthi, K., & Mohanty, K. K. (2018). Nanoparticle-encapsulated acids for stimulation of calcite-rich shales. *SPE/AAPG/SEG Unconventional Resources Technology Conference* (p. 15) Houston, Texas, USA: Unconventional Resources Technology Conference.

Worthen, A. J., Bagaria, H. G., Chen, Y., Bryant, S. L., Huh, C., & Johnston, K. P. (2013). Nanoparticle-stabilized carbon dioxide-in-water foams with fine texture. *Journal of Colloid and Interface Science, 391*, 142–151.

Yang, J., Ji, S., Li, R., Qin, W., & Lu, Y. (2015). Advances of nanotechnologies in oil and gas industries. *Energy Exploration & Exploitation, 33*, 639–657.

Yang, X., Shang, Z., Liu, H., Cai, J., & Jiang, G. (2017). Environmental-friendly salt water mud with nano-SiO_2 in horizontal drilling for shale gas. *Journal of Petrol Science and Engineering, 156*, 408–418.

Yen, M., Lin, P., Dave, B., Chen, D., Groff, S., Hauter, E., ... McInerney, M. (2012). Examination of corrosion on steel structures by innovative nano sol-gel sensors. *CORROSION 2012* (p. 10) Salt Lake City, Utah: NACE International.

Zhang, P., Fan, C., Lu, H., Kan, A. T., & Tomson, M. B. (2011). Synthesis of crystalline-phase silica-based calcium phosphonate nanomaterials and their transport in carbonate and sandstone porous media. *Industrial & Engineering Chemistry Research, 50*, 1819–1830.

Zhang, X., Gong, J., Yang, X., Slupe, B., Jin, J., Wu, N., & Sum, A. K. (2019). Functionalized nanoparticles for the dispersion of gas hydrates in slurry flow. *ACS Omega, 4*, 13496–13508.

Zhang, P., Kan, A. T., Fan, C., Work, S., Lu, H., Yu, J., ... Tomson, M. B. (2010). Silica-templated synthesis of novel zinc-dtpmp nanoparticles, their transport in carbonate and sandstone porous media and scale inhibition. *SPE International Conference on Oilfield Scale* (p. 17) Aberdeen, UK: Society of Petroleum Engineers.

Zhang, P., Kan, A. T., & Tomson, M. B. (2016). Enhanced transport of novel crystalline calcium-phosphonate scale inhibitor nanomaterials and their long term flow back performance in laboratory squeeze simulation tests. *RSC Advances, 6*, 5259–5269.

Zhang, P., Ruan, G., Kan, A. T., & Tomson, M. B. (2016). Functional scale inhibitor nanoparticle capsule delivery vehicles for oilfield mineral scale control. *RSC Advances, 6*, 43016–43027.

Zhang, P., Shen, D., Kan, A. T., & Tomson, M. B. (2016). Phosphino-polycarboxylic acid modified inhibitor nanomaterial for oilfield scale control: transport and inhibitor return in formation media. *RSC Advances, 6*, 59195–59205.

Index

Note: Page numbers followed by "*f*" and "*t*" refer to figures and tables, respectively.

A
Acetylated dextran (AcDEX), 20–21
Acetylene, 158
Active pharmaceutical ingredients (APIs), 290–291
Addition reactions, 233–235
ADME, 312–313
AEAPS. *See* N-(2-aminoethyl)-3-aminopropyltrimethoxysilane (AEAPS)
Affordable, Sensitive, Specific, User-friendly, Rapid and robust, Equipment free, and Deliverable to end-users (ASSURED), 269–270
AHAPS. *See* N-(6-aminohexyl)-3-aminopropyltrimethoxysilane (AHAPS)
Air mass (AM), 171–173
Alkylalkoxysilane hydrolysis, 15–17
Alternating magnetic field (AMF), 341–343
Aluminum (Al), 181–182
Al–insulator–semiconductor memory structures, 12–15
Aminoclay (AC), 221–222
3-aminopropyl trimethoxysilane (APTMS), 332–334
Aminopropyltriethoxysilane, 259, 261–262
3-aminopropyltriethoxysilane (APTES), 213–214
Amorphous carbon (AC), 127
Amorphous SiNPs backbone of graphene nanocomposite (aSBG), 12–15
Amorphous TiO_2 (a-TiO_2), 134
Amplification method, 271
Analyte, 247–248
Anodic aluminum oxide (AAO), 304
Anodization of silicon wafers, 68
Antibody detection, 260
Aryldiazirine, 261–262
Atom transfers radical polymerization (ATRP), 291
Atomic force microscopy (AFM), 73
Atomic layer deposition technique, 71–72

B
Battery
 electrodes, 1
 nanosilicon for, 146–151
Beidellite, 202
Bencheikh's group, 116–118
Bentonite, 202
Biomedical applications, 277
 biosafety of mesoporous silica nanoparticles, 312–314
 blood transport, 278*f*
 drug delivery vessels, 278*f*
 drug loading for MSNs, 290–295
 gene delivery, 295–299
 nucleic acid detection and purification, 305–307
 pharmaceutical applications of mesoporous silica nanoparticles, 287–290
 protein absorption and separation, 299–305
 remarkable trends in mesoporous silica nanoparticles toward cancer therapeutic applications, 307–311
 silica nanoparticles, 280*t*, 285–286
 of silicon-based hybrid nanoparticles, 285*f*
Biomolecules, 200–201
Bioreceptor, 247–248
 attachment of bioreceptor on silicon-based materials, 255–257
 chemical surface modification, 255–257
Biosafety of mesoporous silica nanoparticles, 312–314
Biosensors, 247–248, 271
Biotinylation, 255–257
Bloch subsurface wave (BSSW), 258
Bloch surface wave (BSW), 258
Boron carbide (BC), 12–15
Bosch process, 252–253
Bottom-up approach, 65–66, 149, 169–170
Bovine serum albumin, 262
Bronsted acids, 200–201
Brownian relaxation, 331
Brucite, 202
Buchwald-Hartwig reaction, 211–213
Bulk silicon, 55–56, 65

C
Calcium, 181–182
Camphor sulfonic acid (CSA), 18–20
Cancer therapeutic applications, remarkable trends in mesoporous silica nanoparticles toward, 307–311

363

Cancer therapy, 307–310
Capacity failure mechanism of Si electrode in LIB, 126–127
Capture probe, 247–248
Carbamates, 229–230
Carbodiimide, 261–262
Carbon (C), 151–157
    C-N cross-coupling reactions, 211–213
    C-O cross-coupling reactions, 213–217
    C-S cross-coupling reactions, 213–217
    carbon-coated technology, 127–128
    carbon–carbon coupling reactions, 203–211
    carbon coating of silicon, 157–163
        gas-phase synthesis, 158–159
        liquid process, 159–161
        mechanical milling, 157–158
        Si/C composites, 161–163
    chemistry, 156–157
    coating, 128–129
    covering, 110–111
    dots, 45
    forms, 152–155
    materials, 129–130
    nanomaterials, 46–47, 109
    oxidation, 154
    pitch as precursor, 155–157
    sources, 45
Carbon dioxide (CO_2), 228–229
    conversion, 228–233
Carbon nanofiber (CNF), 12–15
Carbon nanotubes (CNTs), 3–4, 45, 127
Carbon quantum dots (CQD), 175
Carboxyethylsilanetriol sodium salt, 255–257
Catalysis, 1, 65–66, 169
Catalysts, 27, 199–200
Catalytic system, 200–201
Cell detection, 260–261

Chemical bath deposition technique, 91–92
Chemical surface modification, 255–257
Chemical vapor deposition (CVD), 46–47, 169–170, 173–175
3-chloropropyltrimethoxysilane, 210–211
Ciprofloxacin, 58
Cis-diamine platinum dichloride (CDDP), 291
Clustered regularly interspaced short palindromic, 271
Cobalt ferrites ($CoFe_2O_4$), 336–337
Colloidal system, 23–25
Commercial PEDOT: PSS solution, 185
Conductive polymers (CPs), 113–114
Contact inhibition, 279
Convergent reactions, 222–223
Copper (Cu), 130–131, 181–182
Coprecipitation method, 12–15
Coronavirus (CoV), 269–270
Coupling reactions, 203–217
    C-N cross-coupling reactions, 211–213
    C-S and C-O cross-coupling reactions, 213–217
    carbon–carbon coupling reactions, 203–211
COVID-19, 269–270
Cross-dimensional Si/C hybrids, 48–49
Crystalline silicon–calcium phosphonate scale inhibitor nanoparticles (Si-Ca-DTPMP SINPs), 357–358
Crystalline TiO_2 (c-TiO_2), 134
Cyclic carbonates, 229–230
Cyclotrimerization reactions, 236–237

D

Debye–Scherrer method, 73–74, 93–94

Decarboxylation, 255–257
Dialkyl carbonates, 229–230
1,4-diaza-bicyclo (2.2.2) octane-based ammonium salts (DABCO-based ammonium salts), 230–232
Diethylene triamine pentamethylene phosphonate (DTPMP), 355–356
Diethylenetriamine pentamethylene phosphonic acid (DETPMP), 355
Diethylvinylphosphonate (DEVP), 18–20
Diffusion assisted loading (DiSupLo), 291
Dimethyldiethoxysilane (DMDEOS), 15–17
Diodes, 1
DIPEA. See N, N-diisopropylethylamine (DIPEA)
Discrete-dipole approximation (DDA), 327–328
Disulfide bond reducing molecules, 307
Dithiothreitol (DTT), 307
DNA
    detection, 258–259
    purification, 305–307
Docetaxel (Dtxl), 307–310
Doping of boron, 110
Double-layer biosensor, 262–264
Double-walled silicon nanotubes (DWSiNTs), 179–180
Doxorubicin (DOX), 289–290, 294, 307–310
Doxorubicin hydrochloride, 21–22
Drug delivery, 1, 277
Drug loading, 21–22
    drug encapsulation mechanisms, 290–295
    for MSNs, 290–295
Dry fabrication methods, 169–170

DX. *See* Doxorubicin (DOX)
Dynamic light scattering (DLS), 31–33

E

Electrical properties
 of Si/C nanoparticles, 49–55
 of silicon–zinc oxide hybrid nanoparticles, 79–80
Electrocatalysis, 169
Electrochemical double-layer capacitors (EDLCs), 183
Electrochemical impedance spectrum, 53
Electroless etching, 110
Electromagnetic energy, 325–326
Electron beam evaporation, 173–175
Electron microscopy study, 294–295
Electroporation, 297–298
Encapsulation process, 20–21
Energy crisis, 103–104
Energy dispersive X-ray spectroscopy (EDS), 73, 338
Energy storage
 general background and progress, 105–108
 Si-based nanomaterials for lithium storage, 108–115
 supercapacitors, 115–118
 structure–property relationship, 106–108
  crystallinity, 107
  dispersity, 108
  porosity, 107
  size, 106
Enhanced permeation and retention effect (EPR effect), 277
Enhancement factor (EF), 3
Enzyme detection, 259
Enzyme-linked immunosorbent assay (ELISA), 270–271
Escherichia coli, 260–261
Esomeprazole, 218–219

1-ethyl-3-(3-dimethylaminopropyl) carbodiimide/N-hydroxysuccinimide (EDC/NHS), 255–257

F

Fabrication process, 12–15
Field effect glucose biosensor, 250–251
Field effect transistor (FET), 272. *See also* Graphene-based field effect transistor (GFET)
Field emission scanning electron microscopy (FESEM), 93
Fill factor (FF), 171–173, 175
Finely configured Si/C hybrid nanoparticles, 47–48
Finite element method (FEM), 327–328
Finite-difference time-domain technique (FDTD technique), 327–328
Fluorescence of Si/C quantum dots, 56–58
Fluorophore, 259
Fourier-transform infrared spectroscopy (FTIR), 27–29, 53, 73
 of silicon–zinc oxide hybrid structures, 75–77
Fourier-transformed spectroscopy (reflectometric interference) (FT-RIS), 260–261
Full width at half maximum (FWHM), 96–98
Fullerene, 45

G

Gadolinium oxide (Gd_2O_3), 355
Galvanostatic intermittent titration technique, 53
Gas sensors, 65–66
Gas-phase synthesis, 26, 158–159
Gatekeeping, 295

Gene delivery, 295–299
Glaser-type reaction, 208–210
Glutaraldehyde, 259, 261–262
3-glycidoxypropyltri methoxysilane, 255–257
Glycols, 229–230
Gold (III) chloride trihydrate ($HAuCl_4$), 332–334
Gold nanoparticles (AuSiNPs), 332–334
Gold nanorods (GNRs), 332–334
Gold shell-isolated nanoparticles (AuSHINs), 340–341
Grafting, 200–201
Graphene, 3–4, 45, 111–112, 127, 269
Graphene oxide (GO), 25
Graphene-based field effect transistor (GFET), 272
 amplification method, 271
 as biosensors, 272
 immunological assays, 270–271
 nanobiosensors, 271–272
Graphite, 3–4, 145–146, 152
Green Chemistry, 199–201
Green emissions, 96–98

H

4*H*-Chromene derivatives, 223–225
Halloysite, 202
Hazardous contaminants, 58
Heterocatalytic system, 200–201
Hexadecyltrimethyl ammoniumcation (HDTMA), 236–237
Hexamethyldisiloxane (HMDSO), 334–336
Hierarchical Si/C porous structures, 52
High-resolution transmission electron microscope (HR-TEM), 93, 337–338

Hollow mesoporous silica (HmSiO$_2$), 293
Homogeneous catalysts, 200–201
Hot wall reactor (HWR), 26
Hybrid capacitors, 183
Hybrid memory system, 12–15
Hybrid nanomaterials, 247–248
Hybrid nanoparticles, 247–248
Hybrid Si/MoO$_3$ nanostructures, 3
Hybridization, 162
Hydrogen, 169
Hydrogen evolution reaction (HER), 186–188
 silicon as electrocatalyst for, 186–194
Hydrogen silsesquioxane (HSQ), 17–18
Hydroquinone, 58
1-(2-hydroxyethyl)-3-methylimidazolium (ImIL), 206
6-hydroxyhexyl 3-(methylthio)-2-phenyl-3-thioxopropanoate (HMT), 18–20
5-hydroxymethylfurfural (5-HMF), 237
Hyperthermia treatment. *See also* Magnetic hyperthermia
 applications, 339–343
 nanoparticle heating, 326–331
 synthesis of hyperthermia agents, 331–339
Hysteresis loss, 326–327

I

Illite, 202
Imidazolium-based ionic liquid moiety, 237
Immunoglobulin G, 260
Immunological assays, 270–271
In situ
 characterization on of Si/C hybrids properties, 53–55
 NMR spectroscopy, 146

Incipient wetness impregnation (IWI), 290–291
Inductively coupled plasma optical emission spectrometry (ICP-OES), 313
Initial coulombic efficiencies (ICE), 163
Intermetallic silicon composites, 181–182
Ir/Si binary nanowire catalyst, 193–194
Iron, 181–182
Iron oxide nanoclusters (IONC), 341–343
Isocyanatepropyltriethoxysilane (ICPTS), 255–257
Isothermal nucleic acid technique, 271
Itaconic acid (IA), 291

K

Kaolinite, 202
Kenyaite, 202

L

L-lactate dehydrogenase, 259
Lactobacillus acidophilus, 260–261
Langmuir–Blodget technique, 12–15
Laser pyrolysis, 149
Lateral flow immunoassay (LFIA), 270–271
Latex nanospheres, 261–262
Lead sulfide (PbS), 355
Lewis acids, 200–201
Light absorption of silicon quantum dots/graphene, 58–60
Light-emitting diode (LED), 89
Liquid process, 159–161
Lithium
 diffusion, 152
 intercalation, 152
 Si-based nanomaterials for lithium storage, 108–115
 Si/carbon-based nanomaterials for Li storage, 109–112

 Si/metals-based nanomaterials for Li storage, 112–113
 Si/other materials-based nanostructures for Li storage, 113–115
 Si/polymers-based LIB anodes, 113–114
 Si/transition metal nitrides/carbides-based LIB anodes, 114–115
Lithium-ion batteries (LIBs), 3–4, 46, 125, 169, 175–177
 capacity failure mechanism of Si electrode in, 126–127
 commercialization, 151
 silicon for, 177–182
Localized surface plasmon polaritons (LSPPs), 325–326
 LSPP-based heating mechanism, 327–331
Localized surface plasmon resonance (LSPR), 1–2, 271
Loop-mediated isothermal amplification, 271
Lower critical solution temperature (LCST), 293
Luminescent SiNCs/polymer hybrids, 18–20

M

m-maleimidobenzoyl-N-hydroxysuccinimide ester (MBS), 255–257
M13KO7 bacteriophage, 261–262
Maghemite (γ-Fe$_2$O$_3$), 2–3
Magnesium, 181–182
Magnetic hyperthermia. *See also* Plasmonic hyperthermia
 applications, 341–343
 nanoagents, 334–339
Magnetic mesoporous silica system (MMSS), 334–336, 341–343
Magnetic nanoparticles (MNPs), 204–205
 coating with, 26–27

Magnetic resonance imaging (MRI), 1
Magnetic-based heating mechanism, 331
Magnetite (Fe_3O_4), 2–3, 293, 334–336
Magnetoreception, 297–298
Magnetron sputtering technique, 91–92
Malignant mesothelioma (MM), 307–310
Matrix metalloproteinases-2 (MMP-2), 300–302, 301f
Mechanical milling, 157–158
Mercapto-functionalized-41 anchored palladium (0) complex [MCM-41-SH-Pd(0)], 204–205
Mercaptoethanol (ME), 307
3-mercaptopropyltrimethoxysilane (MPTS), 255–257, 307
Mercuric ions, 58
Mesoporous magnetic nanocomposites (Mag-MSN), 204–205
Mesoporous silica nanoparticles (MSNs), 201–202, 285–286
  biosafety of, 312–314
  drug loading for, 290–295
  modified MSNs for protein delivery, 302t
  MSN-based nanodevices, 296t
  pharmaceutical applications of, 287–290
  remarkable trends in mesoporous silica nanoparticles toward cancer therapeutic applications, 307–311
  tailoring, 297f
Mesoporous silica thin film (MSTF), 304
Metal oxide NPs, 202
Metal(s), 202
  assisted chemically etching of silicon wafers, 69
  complexes, 200–201
  ions, 200–201
  metal-assisted chemical etching, 71–72
  metal-based nanomaterials, 112–113
  metal–organic chemical vapour deposition, 91–92
Metal–organic nanostructures (MONs), 286
Microelectromechanical systems (MEMS), 249–250
Microporous silicon synthesis, 181
Microsupercapacitors (μ-SCs), 183–185
Microwave plasma, 169–170
Mixed Si/C nanoparticles, 47
Mobil crystalline materials (MCM-41), 201–202, 287, 300
Modafinil, 218–219
Molecular beam epitaxy, 173–175
Monodispersal, 108
MTD, 313
Multicomponent heterostructured NPs, 12–15
Multicomponent reactions (MCRs), 222–228
Multidrug therapy, 295
Murray place exchange method, 20–21
MXenes, 115

N
N,N-diisopropylethylamine (DIPEA), 208–210
N-(2-aminoethyl)-3-aminopropyltrimethoxysilane (AEAPS), 305–307
N-(6-aminohexyl)-3-aminopropyltrimethoxysilane (AHAPS), 305–307
N-(trimethoxysilylpropyl) ethylene-diamine triacetic acid, 255–257
N-heterocyclic carbene-nickel (NHC-Ni), 213–214
N-Hydroxysuccinimide ester, 255–257
Nacrite, 202
Nano silicon carbon hybrid particles and composites for batteries
  carbon, 151–157
  carbon coating of silicon, 157–163
  nanosilicon for batteries general behavior and SEI, 149–151
  nanosizing, 146–148
  silicon generalities, 146
  various forms of nanosilicon, 148–149
Nanobelts, 11–12
Nanobiosensors, 271–272
Nanoclays, 201–202
Nanoclinoptilolite (NCP), 236–237
Nanocomposites, 11–12
Nanofilms, 149
Nanofluids, 355–356, 358
Nanoforms of silicon, 65
Nanomaterials, 107, 247–248
Nanoparticles (NPs), 1, 65, 277, 293
  heating, 326–331
    LSPP-based heating mechanism, 327–331
    magnetic-based heating mechanism, 331
  nanoparticle-based hyperthermia, 325–326
Nanoplates, 65
Nanopore target sequencing (NTS), 271
Nanoporous Si pillar array (NSPA), 71
Nanoranged particles, 200–201
Nanorods, 11–12, 65
Nanosilica sulfuric acid (SSA), 236
Nanosilicon, 154
Nanosized carbon, 45
Nanosized silica, 201–202
Nanosizing, 146–148

Nanospheres, 11–12
Nanostructured materials, 11–12
Nanostructured silicon, 169
　supercapacitors based on, 183–186
Nanowires, 11–12, 65, 149
Naphthalene di-imide (NDI), 171–173
Near-infrared light (NIR light), 325–326, 332–334
Néel relaxation, 331
Nitrile rubber (NBR), 25
No reported adverse reaction level (NOAEL), 313
Nuclear magnetic resonance spectroscopy (NMR), 53
Nucleic acid detection and purification, 305–307

O

Oilfield scale, 353–354
One-pot method, 221–222
Optical biosensors, 247–248
Optical properties
　of silicon–zinc oxide hybrid nanoparticles, 80–83
　of ultrasmall Si/C nanoparticles, 55–60
Organic functional groups, 53
Organic nanostructure, 52
Organic solvent process (OSM), 290–291
Orthosilicic acid, 1
Oxidation reactions, 217–220

P

Pharmaceutical applications of mesoporous silica nanoparticles, 287–290
Phenylboronic acid (PBA), 300–302
Phenylboronic acid with human serum albumin (PBA-HSA), 300–302
10-phenylphenothiazine (Ph-PTZ), 291
Phosphorus-doped Si nanocrystals, 1–2
Photocarrier behavior of silicon quantum dots/graphene, 58–60
Photocatalysis, 169
Photodetectors, 65–66
Photodynamic therapy, 295
Photoelectrochemical, silicon applications in, 171–177
Photoluminescence (PL), 21–22
　spectroscopy, 80–81
　results of silicon-ZnO hybrid structures, 83
Photon correlation spectroscopy (PCS), 31–32
Photovoltaic applications, 1
Photovoltaic devices, silicon applications in, 171–177
Physical adsorption, 296
Physical vapor deposition techniques, 173–175
Piperidinium benzene-1, 3 disulfonate functionalized $SiO_2@Fe_3O_4$ nanocatalyst (PBDS-SCMNPs), 228
Pitch as precursor, 155–157
Plasma enhanced chemical vapor deposition (PECVD), 249–250
Plasmonic hyperthermia
　applications, 340–341
　nanoagents, 332–334
Plasmonic nanoparticles, 325–326
Poly(3-hexylthiophene) (P3HT), 12–15
Poly(3,4-(ethylenedioxy)thiophene) (PEDOT), 183–185
Poly(9,9-dioctylfluorene-co-fluorenone-co-methylbenzoic ester material, 135–136
Poly(N-isopropylacrylamide) (PNIPAM), 334–336
Poly(vinyl alcohol) (PVA), 15–17
Polyacrylonitrile (PAN), 12–15
Polyaniline (PANI), 18–20, 113–114, 135–136
Polydisprse alkyl-capped SiNCs, 24–25
Polydopamine (PDA), 20–21
Polyethylene glycol (PEG), 334–336
Polyethylene glycol methyl acrylate (PEGMA), 291
Polyethylenedioxythiophene (PEDOT), 113–114
Polyethyleneimine (PEI), 21–22
Polyhedral oligomericsilsesquioxanes (POSS), 201
Polymer grafting, 17–20
Polypyrrole (PPy), 113–114
Polystyrene (PS), 17–18, 304
Polyurethane (PU), 15–17
Pore volumes (PVs), 357–358
Porous silicon (P-Si), 115–116
　biosensors, 247–248, 260
　NPs, 1, 21–22
　porous Si/C hybrid, 48
　structures, 181
Power conversion efficiency (PCE), 12–15, 171–173
Protein
　absorption and separation, 299–305
　detection, 262
Pseudocapacitors, 183
Pt-graphene, 192–193
Pulverization, 157
Pyrimidines, 229–230
Pyrolysis, 156

Q

Quantum dots (QDs), 46, 65

R

Radio frequency (RF), 17–18
　magnetron sputtering method, 91–92
Raman graph of silicon–zinc oxide hybrid structures, 75–77
Raman spectroscopy, 73
Reactive oxygen species (ROS), 332–334

Reduced graphene oxide (rGO), 20–21
Reduction reactions, 220–222
Resonant waveguide biosensor, 259
Reticuloendothelial system (RES), 279
Reverse transcription polymerase chain reaction (RT-PCR), 271
Reversible addition–fragmentation chain transfer (RAFT), 17–18
Reversible hydrogen electrode (RHE), 188–192
Rolling circle amplification, 271

S

Sabatier's principle of catalysis, 193–194
Santa Barbara Amorphous type materials (SBA-15), 201–202, 287–288
Scalable flame technology (FPS), 334–336
Scale formation, 353–354
Scale inhibitors (SIs), 353–354
Scanning electron microscopy (SEM), 73
Scanning transmission electron microscope (STEM), 337–338
Selenium (Se), 332–334
Self-templating mechanism, 108
Sensing, 1
  device, 247–248
Sheet-like zinc oxide/silicon light-emitting diode experimental details
  analysis of samples, 93
  fabrication of samples, 92–93
  results, 93–98
Sheet-like ZnO, 92
Si nanoparticles (Si-NPs), 110–111
Sialic acid (SA), 300–302
Silanol groups, 150
Silica (SiO_2), 201, 286, 332–334

silica-based nanomaterials, 279
Silica nanoparticles (SiNPs), 277, 332–334
  biomedical applications, 280t, 285–286
Silicates, 201
Silicon (Si), 1, 3–4, 73, 104–105, 125, 146, 169–170, 326–327. See also Porous silicon (P-Si)
  capacity failure mechanism of Si electrode in LIB, 126–127
  carbon coating of, 157–163
  chip-based biosensor, 253–254
  classification of Si-based composites
    Si/carbon composites, 127–130
    Si/metal composites, 130–133
    Si/metal oxide composites, 133–135
    Si/polymer composites, 135–137
  for energy applications, 171–194
    silicon applications in photoelectrochemical and photovoltaic devices, 171–177
    silicon as electrocatalyst for HER, 186–194
    silicon for lithium-ion batteries, 177–182
    supercapacitors based on nanostructured silicon, 183–186
  generalities, 146
  microwire array, 188
  nanopillars, 332–334
  oxide groups, 150
  Si/C composites, 127–130, 161–163
  Si/polymers-based LIB anodes, 113–114

Si–Ag system, 12–15
silicon-based biosensor, 247–248
  applications of biosensor, 257–264
  attachment of bioreceptor on silicon-based materials, 255–257
  fabrication of silicon-based hybrid nanoparticles, 248–254
silicon-based composite application in batteries, 125–126
silicon-based hybrid nanomaterials, 279
silicon-based nanomaterials, 247–248
silicon-based nanoparticle, 201–202
silicon-based solar cells, 175–177
silicon-Zn-DTPMP nanofluid scale inhibitor, 355–356, 356f
silicon–metal hybrid nanoparticles, 355–356
silicon–metal oxide hybrid nanostructures, 3
silicon–organic heterojunction solar cells, 171–173
silicon–zinc hybrid nanoparticles, 355–356
wafers
  anodization of, 68
  metal assisted chemically etching of, 69
Silicon carbide (SiC), 12–15, 18–20, 183–185
Silicon dioxide (SiO_2), 201, 251–252
  nanoparticles, 355
Silicon nanoparticles (SiNPs), 11–12, 145–146, 149–150, 353. See also Mesoporous silica nanoparticles (MSNs)
  application of, 353–354, 354t

Silicon nanoparticles into porous CNF (Si/PCNF), 12–15
Silicon nanowire array (SiNA), 27
Silicon nanowires (Si NWs), 12–15, 104–105, 173–175, 340–341
Silicon-based hybrid nanoparticles (Si-HNPs), 1–2, 11–12, 247–248
  as catalyst for organic conversions, 203–238
    addition reactions, 233–235
    carbon dioxide conversion, 228–233
    coupling reactions, 203–217
    miscellaneous reaction, 235–238
    multicomponent reactions, 222–228
    oxidation reactions, 217–220
    reduction reactions, 220–222
  characterization, 27–34
    DLS, 31–33
    FTIR, 27–29
    TEM, 33–34
    XPS, 29–31
  design of Si/C nanoparticles, 46–49
    cross-dimensional Si/C hybrids, 48–49
    finely configured Si/C hybrid nanoparticles, 47–48
    mixed Si/C nanoparticles, 47
  electrical properties of Si/C nanoparticles, 49–55
    influence of chemical components, 52–53
    influence of morphology and structure, 50–52
    in situ characterization on of Si/C hybrids properties, 53–55
  fabrication of, 248–254
  optical properties of ultrasmall Si/C nanoparticles, 55–60
    fluorescence of Si/C quantum dots, 56–58
    light absorption and photocarrier behavior of SiQDs /graphene, 58–60
  synthesis, 12–27
    catalyst, 27
    coating with magnetic nanoparticles, 26–27
    colloidal system, 23–25
    drug loading, 21–22
    encapsulation process, 20–21
    fabrication process, 12–15
    gas-phase synthesis, 26
    polymer grafting, 17–20
    sol–gel process, 15–17
    solution blending, 25
Silicon/carbon quantum dots (SiQDs), 56–58
  fluorescence of, 56–58
  light absorption and photocarrier behavior of SiQDs /graphene, 58–60
Silicone elastomers, 18–20
Silicon–ZnO hybrid nanoparticles, properties of
  applications of silicon–ZnO hybrid nanoparticles, 84
  electrical properties of silicon–ZnO hybrid nanoparticles, 79–80
  optical properties of silicon–ZnO hybrid nanoparticles, 80–83
  photoluminescence spectroscopy results of silicon-ZnO hybrid structures, 83
  UV–Vis spectroscopy results of silicon–ZnO hybrid structures, 81–82
  preparation of silicon–ZnO hybrid structures, 67–72
    anodization of silicon wafers, 68
    metal assisted chemically etching of silicon wafers, 69
    synthesis of silicon–ZnO hybrid structures, 70–72
  structural and morphological properties of silicon–ZnO hybrid structures, 73–79
    FTIR or Raman graph, 75–77
    morphological characterization of silicon–ZnO hybrid structures, 77–79
    X-ray diffraction patterns, 73–75
  ZnO nanoparticles, 65–67
Silsesquioxanes, 202
Silver (Ag), 131–132, 181–182
Silver nanoparticles (AgNPs), 12–15
SiRNA, 299
Small analyte detection, 262–264
Sodium acetate (CH_3COONa), 334–336
Sodium dodecylbenzene sulfonate (SDBS), 355–358
Sodium hydroxide (NaOH), 332–334
Solar cells, 65–66, 169
Solar energy, 171
Sol–gel process, 15–17
Solid electrolyte interface (SEI), 52–53, 177–178
  general behavior and, 149–151
  membrane, 126–127
Solution blending, 25
Sonogashira–Hagihara reactions, 208–210
Sonoporation, 297–298
Spin-coating process, 17–18
Stöber method, 332–334
Sucrose, 160

Sulfonic acid-functionalized silica-coated $CuFe_2O_4$ NPs, 227–228
Sulindac, 218–219
Supercapacitors, 169
　based on nanostructured silicon, 183–186
　Si-based nanomaterials for, 115–118
Surface-enhanced Raman scattering (SERS), 3, 71
Surface-initiated group-transfer polymerization process (SIGTP process), 18–20

T
Talc, 202
Tetrabutyl ammonium bromide (TBAB), 229–230
Tetraethyl orthosilicate (TEOS), 15–17, 332–334
Tetrahydrofuran (THF), 334–336
Tetrakis (hydroxymethyl) phosphonium chloride (THPC), 332–334
Thermal annealing of silicon oxide based polymers, 169–170
Three dimension (3D)
　network architecture, 135–136
　ordered structure, 108
　PANI structure, 113–114
Tin-doped silicon nanowire, 181–182
TiO_2 NPs, 12–15
Top-down approach, 65–66, 169–170
Transition metal (TM), 114–115, 133–134
Transition metal oxides (TMO), 133–134
Transmission electron microscopy (TEM), 33–34, 73, 106, 313, 358
Transparent conducting oxides (TCOs), 90
Transparent solar cells (TSCs), 175–177
Triethanolamine (TEA), 206
Trimethylsilylcyanides (TMSCN), 226–227
2,4,6-trinitrotoluene (TNT), 3
Tunable silica nanomaterials, 285–286

U
Ultrasmall carbon particles, 46–47
Ultrasmall Si/C nanoparticles, optical properties of, 55–60
Ultrasonic spray pyrolysis, 90–91
UV
　lithography, 71
　UV-lasers, 65–66
　UV–Vis spectroscopy, 80–81
　results of silicon–ZnO hybrid structures, 81–82

V
Vanadium-incorporated silica-based complex, 219–220
Vapour–liquid–solid (VLS), 118
　development mechanism, 118
　methods, 91–92, 173–175
Vinylaniline-terminated hybrid composite PSiNPs (VANi-PSiNPs), 21–22
Virus detection, 261–262

W
World Health Organization (WHO), 269–270

X
X-ray diffraction (XRD), 73–75, 93
X-ray photoelectron spectroscopy (XPS), 29–31

Z
Zero premature exposure, 295
Zinc chloride ($ZnCl_2$), 355–356
Zinc oxide (ZnO), 65, 90
　nanoparticles, 65–67
Zinc sulfide (ZnS), 355
Zirconium tetra-kis(2,4-pentanedionate) (ZTP), 15–17

Printed in the United States
by Baker & Taylor Publisher Services